AF352832

Locally Available Energy Sources and Sustainability

Locally Available Energy Sources and Sustainability

Special Issue Editors

Antonio Colmenar Santos
David Borge Diez
Enrique Rosales Asensio

MDPI • Basel • Beijing • Wuhan • Barcelona • Belgrade

Special Issue Editors
Antonio Colmenar Santos
National University of
Distance Education
Spain

David Borge Diez
University of Léon
Spain

Enrique Rosales Asensio
Universidad de Las Palmas de
Gran Canaria
Spain

Editorial Office
MDPI
St. Alban-Anlage 66
4052 Basel, Switzerland

This is a reprint of articles from the Special Issue published online in the open access journal *Sustainability* (ISSN 2071-1050) in 2019 (available at: https://www.mdpi.com/journal/sustainability/special_issues/Locally_Available_Energy_Sources).

For citation purposes, cite each article independently as indicated on the article page online and as indicated below:

LastName, A.A.; LastName, B.B.; LastName, C.C. Article Title. *Journal Name* **Year**, *Article Number*, Page Range.

ISBN 978-3-03928-993-6 (Pbk)
ISBN 978-3-03928-994-3 (PDF)

© 2020 by the authors. Articles in this book are Open Access and distributed under the Creative Commons Attribution (CC BY) license, which allows users to download, copy and build upon published articles, as long as the author and publisher are properly credited, which ensures maximum dissemination and a wider impact of our publications.
The book as a whole is distributed by MDPI under the terms and conditions of the Creative Commons license CC BY-NC-ND.

Contents

About the Special Issue Editors

Antonio Colmenar Santos has been a senior lecturer in the field of Electrical Engineering at the Department of Electrical, Electronic and Control Engineering at the National Distance Education University (UNED) since June 2014. Dr. Colmenar-Santos was an adjunct lecturer at both the Department of Electronic Technology at the University of Alcalá and at the Department of Electric, Electronic and Control Engineering at UNED. He has also worked as a consultant for the INTECNA project (Nicaragua). He has been part of the Spanish section of the International Solar Energy Society (ISES) and of the Association for the Advancement of Computing in Education (AACE), working in a number of projects related to renewable energies and multimedia systems applied to teaching. He was the coordinator of both the virtualisation and telematic Services at ETSII-UNED, and deputy head teacher and the head of the Department of Electrical, Electronics and Control Engineering at UNED. He is the author of more than 60 papers published in respected journals (http://goo.gl/YqvYLk) and has participated in more than 100 national and international conferences.

David Borge Diez has a Ph.D. in Industrial Engineering and an M.Sc. in Industrial Engineering, both from the School of Industrial Engineering at the National Distance Education University (UNED). He is currently a lecturer and researcher at the Department of Electrical, Systems and Control Engineering at the University of León, Spain. He has been involved in many national and international research projects investigating energy efficiency and renewable energies. He has also worked in Spanish and international engineering companies in the field of energy efficiency and renewable energy for over eight years. He has authored more than 40 publications in international peer-reviewed research journals and participated in numerous international conferences.

Enrique Rosales Asensio is an industrial engineer with postgraduate degrees in electrical engineering, business administration, and quality, health, safety and environment management systems. He has been a lecturer at the Department of Electrical, Systems and Control Engineering at the University of León, and a senior researcher at the University of La Laguna, where he has been involved in water desalination project in which the resulting surplus electricity and water would be sold. He has also worked as a plant engineer for a company that focuses on the design, development and manufacture of waste-heat-recovery technology for large reciprocating engines; and as a project manager in a world-leading research centre. Currently he is an associate professor at the Department of Electrical Engineering at the University of Las Palmas de Gran Canaria.

Preface to "Locally Available Energy Sources and Sustainability"

Renewable energy is electricity generated by fuel sources that restore themselves over a short period of time and do not diminish. Although some renewable energy technologies impact n the environment, renewables are considered environmentally preferable to conventional sources and, when replacing fossil fuels, have significant potential to reduce greenhouse gas emissions. This book focuses on the environmental and economic benefits of using renewable energy, which include: (i) generating energy that produces no greenhouse gas emissions from fossil fuels and reduces some types of air pollution, (ii) diversifying energy supply and reducing dependence on imported fuels, and (iii) creating economic development and jobs in manufacturing, installation, and more.

Local governments can dramatically reduce their carbon footprint by purchasing or directly generating electricity from clean renewable sources.

The most common renewable power technologies include: solar (photovoltaic (PV), solar thermal), wind, biogas (e.g., landfill gas, wastewater treatment digester gas), geothermal, biomass, low-impact hydroelectricity, and emerging technologies such as wave and tidal power.

Local governments can lead by example by generating energy on site, purchasing green power, or purchasing renewable energy. Using a combination of renewable energy options can help to meet local government goals, especially in some regions where availability and quality of renewable resources vary. Options for using renewable energy include: generating renewable energy on site, using a system or device at the location where the power is used (e.g., PV panels on a state building, geothermal heat pumps, biomass-fueled combined heat and power), and purchasing renewable energy from an electric utility through a green pricing or green marketing program, where buyers pay a small premium in exchange for electricity generated locally from green power resources.

Antonio Colmenar Santos, David Borge Diez, Enrique Rosales Asensio
Special Issue Editors

Article

The Practice and Potential of Renewable Energy Localisation: Results from a UK Field Trial

Peter Boait [1,*], J. Richard Snape [1], Robin Morris [2], Jo Hamilton [3] and Sarah Darby [4]

[1] Institute of Energy and Sustainable Development, De Montfort University, Queens Building, The Gateway, Leicester LE1 9BH, UK; jsnape@dmu.ac.uk

[2] Energy Local (Development) Ltd., Crickhowell Resource and Information Centre, Beaufort Street, Crickhowell, Powys NP8 1BN, UK; robin@energylocal.co.uk

[3] Department of Geography and Environmental Science, School of Archaeology, Geography and Environmental Science, The University of Reading, Whiteknights, P.O. Box 227, Reading RG6 6AB, UK; e.j.hamilton@pgr.reading.ac.uk

[4] Environmental Change Institute, School of Geography and the Environment, University of Oxford, South Parks Road, Oxford OX1 3QY, UK; sarah.darby@ouce.ox.ac.uk

* Correspondence: p.boait@dmu.ac.uk; Tel.: +44-774-064-4211

Received: 13 December 2018; Accepted: 1 January 2019; Published: 4 January 2019

Abstract: The adaptation of electricity demand to match the non-despatchable nature of renewable generation is one of the key challenges of the energy transition. We describe a UK field trial in 48 homes of an approach to this problem aimed at directly matching local supply and demand. This combined a community-based business model with social engagement and demand response technology employing both thermal and electrical energy storage. A proportion of these homes (14) were equipped with rooftop photovoltaics (PV) amounting to a total of 45 kWp; the business model enabled the remaining 34 homes to consume the electricity exported from the PV-equipped dwellings at a favourably low tariff in the context of a time-of-day tariff scheme. We report on the useful financial return achieved by all participants, their overall experience of the trial, and the proportion of local generation consumed locally. The energy storage devices were controlled, with user oversight, to respond automatically to signals indicating the availability of low cost electricity either from the photovoltaics or the time of day grid tariff. A substantial response was observed in the resulting demand profile from these controls, less so from demand scheduling methods which required regular user configuration. Finally results are reported from a follow-up fully commercial implementation of the concept showing the viability of the business model. We conclude that the sustainability of the transition to renewable energy can be strengthened with a community-oriented approach as demonstrated in the trial that supports users through technological change and improves return on investment by matching local generation and consumption.

Keywords: community energy; energy storage; time of use tariff; home battery; demand response; renewable energy; business model

---

1. Introduction

Community energy initiatives are widely recognised as a valid and useful response to the sustainability challenges of climate change, energy security, and energy affordability. The UK government published a Community Energy Strategy in 2014 [1], updated 2015 [2], aimed at encouraging both supply and demand side projects. Municipal and co-operative ownership models providing renewable generation capacity are playing a major role in Germany's "Energiewende" [3], while the USA's 900 rural electricity co-operatives [4], founded as a response to economic depression in the 1930s, are evolving to promote energy efficiency and adopt low carbon generation. Community energy schemes can have many organizational forms based on community of place or interest, but surveys and reviews such as [5–8] identify as typical benefits their ability to engage consumer participation in the systemic changes taking place and a contribution to societal cohesion through shared goals and a fair and transparent allocation of financial costs and returns. Another benefit found by [9] is that community members have a more favourable attitude overall to local renewable energy installations, mitigating the "not in my backyard" attitude that is otherwise common. However these studies also indicate that, at least in the UK, schemes are often financially fragile and depend on enthusiastic volunteers, so need policy and regulatory support to flourish.

An opportunity to strengthen the economic basis for community energy arises from the rising demand for electricity expected as electrification of transport and heating takes place. This prospect was reinforced in the UK by the government's publication of a policy to ban the sale of most petrol and diesel vehicles by 2040 [10], a goal also set by the government of France [11]. The UK's electricity system operator, National Grid, predicted in their 2017 Future Energy Scenarios report [12] that with consumer engagement in end-use energy efficiency and demand response for system efficiency, peak demand by 2050 is limited to 74 GW (from a baseline of 62 GW), but rises to 85 GW without it. So there should be scope for rewarding communities for their contribution to a more efficient solution, which earlier studies such as [13] have valued as worth up to £30 Bn in deferred or avoided network reinforcement costs. Reflecting this potential, the 2018 Future Energy Scenarios report from National Grid [14] includes "Community Renewables" as one of the scenarios that can deliver UK commitments to the Paris Agreement.

The community contribution can be seen as the product of service expectations, activities and technologies within a given community at a given time: a 'demand response space' [15]. There may well be greater opportunities for developing demand response at community scale rather than focusing on individual customers. These opportunities could arise from norms and practices developed through social learning about new technologies and processes [16], from trust-building [17]; and from the diversity of activities, skills and technologies to be found within communities [18].

In this paper we report the results from trialing a combination of business model and "smart home" technology designed for communities of place whose common factor is that they reside on the same segment of the local electricity distribution network—e.g., they might share the same low voltage (LV) network. An assumption of the model is that there is some distributed low carbon electricity generation on the shared LV network. This can take any of the common forms such as solar photovoltaics (PV), wind generation, combined heat and power (CHP) or micro hydro. The technology and the incentives from the business model are then framed to work synergistically with community activities to empower participants to reduce their cost of electricity through three mechanisms in order of priority:

- adapting demand to make use of the local generation wherever possible;
- avoiding use of non-local electricity at high cost times such as early evening;
- reducing overall consumption of electricity.

A benefit of the proposed business model is that it overcomes a legal constraint on financing local generation by forming a community co-operative. UK financial regulation requires that the investing members of a co-operative must either be workers in, or consumers of, the commercial product of the enterprise. This is to avoid the regulatory concessions available to co-operatives being exploited by purely speculative investment offers. Where the whole output of a generator is sold to an electricity supplier through a power purchase agreement, under a recent UK regulatory clarification [19] the investors in the generator cannot be considered consumers. But as this model allows consumers to purchase locally-generated electricity they can form a co-operative to fund the generator, which has advantages over other forms of legal entity such as a Community Interest Company (CIC) in allowing more flexibility in the use of profits.

There have been many trials of time-of-day tariffs and use of technology to influence patterns of residential electricity consumption, as for example summarised in [20,21]. More recently the potential of energy storage to contribute to demand flexibility has been recognized [22]. Drawing on lessons from that experience, the present trial incorporated a comprehensive combination of features and demand response measures not previously tested in the UK. These were:

- a time-of-use tariff with a static baseline and a day-ahead dynamic adjustment reflecting the predicted availability of locally-generated electricity from PV panels owned by some participants;
- a web-based display of the current tariff and consumption on user's smart phones, tablets, and computers;
- technology to automatically schedule loads at an optimum time with respect to the tariff while prioritizing user needs and preferences;
- exploitation of domestic energy storage in batteries and thermal storage heaters;
- regular feedback on the financial savings achieved by individual users and the participant group as a whole;
- a sustained program of engagement aimed at retaining user interest and obtaining their feedback.

The goals of the project (called CEGADS) undertaking this trial were to demonstrate the viability of the business model, test the acceptability of this level of innovation to consumers, and evaluate the amount of demand-side response to the measures deployed. The remainder of the paper is structured as follows. In Section 2 we describe the participant community, the business model, and the technology employed. Section 3 provides the results obtained, in respect of the use of local generation, financial outcome, demand side response, and the experience of participants. Section 4 describes briefly an agent-based modelling study of the scheme implemented, and a follow-up project implementing the business model now in fully commercial operation. The overall implications of the findings are discussed in Section 5 followed by conclusions.

2. The CEGADS Trial

2.1. The Participants and Business Model

The project name CEGADS stands for Community Electricity Generation, Aggregation, and Demand Shaping, indicating the key features involved. The project is also known by the acronym SWELL, referring to the Energy Local model in the cluster of Oxfordshire villages Shrivenham, Watchfield, and Longcot, in which the trial took place and the 48 participating households were recruited. Many of the residents had already subscribed to the local Westmill energy co-operatives [23,24] operating substantial wind and solar farms, but these generators were subject to wholesale power purchase agreements (entered into prior to the regulatory clarification at [19] summarised above), and being connected at 33 kV were not accessible to the present scheme. However the charitable trust associated with these co-operatives was thereby able to facilitate recruitment for this project from an informed community. Metering of electricity consumption and generation at one-minute intervals was installed in each household, along with the display and control technology. This equipment was installed at no cost to users, as were the batteries described later.

The generation for CEGADS was provided by roof-mounted PV panels already owned by 14 participants with a total capacity of 45 kWp. The export electricity from these panels (i.e., the generated electricity not consumed within the household) was metered and aggregated to form a resource which was considered available for supply at a favourable tariff (£0.065/kWh) to the remaining non-generating participants. The allocation of this export energy aggregate A in each half-hour was computed by finding iteratively a "fill level" L such that for each of n consumers with demand e_i in the half hour greater than L, L kWh would be considered supplied from A, and for those remaining m consumers with demand e_j less than L, their demand would be fully met from A, with L also satisfying:

$$A = nL + \sum_{j=1}^{j=m} e_j \tag{1}$$

where A was large enough to more than supply all non-generating consumers then the residue was considered community export. This method of fair allocation of local generation is a key feature of the business model, which also allows the community export to be sold to an electricity supplier under a power purchase agreement. To enable participants in the trial to retain their existing electricity supplier and tariffs, the time-of-use tariff applied to electricity not generated locally was implemented as an incentive scheme where the difference between the actual cost of electricity to participants and the cost they would have incurred under the trial tariff is given to them in the form of credit vouchers exchangeable for goods at a supermarket chain associated with the electricity supplier supporting the project. The commercially-realistic (for 2016) time of use tariff rates offered are shown in Figure 1. Generators were credited with £0.065 for each export kWh matched with consumption, and £0.055 for each kWh not matched so considered as taken up by a power purchase agreement (PPA).

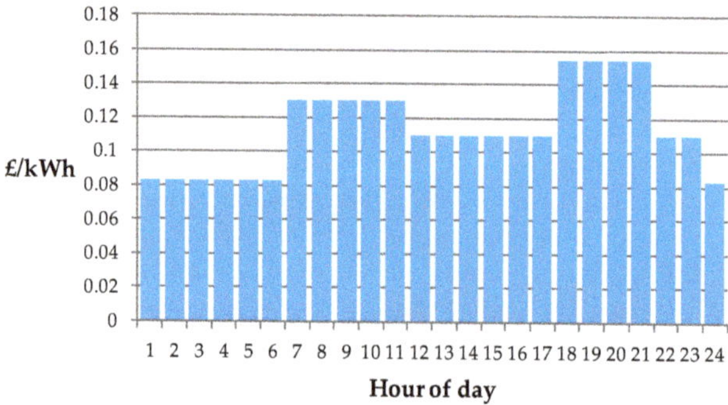

Figure 1. Time-of-use electricity tariff rates.

2.2. The Metering and Demand Response System

To execute the metering required for this scheme and enable the participants to make best use of the local generation and time-of-use tariffs a smart metering and control unit was installed in each household. Branded "Hestia", this unit provided from an internal web server a display of the tariff rates on any convenient IT device connected to the household broadband, but modified with a dip in the displayed rates during the middle of the day that reflected approximately the amount of local PV generation predicted to be available based on the overnight local weather forecast. It also provided displays of electricity consumption and generation over the last 24 h for the household, the participant community as a whole, the aggregate PV generation, and the PV generation for the household for those so equipped. To provide the data for these displays, metering data at one minute intervals was collected and processed in a central database using a commercial cloud service. A simplified view of the system is shown in Figure 2.

The Hestia control unit also performed automatic demand response for controllable appliances as illustrated in Figure 2. Six of the participating dwellings had space heating provided by electrically-heated thermal storage heaters and hot water from an immersion-heated tank. In aggregate these appliances provided about 60 kWh of thermal storage in each of the six homes. Charging of these useful thermal energy stores was controlled such that user comfort requirements as expressed on the Hestia user interface were prioritized, but was otherwise optimized against a tariff-dependent signal from the database server that ensured cost effective use of local generation and the time-of-day tariff while preventing peaks in aggregate demand at tariff boundaries by randomizing dispatch of loads.

This signaling and optimization methodology is fully described in [25,26] and the peaking risk that is mitigated, which has been identified in many simulation studies, is identified for example in [27,28].

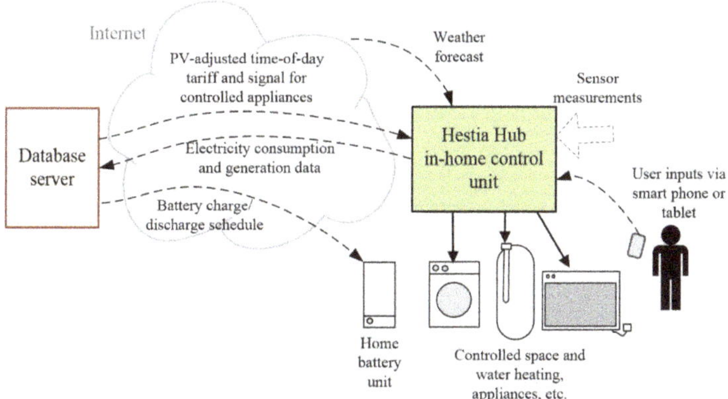

Figure 2. CEGADS system diagram.

An example page from the Hestia Hub display is shown in Figure 3a. All of the participants were given a smart plug as shown in Figure 3b for which the on/off status could be radio-controlled via a user interface provided by the Hestia unit. This allowed users to set a time window within which an appliance powered via the smart plug should operate, and the required operating duration. If some scheduling flexibility was available from the difference between the time window and the operating duration, the Hestia selected an optimized dispatch time using the demand response signal.

Nine of the participant households were equipped with a 2 kWh home battery unit (Figure 3c). The nine were deliberately chosen to exclude households with PV panels or thermal storage. These lithium-ion batteries were controlled to charge during low tariff periods and discharge during the early evening high tariff rate period, with the objective of improving the benefit these households obtained from the tariff scheme.

To summarise, the trial involved expanding the demand response potential in three villages by extending people's ideas of what community energy could do for them, by developing new activity in relation to the cooperative business model and tariff, and introducing technology in the shape of the Hestia control units, database server, smart plugs, batteries and display capability. In doing this, it was building on trust and knowledge that had been established through everyday social interactions and (for some) involvement in a local energy cooperative.

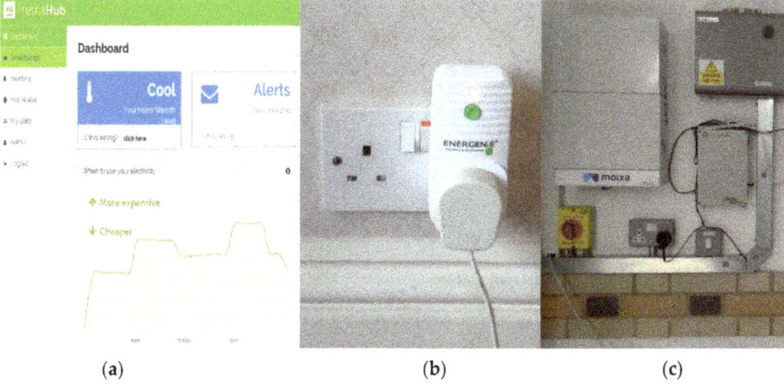

(**a**)　　　　(**b**)　　　　(**c**)

Figure 3. (**a**) Hestia Hub display. (**b**) Smart plug. (**c**) Home battery and heating control units.

3. Results from the Trial

3.1. Utilisation of Local Generation

Over the year of trial operation, out of the total PV generation of 43,406 kWh from the 14 generators, 18,307 kWh were used within the generating households, and 25,908 kWh were available to share with other participants. Of this available total, 22,154 kWh were matched with consumption using the algorithm described earlier, and the balance of 2944 kWh was allocated to the PPA. Figure 4 illustrates this outcome on a monthly basis. The generation tariffs gave an improved financial return simulated through the credit vouchers of about 80% to generators (£719 in addition to £868 from 50% deemed export feed-in tariff at £0.040/kWh making a total of £1587). The generation matched with consumption represented about 9.5% of the total electricity consumed (c. 233 MWh) by all the participants during the year.

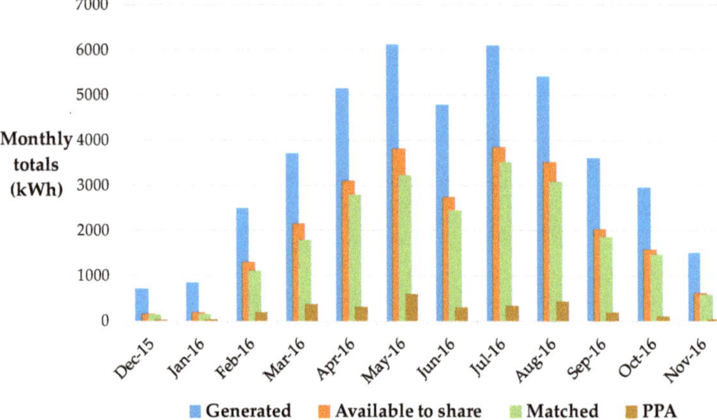

Figure 4. Allocation of PV generation by month.

The savings with respect to their existing tariffs that accrued to all participants are shown in Figure 5. The "tariff" plot shows the savings from the baseline time-of-day tariff. "Local use" shows the savings to participants who consumed the matched generation shown in Figure 4. "PV shared" shows the additional return to the PV generators.

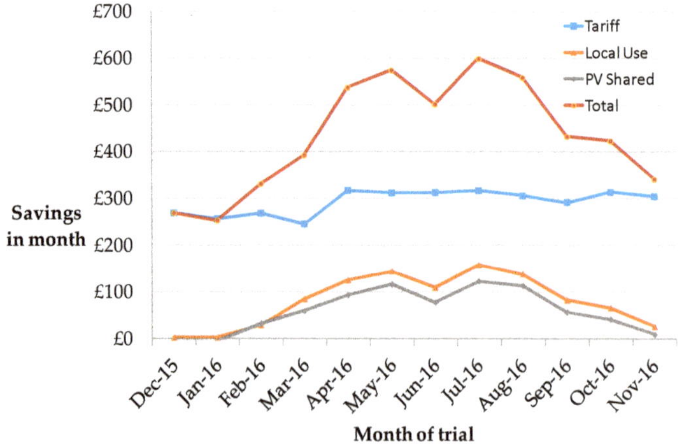

Figure 5. Financial savings to participants from the project.

3.2. Demand Side Response

The overall demand side response is illustrated in Figure 6a–d by comparing aggregate demand profiles at the start and end of the trial. The response derived from four sources:

- automatic "smart" control of water and space heating in six electrically-heated homes;
- time-shifting of supply via the home batteries installed in nine homes;
- semi-automatic time-shifting of demand using the smart plugs in all homes;
- decisions made by any of the participants to manually control any appliance taking account of the incentives offered.

The comparison between average consumption profiles in Figure 6 is influenced by the fact that December 2016 was much colder (295 degree-days) than December 2015 (154 degree-days). The six electrically-heated homes presented a special case in that they were already using a time-dependent tariff known as Economy 7. This comprises a low rate for 7 h overnight of about £0.07/kWh and a higher day rate of about £0.016 typically used, as in the present case, with thermal storage heating and domestic hot water tanks that can be charged at the low rate. So the automatic controls were given a signal which moved some of the heating demand into the middle of the day to take advantage of the local generation and lower mid-day tariff. The controls also ensured a more precise matching of stored thermal energy to the weather-dependent heating demand.

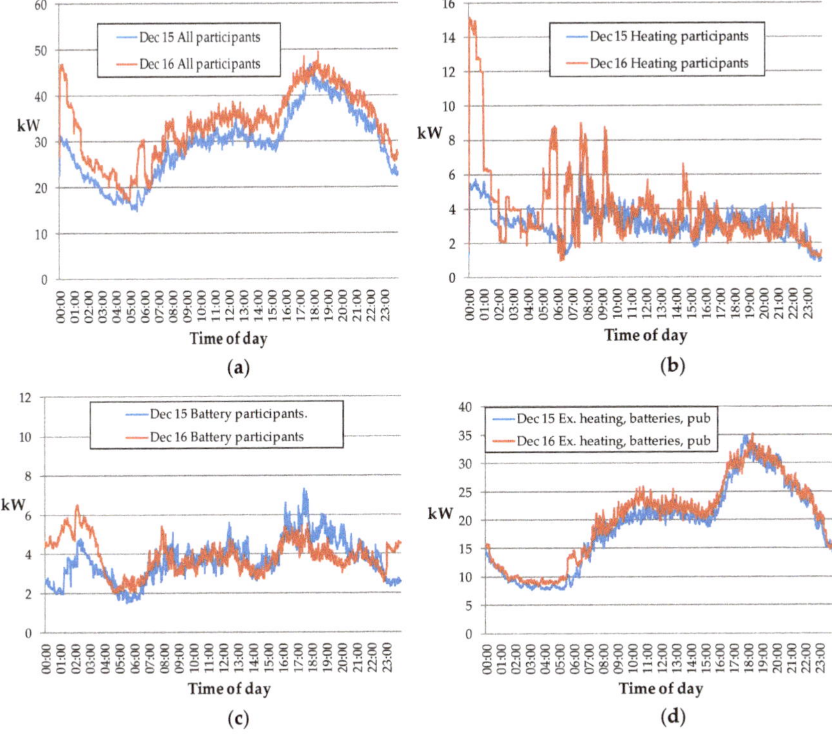

Figure 6. Comparison of average daily aggregate electricity use profile in December 2015 with December 2016, for different participant groups: (**a**) All participants (**b**) Participants with electric heating (**c**) Participants with batteries (**d**) All participants excluding those with batteries, electric heating and the pub.

The resulting shift in distribution of demand is illustrated in Figure 6b which shows the increased heating demand during the day and also a reduction of 16% in consumption during the peak tariff hours despite the colder weather in 2016. A comprehensive report focused on the performance of the heating controls is provided in [25].

The operation of eight batteries (one had to be removed before December 16) can be seen in Figure 6c with charging commencing at the start of the low tariff rate at 23:00 and continuing overnight. The reduction in demand by 20% during the evening high tariff period is also evident. It was found desirable to configure the batteries with a maximum discharge rate of 0.25 kW to ensure that the battery output always offset local consumption and was not exported, for which no reward could be offered. One of the participants with a battery also acquired an electric car during the trial year which contributed to the increased overnight demand seen in Figure 6c.

The relatively limited aggregate impact of the smart plugs and manual time-shifting can be seen in Figure 6d, which excludes the participants shown in Figure 6b,c and also the village pub (i.e., bar) which had a significant increase in power consumption during the year for commercial reasons unrelated to this trial. The participants in 6d had gas central heating so consumption was not greatly affected by the weather difference in the comparison. The increased overnight and mid-day consumption can be seen and also a slight reduction during the peak tariff time. The peak before 06:00 is believed to be wet appliance operation scheduled at the end of the low tariff period.

To examine the range of individual household responses, the average demand in each of the six tariff periods shown in Figure 1 was calculated for each household for October–December 2015 and for the same period in 2016. The correlation between changes in demand in each tariff period, and the tariff rate was then tested, with the hypothesis that demand would have changed over the year in inverse proportion to the tariff as consumers became accustomed to a time-of-day tariff and adjusted their demand accordingly. The results for different participant groups are shown in Table 1. A participant was counted as a "responder" in the table if a negative correlation was observed between change in demand and tariff rate with an R^2 value greater than 0.1. The much greater proportion of responders among participants with some additional technology that reinforces their engagement is evident. Note all the groups in Table 1 are independent i.e., there is no overlap of membership.

Table 1. Correlation of change in demand over a year with time-of-day tariff.

Group Attribute	Number in Group	Responders	R^2 Range for Responders	Responders as % of Group
Controlled electric heating	6	5	0.15–0.26	83%
PV generator	14	11	0.13–0.7	79%
Battery storage	8	7	0.23–0.82	88%
All other participants	19	4	0.16–0.67	21%

3.3. User Experience

Most participants in the CEGADS project were interviewed in three rounds of surveys; (by telephone, and online using Survey Monkey) around the start in late 2015 early 2016, during summer 2016, and spring 2017. They began with largely positive attitudes, with responses to questions concerning demand response and time of use tariffs as shown in Figure 7a,b. In round 1 of the interviews, most participants were positive about being able to switch the time of use of some of their electricity although the extent to which they could was determined by patterns of household occupation (i.e., if they were in the home during the daytime). Most participants felt able to switch devices such as white goods such as washing machines, tumble driers, and dishwashers and battery charging (e.g., for computers) and where time and space flexibility permitted (e.g., either being able to use them, or program them to run, in off peak hours).

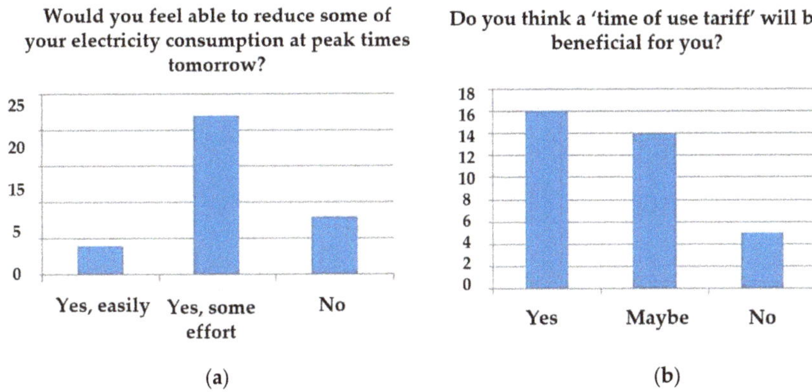

Figure 7. Initial perceptions of (**a**) demand response possibility and (**b**) a time-of-use tariff.

By the second round of interviews practices relating to the Hestia Hub technology provided were evident, with a range of usage levels as shown in Figure 8, and comments such as *"when I first started out I was looking at it daily or more frequently, but now … I'm more familiar with it"*. Overall, 13 of the 37 (35%) respondents reported looking at their Hestia Hub at least once a week. This is roughly consistent with findings relating to in-home display usage by smart-metered customers in Great Britain as a whole, where 44% of householders reported that they were consulting their display at least once a week, between seven and 29 months after installation [20].

However, on the broad question as to whether the project had influenced day to day habits of electricity use, 31 (out of 39) said *"yes"*. Of those who responded *"yes"*, many reported shifting of activities, such as using the washing machine and dish washer at different times, or changing cooking practices, for example: *"sometimes opting to microwave or grill instead of using oven"*. Of the 6 who responded no, some were already producing electricity through solar PV, some considered they had already made changes in their consumption, and some had family routines which they didn't want to shift.

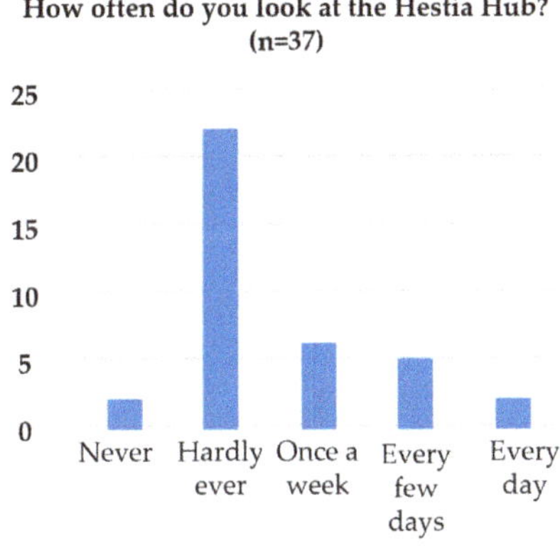

Figure 8. Engagement with smart metering and control device after six months.

The final survey revealed many practical changes in household practices that participants had taken, such as *"Weather—I was advised if I could use my washing machine when the sun was at its highest, that was the best time to do it"*; and *"Put the electric towel rail on a timer plug socket. It only operates sporadically in the peaks"*. The motivation for demand response decisions was drawn from a wide range of factors as indicated in Figure 9, showing that the different channels used by the project to communicate with participants all had a role in the results obtained. The variety of integrated approaches enabled participants to learn and incorporate their learning into new routines in different ways, through different means, and at different times, for example: *"when you have the reports that came through and here you have the cheque for your money and your report on your energy use, you could practically see how it all fitted together"*. Of 21 participants who said they had shifted usage during the trial, 17 said they would continue to do so, while 16 respondents who had reduced the amount of electricity usage during the trial were planning to maintain their reduction.

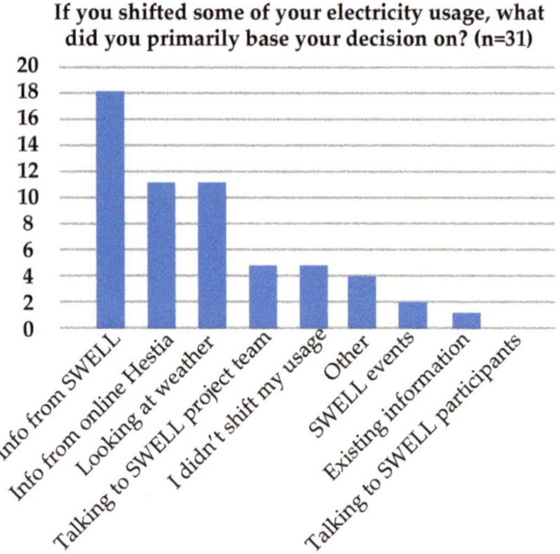

Figure 9. Decision basis for demand response actions.

4. Follow-On System Modelling and Commercial Implementation

4.1. Agent-Based Modelling

To investigate the wider applicability of this business and technical approach to localisation of energy use, an agent-based model (ABM) was constructed embodying both the technical features and the economic and social incentives of the real-life CEGADS trial. The model used the CASCADE framework [29] in which each agent is a household for which their energy-using appliances, energy flows, costs, and decisions are simulated taking account of their physical and social environment. The environmental factors included weather, tariffs, and the feedback and encouragement provided by participation in the project. The governing attributes of each agent (such as the number of household occupants and size of dwelling) were given values selected randomly from an appropriate range and distribution. Each run of the model (typically simulating a year's operation) therefore gave different outcomes, and as is conventional for an ABM, interpretation of results is based on the range of outcomes.

The initial correspondence between empirical and modelled results indicated that the model was useful. A series of tests were then undertaken to test how robust the positive results of the empirical trial were to changes in the scenario. Firstly, different weather files were used to investigate how dependent the overall results and benefits to individuals were on weather. This showed that the important characteristic of no participant being expected to lose out financially was preserved in differing weather conditions. Next, several communities were programmed to co-exist in the model and the smart signals sent to consumers in each model were examined after the model had evolved. It was noted that the signals exhibited similar characteristics, but that they evolved in a way that was specific to the community—indicating that the model was transferrable, but that the specific evolution in response to the demographic and technology mix within the community would differ, resulting in different levels and patterns of demand response.

4.2. Commercial Implementation

The success of the CEGADS trial has led to a first fully commercial implementation of the concept for a community in the small town of Bethesda, North Wales [30]. This is based around a 100 kW micro hydro generator. 100 consumers have been recruited, who pay £0.07/kWh for matched use of local generation, and a small charge for membership of the co-operative club. Any electricity not supplied by the hydro is charged according to a time-of-day tariff similar to that in Figure 1. Because the output of this generator varies seasonally, power availability is signalled to users. Figure 10 shows the proportion of each user's power matched to low cost local generation, with an average of 65% over the year of operation. The overall shape of the graph reflects the availability of hydro generation, including the impact in May 2017 of a period of dry weather.

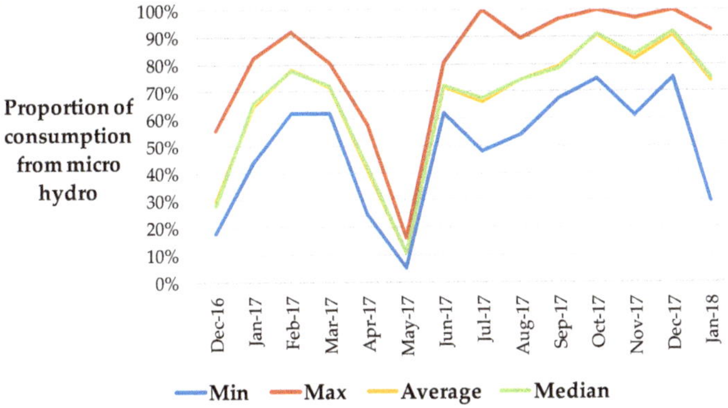

Figure 10. User demand matched to local micro-hydro.

The social enterprise promoting this model, Energy Local, is now developing a "starter pack" of processes and documentation [31] with follow-up support, allowing social enterprise clubs to be formed and the model implemented wherever appropriate generation, network configuration and community enthusiasm exist. New micro-hydro clubs are in progress elsewhere in Wales, and PV-based clubs in London and Gloucester.

5. Discussion

A limitation of this study is that there is no unambiguous counterfactual against which the impact on electricity use of the full range of interventions can be assessed. The process of recruiting participants and installing equipment in their homes inevitably involved engaging them with the objectives of the study. So the comparison of electricity demand measurements from the end of the study with those at the start cannot fully demonstrate the behavioural shifts that may have occurred, but they do reliably show the effect of the heating control and battery technologies. With further development, the ABM modelling techniques tested as part of this study could provide policymakers with additional predictive insight on the impact of wide-scale adoption of this kind of energy localisation scheme.

The increased demand response from those participants who had a significant supporting technology is very evident from Table 1. The smart plugs that were given to all participants clearly did not have the same impact as the other technologies. The increased response from participants with controlled electric heating and PV is notable since both started the trial with an incentive to attend to the timing of electricity demand, in the heating case through their longstanding use of the Economy 7 tariff, and in the PV case to make use of their own generation. So it must be concluded that their initial sensitivity to timing was further reinforced through the various inputs categorized in Figure 7, and for the heating users by the control signal alignment with the tariff. For the battery-equipped participants, the response seen fully reflects the automatic operation of the technology because the batteries were installed in January 2016. For the remaining participants it is evident from the data that only a few enthusiasts were able to sustain an increasing level of demand response that could be detected in the start-to-finish comparison. This does not of course preclude that many of this group may have taken demand response actions on an intermittent basis as suggested by the interviews. These results are very consistent with previous studies such as the review of 21 trials by Frontier Economics [32] which has a key finding "Interventions to automate responses deliver the greatest and most sustained household shifts in demand where consumers have certain flexible loads".

The financial benefit of the business model is clear from Figure 5, amounting to an average of £109 to each participant over a year, on average consumption of 4854 kWh that would cost £688 at the benchmark fixed rate used as a comparator of £0.135/kWh and standing charge of £0.09/day. These substantial savings illustrate the value that will be made accessible by the UK national smart meter rollout as long as regulation, and suitable smart meter data processing capability, facilitates this form of community energy scheme and tariff structure. The way in which the benefit of the PV generation is spread across the community makes this model particularly attractive for social housing, where otherwise the restriction of local generation to a subset of dwellings with favourable roof orientation can lead to perception of unfairness [33].

The recent UK regulatory review [22] has identified the issue that consumers who benefit from local generation, as in the present scheme, pay less towards the distribution and transmission infrastructure through per-kWh tariffs, but the burden they place on that infrastructure is determined by their peak network demand. This is ultimately likely to lead to tariff structures that are less dependent on the volume of electricity consumed and have some dependency on peak demand, such as those proposed by Simshauser [34] and Nijhuis et al. [35]. The battery and smart heating control technologies demonstrated in this project directly address this issue by substantially reducing peak demand.

This trial also provides a more positive perspective on the economics of home batteries than other trials such as Uddin et al. [36] who tested exactly the same product as used in the present project and concluded there was no saving from increased self-consumption of PV generation and a substantial loss to the user from battery depreciation. It is worth noting that the trial reported by Uddin et al. was purely techno-economic, with empirical testing in a single household as the basis for economic modelling: there was no community or social dimension. In the case of the CEGADS/SWELL trial, the battery users made savings at an average rate of about £23 per annum from tariff arbitrage, by avoiding 1 kWh per day in the evening peak rate period and drawing the corresponding charge overnight. Clearly this alone would not justify the investment, or cover the depreciation, but in combination with other grid services such as Short Term Operating Reserve, for which the UK system operator National Grid currently pays about 12 p/kWh [37] adding c. £10 per annum to the return, a path to viability, as battery costs fall, can be seen. Another role for batteries relevant to community energy is to facilitate connection of generation capacity. In [38] Idlbi et al. elaborate a case study in which 0.5 MWh of lithium battery capacity enables connection of 3 MWp PV generation to a network that would otherwise require reinforcement at much greater cost to maintain voltage compliance. However, batteries are not essential to the value of the Energy Local model, as shown by the results from the Bethesda follow-up project, which did not deploy them.

6. Conclusions

This trial, combined with the subsequent commercial implementation of the business model, has demonstrated that valuable technical, economic, and social outcomes can be achieved by localisation of electricity generation and consumption within a community-of-place-based organisational framework. The generally positive user experience described in Section 3 reflects the support given to participants by the project team allowing them to understand and engage with the novel technology and tariffs. The business model delivered useful financial savings for consumers which have been shown to be repeatable elsewhere, while the technology successfully demonstrated the use of both electrical and thermal energy storage to reshape the daily profile of electricity demand, in response to technical and financial signals, such that peak demand is reduced and local consumption of local generation is increased. This demand response was stronger for the automated mechanisms managing energy storage than from the simpler devices providing appliance scheduling which required repeated user configuration. These results show that the sustainability of the transition to renewable energy can be strengthened with a community-oriented approach that supports users through technological change and improves the return on investment by localising generation and consumption. The enterprises and institutions taking part in this trial are now seeking to build on this experience by evolving the business model and technology so that they can be widely deployed, and are also motivating adjustments to the regulatory environment that facilitate local initiatives and consumer engagement.

Author Contributions: Formal analysis, P.B., J.R.S., R.M. and J.H.; Investigation, P.B., J.R.S., R.M., J.H. and S.D.; Writing—original draft, P.B., J.R.S., R.M., J.H. and S.D.

Funding: As detailed under Acknowledgements and Conflicts of Interest below.

Acknowledgments: The authors would like to thank the Engineering and Physical Sciences Research Council (EPSRC) and Innovate UK for providing the financial support for this study as part of the CEGADS project (EP/M507209/1 and EP/M507210/1), including funds for covering the costs to publish in open access. The project implementation was also part funded by Energy Local (Development) Ltd., Exergy Devices Ltd., Moixa Technology Ltd., Westmill Sustainable Energy Trust, and Co-Operative Energy who provided the supermarket vouchers to participants reflecting their benefit from the simulated tariffs.

Conflicts of Interest: Two authors (Boait and Morris) have roles in the enterprises acknowledged above that provided match funding and technology for the project. None of the other authors has a conflict of interest which could inappropriately influence this work. The results reported were generated and reviewed by all the authors. This work (i.e., the data analysis and drafting of the manuscript) was funded by the UK Engineering and Physical Sciences Research Council and Innovate UK. They had no role in the collection of data or in its analysis and interpretation in the paper and none in the decision to submit to this journal.

References

1. Department of Energy and Climate Change, Community Energy Strategy. 2014. Available online: https://www.gov.uk/government/publications/community-energy-strategy (accessed on 29 December 2018).
2. Department of Energy and Climate Change, Community Energy Strategy Update. 2015. Available online: https://www.gov.uk/government/publications/community-energy-strategy-update (accessed on 29 December 2018).
3. Romero-Rubio, C.; Diaz, J.R. Sustainable energy communities: A study contrasting Spain and Germany. *Energy Policy* **2015**, *85*, 397–409. [CrossRef]
4. America's Electric Co-operatives. Available online: https://www.electric.coop/ (accessed on 29 December 2018).
5. Walker, G.; Devine-Wright, P. Community renewable energy: What should it mean? *Energy Policy* **2008**, *36*, 497–500. [CrossRef]
6. Seyfang, G.; Hielscher, S.; Hargreaves, T.; Martiskainen, M.; Smith, A. A grassroots sustainable energy niche? Reflections on community energy in the UK. *Environ. Innov. Soc. Trans.* **2014**, *13*, 21–44. [CrossRef]
7. Seyfang, G.; Park, J.J.; Smith, A. A thousand flowers blooming? An examination of community energy in the UK. *Energy Policy* **2013**, *61*, 977–989. [CrossRef]
8. Klein, S.J.W.; Coffey, S. Building a sustainable energy future, one community at a time. *Renew. Sustain. Energy Rev.* **2016**, *60*, 867–880. [CrossRef]
9. Bauwens, T.; Devine-Wright, P. Positive energies? An empirical study of community energy participation and attitudes to renewable energy. *Energy Policy* **2018**, *118*, 612–625. [CrossRef]
10. Department for Environment, Food, and Rural Affairs. UK Plan for Tackling Roadside Nitrogen Dioxide Concentrations. 2017. Available online: https://www.gov.uk/government/publications/air-quality-plan-for-nitrogen-dioxide-no2-in-uk-2017 (accessed on 29 December 2018).

11. Ministére de la Transition Ecologique et Solidaire. Plan Climat, 1 Planéte, 1 Plan. 2017. Available online: https://www.ecologique-solidaire.gouv.fr/sites/default/files/2017.07.06%20-%20Plan%20Climat_0.pdf (accessed on 29 December 2018).

12. National Grid. Future Energy Scenarios July 2017. Available online: http://fes.nationalgrid.com/media/1253/final-fes-2017-updated-interactive-pdf-44-amended.pdf (accessed on 29 December 2018).

13. Pudjianto, D.; Djapic, P.; Aunedi, M.; Gan, C.K.; Strbac, G.; Huang, S.; Infield, D. Smart control for minimizing distribution network reinforcement cost due to electrification. *Energy Policy* **2013**, *52*, 76–84. [CrossRef]

14. National Grid. Future Energy Scenarios. 2018. Available online: http://fes.nationalgrid.com/fes-document (accessed on 29 December 2018).

15. McKenna, E.; Higginson, S.; Grunewald, P.; Darby, S.J. Simulating residential demand response: Improving socio-technical assumptions in activity-based models of energy demand. *Energy Effic.* **2017**, *7*, 1583–1597. [CrossRef]

16. Ornetzeder, M.; Rohracher, H. User-led innovations and participation processes: Lessons from sustainable energy technologies. *Energy Policy* **2006**, *34*, 138–150. [CrossRef]

17. Kalkbrenner, B.J.; Roosen, J. Citizens' willingness to participate in local renewable energy projects: The role of community and trust in Germany. *Energy Res. Soc. Sci.* **2016**, *13*, 60–70. [CrossRef]

18. Burchell, K.; Rettie, R.; Roberts, T.C. What is energy know-how, and how can it be shared and acquired? In Proceedings of the ECEEE Summer Study: First Fuel Now, Toulon, France, 1–6 June 2015; pp. 1979–1988.

19. Financial Conduct Authority. Guidance on the FCA's Registration Function under the Co-Operative and Community Benefit Societies Act 2014. 2015. Available online: https://www.fca.org.uk/publication/finalised-guidance/fg15-12.pdf (accessed on 29 December 2018).

20. Ipsos MORI. *Smart Metering Early Learning Project: Consumer Survey and Qualitative Research*; Report Prepared by IpsosMORI; Department of Energy and Climate Change: London, UK, 2015. Available online: https://assets.publishing.service.gov.uk/government/uploads/system/uploads/attachment_data/file/407543/3_Smart_Metering_Early_Learning_Project_-_Consumer_survey_and_qual_research_-_Main_report_FINAL_CORRECTED.pdf (accessed on 29 December 2018).

21. Faruqui, A.; Sergici, S. Household response to dynamic pricing of electricity: A survey of 15 experiments. *J. Regul. Econ.* **2010**, *38*, 193–225. [CrossRef]

22. Ofgem. Upgrading our Energy System—Smart Systems and Flexibility Plan July 2017. Available online: https://www.gov.uk/government/uploads/system/uploads/attachment_data/file/633442/upgrading-our-energy-system-july-2017.pdf (accessed on 29 December 2018).

23. Westmill Wind Farm Co-operative. Available online: http://www.westmill.coop/ (accessed on 29 December 2018).

24. Westmill Solar Co-operative. Available online: http://westmillsolar.coop/ (accessed on 29 December 2018).

25. Boait, P.; Snape, J.R.; Darby, S.J.; Hamilton, J.; Morris, R. Making legacy thermal storage heating fit for the smart grid. *Energy Build.* **2017**, *138*, 630–640. [CrossRef]

26. Boait, P.; Ardestani, B.; Snape, J.R. Accommodating renewable generation through an aggregator-focussed method for inducing demand side response from electricity consumers. *IET Renew. Power Gener.* **2013**, *7*, 689–699. [CrossRef]

27. Ramchurn, S.D.; Vytelingum, P.; Rogers, A.; Jennings, N.R. Agent-Based Control for Decentralised Demand Side Management in the Smart Grid. In Proceedings of the 10th International Conference on Autonomous Agents and Multiagent Systems (AAMAS 2011), Taipei, Taiwan, 2–6 May 2011; pp. 5–12.

28. Rastegar, M.; Fotuhi-Firuzabad, M.; Aminifar, F. Load commitment in a smart home. *Appl. Energy* **2012**, *96*, 45–54. [CrossRef]

29. Rylatt, M.; Gammon, R.; Boait, P.; Varga, L.; Allen, P.; Savill, M.; Snape, J.R.; Lemon, M.; Ardestani, B.; Pakka, V. Cascade: An agent based framework for modeling the dynamics of smart electricity systems. *Emerg. Complex. Organ.* **2013**, *15*, 1–13.

30. BBC News. Bethesda Energy Club Shares Hydro Power in UK First. 2016. Available online: http://www.bbc.co.uk/news/uk-wales-38236414 (accessed on 29 December 2018).

31. Energy Local. Making Energy Work for You. 2018. Available online: http://www.energylocal.co.uk/ (accessed on 29 December 2018).

32. Frontier Economics and Sustainability First. Demand Side Response in the Domestic Sector—A Literature Review of Major Trials. 2012. Available online: https://assets.publishing.service.gov.uk/government/uploads/system/uploads/attachment_data/file/48552/5756-demand-side-response-in-the-domestic-sector-a-lit.pdf (accessed on 22 November 2018).

33. McCabe, A.; Pojani, D.; Broese van Groenou, A. Social housing and renewable energy: Community energy in a supporting role. *Energy Res. Soc. Sci.* **2018**, *38*, 110–113. [CrossRef]

34. Simshauser, P. Distribution network prices and solar PV: Resolving rate instability and wealth transfer through demand tariffs. *Energy Econ.* **2016**, *34*, 108–122. [CrossRef]

35. Nijhuis, N.; Gibescu, M.; Cobben, J.F.G. Analysis of reflectivity & predictability of electricity network tariff structures for household consumers. *Energy Policy* **2017**, *109*, 631–641.

36. Uddin, K.; Gough, R.; Radcliffe, J.; Marco, J.; Jennings, P. Techno-economic analysis of the viability of residential photovoltaic systems using lithium-ion batteries for energy storage in the United Kingdom. *Appl. Energy* **2017**, *206*, 12–21. [CrossRef]

37. National Grid Electricity System Operator, 2018, Short Term Operating Reserve (STOR). Available online: https://www.nationalgrideso.com/balancing-services/reserve-services/short-term-operating-reserve-stor (accessed on 29 December 2018).

38. Idlbi, B.; Appen, J.; Kneiske, T.; Braun, M. Cost-benefit analysis of battery storage system for voltage compliance in distribution grids with high distributed generation. *Energy Procedia* **2016**, *99*, 215–228. [CrossRef]

© 2019 by the authors. Licensee MDPI, Basel, Switzerland. This article is an open access article distributed under the terms and conditions of the Creative Commons Attribution (CC BY) license (http://creativecommons.org/licenses/by/4.0/).

 *sustainability*

Article

Comparative Analysis of the Energy and CO_2 Emissions Performance and Technology Gaps in the Agglomerated Cities of China and South Korea

Na Wang * and Yongrok Choi *

Global E-governance Program, Inha University, Incheon 402751, Korea
* Correspondence: 22141792@inha.edu (N.W.); yrchoi@inha.ac.kr (Y.C.)

Received: 4 December 2018; Accepted: 12 January 2019; Published: 17 January 2019

Abstract: This paper presents a comparative analysis of the technology gap, energy efficiency, and CO_2 emission performance of the agglomerated cities in Eastern and Central China and South Korea under economic heterogeneity. The potential reductions of energy and CO_2 emission are estimated from agglomerated city perspectives. The global meta-frontier non-radial direction distance function is used to conduct an empirical analysis of agglomerated cities among Eastern, Central China and South Korea. The results show the potential reduction of 7.58 billion tons of CO_2 emissions in Korea and another potential reduction of 1930.62 toe energy for the research period in China, if Korea and China proactively collaborate with each other. The empirical results conclude several unique findings and their implications. First, there are significant differences between the Chinese and Korean cities, in energy efficiency, CO_2 emission performance, and meta-technology gaps. Korean cities play a leading role at benchmarking efficiency level with meta-frontier technology. Second, there is no significant difference between total-factor and single-factor performance indexes in the Korean cities, because South Korea requires large capital stocks to replace energy in the production process. However, the opposite is true for Eastern and Central China cities. Finally, there is huge potential for the Chinese cities to reduce energy and CO_2 emissions by "catching up" internally as well as by the collaborative efforts with Korean cities.

Keywords: global meta-frontier non-radial direction distance function; energy efficiency; CO_2 emission performance; benchmark; potential CO_2 emission and energy reduction

---

1. Introduction

Since 2010, China is the world's largest energy consumer. According to BP (2018) [1], China's primary energy consumption reached 3132.2 million toe in 2017, accounting for 23.2% of the world's total. Wasteful use of energy resources not only caused energy scarcity, but also led to severe air pollution in China as well as in South Korea. With the increasing concern about climate change mainly arising from CO_2 emissions, the Chinese government has strived to manage its energy consumption and greenhouse gas emissions. Energy efficiency is widely regarded as the most cost-effective way of dealing with energy challenges and environmental deterioration [2]. The Chinese government has taken measures to improve energy efficiency by controlling or slowing down the growth of its energy consumption. In its 11th Five-Year Plan, the Chinese government set out the target for reducing its energy intensity by 20% and decreasing its CO_2 emissions per unit GDP by 60% to 65% of 2005 levels by 2030 [3].

As a neighboring country, South Korea (hereinafter, Korea), with 0.68% of the world's population, consumes 2.8% of the world's oil fuels (ranked 8th), 2.3% of coal (ranked 6th), and 1.3% of natural gas (ranked 18th) in 2017. Korea was ranked the 7th largest CO_2 emitter in the world in 2017. Hence, the Korean government set its ambitious target to reduce its business as usual (BAU) levels to 37%

by 2030 under the 2015 Paris Agreement. The Korean government began to gradually implement a nationwide regulatory system from 2010, starting with the target management system (TMS) to achieve the goal of reducing its BAU levels. TMS is a cap-based regulation mechanism for greenhouse gas emissions. TMS was imposed on 470 Korean companies and institutions in early 2010. TMS includes command-and-control policies that directly regulate greenhouse gas emissions and energy efficiency. TMS-bound entities are divided into seven sectors in the national economic system [4]. Under TMS, each company should maintain its assigned emissions. Every year, each company has to report to the central government whether it met its emissions targets. If the companies fail, they are penalized severely. As shown in Table 1, the Korean government has taken a phased approach starting with the TMS in 2012 to the emissions trading scheme (ETS) in 2015.

Table 1. Comparison of National Targets and policies.

	China	Korea
Policy Paradigm	Green and low-carbon development	Low carbon, green growth
National target	60% to 65% per unit GDP in carbon intensity by 2030 from 2005 levels	37% reduction below the 2010 BAU levels by 2030
Detailed emissions target	Provincial emissions reduction targets	Emissions target system for large firms
Carbon ETS	Pilot start in seven provinces and cities in 2013, nationwide system launched in 2017, phased approach till 2020	Start pilot regions in 2013, unified nationwide system by 2015

Notes: Cities have high population densities and economic development levels and are seen as focal centers for implementing climate change adaptation and carbon reduction policies [5–9].

The 12th TFP report (The Annual Meeting Report of Chinese Parliament and Communist Party of China (http://www.gov.cn/2011lh, accessed on 12 December 2018) proposes a regionally differentiated development strategy to be implemented to build the pattern of coordinated urban development in small and medium-sized cities, and in small towns with urban agglomerations. As the "main form" of national planning and the new-type of agglomerations, the urbanization of these cities has become the most pressing issue in China from 2014, especially from the environmental perspective. These cities also need to allocate corresponding responsibilities to achieve China's national emissions reduction targets. The "central rising plan" adopted in 2004 provides a good opportunity for the sustainable development of environmentally harmonized cities. Moreover, for an effective transformation of industrial structure, the central region in China has the daunting task to catch up with the eastern region. The regional characteristics of China's energy consumption show different patterns due to the imbalances in economic development and resource allocation levels and so the cities in the central region need to create more complex, yet feasible, solutions for overcoming these challenges [10]. It would be a great opportunity for these cities to learn more not only from the experiences of the eastern regions, but also from the Korean cases. So far, there has not been any comparative analysis in the field of sustainable city development, especially involving China and Korea because of a lack of regional cooperation between Korea and China.

On 30 April 2015, the Promotion Conference on Green and Low-Carbon Cooperation Program was initiated by Chinese and Korean non-governmental organizations for a cooperation mechanism for greenhouse gas emission reduction and green conversations in daily life, but there have been no detailed follow-up activities to draw the attention of either government. This research compares the city-level environmental efficiency to fill the missing link on regional cooperation in the field of sustainable development in the two countries. Here, we would argue that China may be too large a country to analyze its national policy effectiveness, especially compared to Korea, and so, we will concentrate our empirical analysis on the northeast Asian regions that includes Central and Eastern China and Korea. This is also more suitable for our research, because of the similar sizes of three subsets of both countries in northeast Asia. The CO_2 emissions data for the city level is not yet available to the public in Korea. To the best of our knowledge, no literature has used city-level data to compare energy efficiency, CO_2 emissions performance, and technology gaps in agglomerated cities in Eastern and Central China and Korea.

There are two lines of research closely related to this. The first line is in the field of emphasizing issues on energy and CO_2 emissions, particularly in China and Korea. Such research can provide valuable information to policymakers when designing effective climate policy tools [11]. Lin and Du (2013) [12] had evaluated the energy efficiency of 30 provinces in China. They argued that China's energy efficiency tends to be underestimated when the technological gaps between groups are ignored. Bian et al. (2013) [13] had used non-radial data envelopment analysis (DEA) model to evaluate energy efficiency of regions in China. Wang and Wei (2014) [14] believe that it is very important to assess energy efficiency and carbon emissions performance at the regional level, given the significant regional differences in China. Wang et al. (2014) [15] have discussed China's energy efficiency from both a static and dynamic perspective, using provincial panel data from 2001 to 2010. Yao et al. (2015) [16] used the meta-frontier non-radial directional distance function to analyze regional energy efficiency, carbon emissions performance, and technology gaps in China, using province-equivalent data for 2011. Lee and Choi (2018) [17] have used panel data from 2009 to 2013 to examine the greenhouse gas (GHG) performance of Korea from the local governments perspective.

The other line of research involves measurement-oriented issues such as DEA based on a non-parametric approach. It is used to calculate input-output relative efficiency of the decision-making unit (DMU) using Farrell's (1957) [18] linear programming analysis. In 1978, Charnes, Cooper, and Rhodes [19] proposed a DEA model with constant scale of returns (CCR). Since then, it has been improved, and a variety of models, such as the BCC [20] and SBM (slacks-based measure) [21] have been developed. The DEA model has a lot of advantages; for example, it does not need to assume a production functional form, is not affected by index dimension, and does not need to estimate parameters. The DEA can consider multiple inputs and outputs at the same time, achieve objective empowerment, and consider undesirable environmental pollution output. Färe et al. (1994) [22] first introduced the concept of undesirable output and incorporated it in the DEA model. Chung (1996) [23] developed the directional distance function to solve for the undesirable output. Färe et al. (2007) [24] estimated the technical efficiency of coal-fired power plants by using the environmental directional distance function. This method was used to analyze the technical efficiency in other fields as well as demonstrated by Oggioni (2011) [25] and Riccardi et al. (2012) [26]. Fukuyama and Weber (2009) [27] proposed a technical inefficiency measurement method based on slacks-based directional distance function, which solves the technical restrictions problem. Barros et al. (2012) [28] proposed a Russell type of directional distance function (DDF), that included desirable and undesirable output. Lin and Du (2013) [12] introduced the parameter element frontier method based on the Shephard type of energy distance function. They argued that this method can consider group heterogeneity. Tian et al. (2011) [29] used a sequential reference method to measure a productivity growth index, such as the Malmquist–Luenberger index, but their method was deficient in some respects. For example, there might have been an infeasible solution because one observation was evaluated based on the environmental technology of another group (different years) [30]. The Malmquist–Luenberger index reflects the growth in total factor rate and not total factor productivity, while in our research we want to analyze the latter. The global benchmark technology [31] can solve the aforementioned problems because it incorporates all contemporary technologies during the study period. However, Pastor did not consider undesirable output. Based on these two distinct lines of content and methodology, this paper mainly contributes to the following two aspects of existing research.

First, we analyze the impact of the technology gap on agglomerated cities among Eastern and Central China and Korea taking energy efficiency and carbon emissions performance into consideration of heterogeneity across cities. The study will focus on the benchmark cities in both countries for future cooperation in the region. In this regard, so far, few studies have been conducted that take into consideration heterogeneity issues across the border. Moreover, there could be an infeasible solution when all the contemporary technologies during the study period are included [30]. Combined with regional heterogeneity, the study will conduct a global meta-frontier non-radial directional distance function (GMNDDF) on agglomerated cities among Eastern and Central China and Korea taking energy efficiency and carbon emissions performance from a regional perspective.

Second, we further identify the sources of energy conservation and CO_2 emissions reduction potential of these three groups. This group comparison provides new insights by focusing on the following issues: Are there significant differences in energy efficiency and CO_2 emissions performance across agglomerated cities and across Eastern and Central China and Korea? What is the potential for energy (CO_2 emissions) reductions when agglomerated cities with low energy (CO_2 emissions) performance "catch up" with high energy (CO_2 emissions) performance cities under the global (Meta) frontier? To answer these questions, this study adopts the GMNDDF model for empirical analysis.

The structure of this paper is as follows. Section 2 will explain the methodology. In Section 3, the results of the empirical study and implications will be discussed. Section 4 will provide the conclusions along with the policy implications.

2. Methodology

In this section, we introduce the GMNDDF model to measure the energy and CO_2 emission performance indices. Following Zhou et al. (2012) [32] and Choi et al. (2013) [33], the group-frontier reflects the actual technology level in the group, and the meta-frontier reflects the all the possible potential technology levels across the different decision-making units (DMUs) in the region. For example, when evaluating the efficiency of 16 cities in Korea from 2009 to 2013, relevant data would constitute the frontier surface A, and the efficiency of each DMU will be measured relative to the distance from benchmark A. When evaluating the efficiency of 23 cities from Eastern China from 2009 to 2013, the relevant data would constitute a new frontier surface B, and the efficiency of each DMU would be measured relative to the distance from benchmark B. It will be the same for 18 cities from Central China from 2009 to 2013. Generally, A, B, and C are not same in their production technology level, so the efficiency in each region will not be comparable. That is the basic motivation of group heterogeneity in a group-frontier model. Nonetheless, if intertemporal comparisons are to be made, the efficiency of the all groups must be based on the same meta-frontier. The GMNDDF in this paper refers to the data for all sample cities from 2009 to 2013 as different DMU, and the 285 sample data points are mixed together to form a common meta-frontier and then measure the efficiency. Later, we decompose it with meta-frontier energy and CO_2 emissions performance.

2.1. Environmental Production Technology

For estimation of the non-radial DDF, the term "environmental production technology" should be defined. Suppose, $j = 1, 2, \ldots \ldots, N$ city units, and each city uses three input vectors $x \in \Re_+^M$, namely capital(K), labor(L), and energy consumption(E). These inputs generate the desirable outputs of GRDP(y) vector $y \in \Re_+^S$ and undesirable outputs CO_2 emissions vector $b \in \Re_+^J$. This production technology can be expressed as follows:

$$T = \{(x, y, b) : x \ can \ produce(y, b)\}, \tag{1}$$

Based on Färe and Grosskopf (2005) [34], the standard axioms of production theory applicable here are as follows: T is always assumed to be satisfied, "finite amounts of input can produce only finite amounts of outputs," and "inactivity is always possible." In addition, for a reasonable model of joint-production technologies, this study assumes weak disposability and null-jointness between desirable and undesirable outputs and strong or free disposability between production inputs and desirable outputs [32,33,35]. Technically, the production technologies can be expressed as follows:

$$i. If \ (x, y, b) \in T \ and \ 0 \leq \theta \leq 1, \ then \ (x, \theta y, \theta b) \in T,$$

$$\text{ii.} \, If \, (x, y, b) \in T \text{ and } b = 0, \text{ then } y = 0.$$

For the environmental production technologies, the weak-disposability assumption implies that reducing CO_2 emissions are costly in terms of proportional reductions in desirable product too. Following Färe et al. (2007), Zhou et al. (2012), and Choi et al. (2013) [32,33,35], we adopt nonparametric DEA piecewise linear production frontiers, where we can express environmental technology T for N DMU generators exhibiting constant returns to scale as:

$$T = \left\{ (x, y, b) : \sum_{n=1}^{N} z_n x_{mn} \leq x_m, m = 1, \ldots, M \right\}$$
$$s.t. \begin{cases} \sum_{n=1}^{N} z_n y_{sn} \geq y_s, s = 1, \ldots, S \\ \sum_{n=1}^{N} z_n b_{jn} = b_j, j = 1, \ldots, J \end{cases}$$
$$z_n \geq 0, n = 1, \ldots, N \tag{2}$$

where M denotes the number of inputs, S and J denote the number of desirable outputs and undesirable outputs, respectively.

We define the global environmental production technology as $T^G = T^1 \cup T^2 \cup \ldots \cup T^T$ following Zhang (2017) [36], where T denotes time. This global frontier consists of a single technique of observations, including the entire period of observation. The global environmental technology can be expressed as follows:

$$GT = \left\{ (x, y, b) : \sum_{t=1}^{T} \sum_{n=1}^{N} z_n x_{mn}^t \leq x_m, m = 1, \ldots, M \right\}$$
$$s.t. \begin{cases} \sum_{t=1}^{T} \sum_{n=1}^{N} z_n y_{sn}^t \geq y_s, s = 1, \ldots, S \\ \sum_{t=1}^{T} \sum_{n=1}^{N} z_n b_{jn}^t = b_j, j = 1, \ldots, J \end{cases}$$
$$z_n \geq 0, n = 1, \ldots, N, t = 1, \ldots T \tag{3}$$

2.2. Non-Radial Directional Distance Functions

Zhou et al. (2012) [32] and Choi et al. (2013) [33] define a non-radial directional distance function (NDDF) as follows:

$$\vec{D}(x, y, b; g) = \sup \left\{ W^T \beta : ((x, y, b) + g \times diag(\beta)) \in T \right\}, \tag{4}$$

where $W = (w_m^x w_s^y w_j^b)^T$ stands for a normalized weight vector relevant to production inputs and outputs. $g = (-g_x, g_y, -g_b,)$ stands for an explicit directional vector. Additionally, $\beta = (\beta_m^x \beta_s^y \beta_j^b)^T \geq 0$ stands for the vector of scaling factors. We can compute the value of $\vec{D}(x, y, b; g)$ as follows:

$$\vec{D}^r (x, y, b; g) = \max w_m^x \beta_m^x + w_s^y \beta_s^y + w_j^b \beta_j^b$$
$$s.t. \begin{cases} \sum_{t=1}^{T} \sum_{n=1}^{N} z_n x_{mn}^t \leq x_m - \beta_m^x g_{xm}, m = 1, \ldots, M \\ \sum_{t=1}^{T} \sum_{n=1}^{N} z_n y_{sn}^t \geq y_s + \beta_s^y g_{ys}, s = 1, \ldots, S \\ \sum_{t=1}^{T} \sum_{n=1}^{N} z_n b_{jn}^t = b_j - \beta_j^b g_{bj}, j = 1, \ldots, J \end{cases}$$
$$z_n \geq 0, n = 1, \ldots, N; \, t = 1, 2, \ldots, T. \tag{5}$$
$$\beta_m^x, \beta_s^y, \beta_j^b \geq 0$$

2.3. Meta-Frontier and Group-Frontier Technologies

To investigate technology gaps among agglomerated cities in Central and Eastern China and Korea, we will combine the idea of an NDDF in Zhou et al. (2012) [32] with meta-frontier technologies in Chiu et al. (2012) [37] to develop several indices for meta-frontier CO_2 emissions and energy performance and to investigate the group heterogeneity among cities. We define two types of frontier technologies: Group-frontier that reflects the actual technology level and the meta-frontier that reflects the potential technology level.

Suppose there are H groups. Following Battese and Rao (2004) [38] and O'Donnell et al. (2008) [39], we define the non-radial directional distance function of group h as follows:

$$T_h = \{(x, y, b) : x \text{ can produce } (y, b)\}, \ h = 1, \ldots, H. \tag{6}$$

In contrast to the group-frontier, the meta-frontier technologies are constructed based on all observations among groups. Therefore, we can consider the NDDF of meta-frontier technology as:

$$\overrightarrow{D}^m (x, y, b; g) = \sup \left\{ W^T \beta : ((x, y, b) + g \times diag(\beta)) \in T_h \right\}, \ h = 1, 2, \ldots, H. \tag{7}$$

Following the method used in Chiu et al. (2012) [37], we can compute $\overrightarrow{D}^m (\bullet)$ after solving the following global meta-frontier non-radial directional distance function:

$$\overrightarrow{D}^m (x, y, b; g) = \max w_m^x \beta_m^x + w_s^y \beta_s^y + w_j^b \beta_j^b$$

$$s.t. \begin{cases} \sum_{h=1}^{H} \sum_{t=1}^{T} \sum_{n_h=1}^{N_h} z_n^h x_{mn}^t \leq x_m^t - \beta_m^x g_{xm}, m = 1, \ldots, M \\ \sum_{h=1}^{H} \sum_{t=1}^{T} \sum_{n_h=1}^{N_h} z_n^h y_{sn}^t \geq y_s^t + \beta_s^y g_{ys}, s = 1, \ldots, S \\ \sum_{h=1}^{H} \sum_{t=1}^{T} \sum_{n_h=1}^{N_h} z_n^h b_{jn}^t = b_j^t - \beta_j^b g_{bj}, j = 1, \ldots, J \end{cases} \tag{8}$$

$$z_n^h \geq 0, n_h = 1, \ldots, N_h; h = 1, \ldots, H; t = 1, 2, \ldots, T.$$
$$\beta_m^x, \beta_s^y, \beta_j^b \geq 0$$

where N_h represents the number of observations in a specific group, and z_n^h represents the intensity variables of meta-frontier technology. As proposed by O'Donnell et al. (2008) [39], to make the meta-frontier smoother, we use the convexity constraint of Variable Returns to Scale (VRS). Therefore, i.e., $\sum_{h=1}^{H} \sum_{n=1}^{N_h} \lambda_n^h = 1$ should be imposed on Equation (8).

2.4. Energy and CO$_2$ Emissions Performance Indices

To derive energy and CO_2 emissions performance indices, we must define the input and output variables explicitly. In this study, the input vector x contains capital (K), labor (L), energy (E), desirable output y that refers to real GRDP (Y), adjusted to 2009 prices, and undesirable output b that is CO_2 emissions (C). The weighted vector for the three inputs and two outputs can be set as (1/9, 1/9, 1/9, 1/3, and 1/3). Moreover, to focus on the efficiency of energy input, we obtain single factor performance indices, the weighted vector can be set as (0, 0, 1/3, 1/3, 1/3).

As there are three inputs, one desirable and one undesirable output, the directional vector is specified as $g = (-x, y, -b) = (-K, -L, -E, Y, -C)$.

For our research, following Choi et al. (2013) [33] and Yao et al. (2015) [16], we have four types of indices measuring CO_2 emissions and energy efficiency performance, including energy potential reduction index (EPRI), total-factor energy efficiency performance index (TEPI), CO_2 emission potential reduction index, and total-factor carbon emission performance index (TCPI). The values of TEPI and TCPI indices range from 0 to 1. A higher value indicates better performance.

Suppose β_E^* and β_Y^* are the optimal solutions to Equation (8) corresponding to the energy input and the GRDP output in the above model. Here energy potential reduction index (EPRI) can be defined as follows:

$$EPRI = \beta_E^* E \tag{9}$$

Following O'Donnell et al. (2008) [39] and Yao et al. (2015) [16], the total-factor energy efficiency performance index (TEPI) can be defined as:

$$TEPI = \frac{Y/E}{(Y + \beta_Y^* Y)/(E - \beta_E^* E)} = \frac{1 - \beta_E^*}{1 + \beta_Y^*}. \tag{10}$$

Similarly, we can further define a CO_2 emission potential reduction index (CPRI) as:

$$CPRI = \beta_C^* C. \tag{11}$$

Next, the total-factor carbon emission performance index (TCPI) can be defined as follows:

$$TCPI = \frac{(C - \beta_C^* C)/(Y + \beta_Y^* Y)}{C/Y} = \frac{1 - \beta_C^*}{1 + \beta_Y^*}. \tag{12}$$

By solving the above equations, we can compute the energy efficiency and carbon emission performance indices under different frontier technologies, as shown in Appendix A nomenclature.

2.5. Decomposition of Meta-Frontier Energy and CO_2 Emission Performance

Based on the method of O'Donnell et al. (2008) [39], this study decomposes technical efficiency under meta-frontier technologies into within-group energy (or CO_2 emission) and meta-technology (also the technology gap ratio) performances. In this regard, the within-group energy (or CO_2 emission) performance reflects the frontier technologies in a specific set of relative efficiency of observed within-group technology. The meta-technology ratio shows relative gaps between group-frontier and meta-frontier. In Figure 1, we consider a simple example:

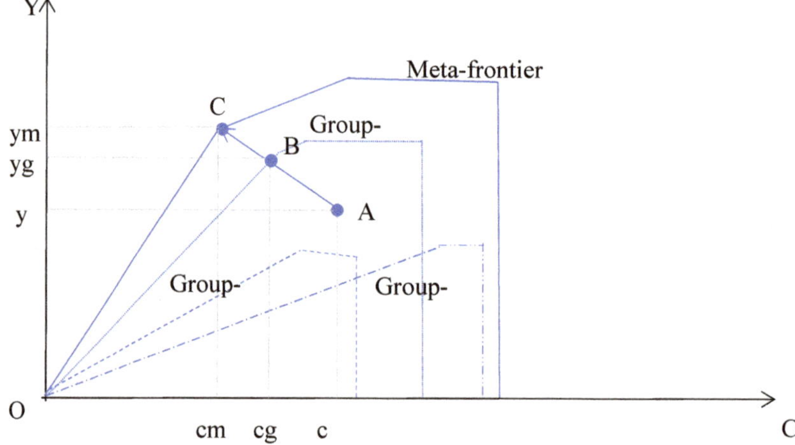

Figure 1. Meta-frontier CO_2 emission performance index and its decomposition.

According to Figure 1, city A belongs to group-frontier 1 and its GRDP is Y with CO_2 emissions. Under the group frontier technologies, point B gives the target values for GRDP and CO_2 emissions as "yg" and "cg." The Group-Frontier Total-Factor Carbon Emission Performance Index (GTCPI) is the ratio of potential CO_2 strength to actual CO_2 strength and thus, it can be measured the relative distance (ocg/oyg)/(oc/oy) in Figure 1. For the meta-frontier technologies, the Meta-Frontier Total-Factor Carbon Emission Performance Index (MTCPI) can be measured by the ratio (ocm/oym)/(oc/oy). In this way, we can get the technology gap between a specific meta-frontier and group-frontier technology in terms of CO_2 emission performance. The meta-technology of CO_2 performance MTCPI can be measured with the ratio (ocm/ocg)/(oyg/oym) and it can be written as follows:

$$MTCPI = \frac{ocm/oym}{oc/oy} = \left(\frac{ocg/oyg}{oc/oy} \right) \cdot \left(\frac{ocm}{ocg} \cdot \frac{oyg}{oym} \right) = GTCPI \times MTRCI. \tag{13}$$

Similarly, the decomposition can be expressed as follows:

$$MTEPI = GTEPI \times MTREI. \tag{14}$$

In the illustration in Figure 2, let us assume that there is a city in a specific group 1 that operates on point A. If a city is on point B, it means it is operating with group-frontier technology. Further, if the city can operate with meta-frontier technology, it will be on point C. The values of the Meta-Frontier Energy Potential Reduction Index (MEPRI) or Meta-Frontier Carbon Emission Potential Reduction Index (MCPRI) indicate the technology gap between group-frontier technology and meta-frontier technology. Higher values correspond to a smaller technology gap between group-frontier and meta-frontier technologies.

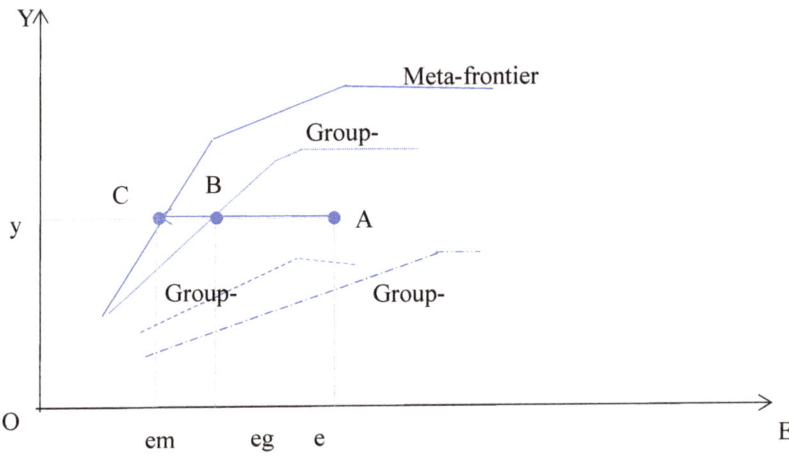

Figure 2. Meta-frontier and group frontier technologies.

In this case, the potential gaps in energy reduction (PGERI) between group-frontier and meta-frontier can be defined as follows [16]:

$$PGERI = MEPRI - GEPRI. \tag{15}$$

Similarly, the potential gaps in carbon emission abatement (PGCRI) between group-frontier and meta-frontier can be defined as:

$$PGCRI = MCPRI - GCPRI. \tag{16}$$

3. Empirical Analysis

3.1. Data Collection

Based on the methodology described in Section 2 to examine energy and CO_2 performance, the data was collected from 16 cities in Korea, 23 cities from the eastern provinces of China, including Shandong, Jingjinji Metropolitan, Yangtze river delta, and Pearl river delta, and 18 cities from central China, including Taiyuan, Zhongyuan, Wanjiang, City Cluster surrounding Poyang Lake, Wuhan, and Greater Changsha Metropolitan cities. Figure 3 shows the location of China and Korea Cites and its energy use amount over five years.

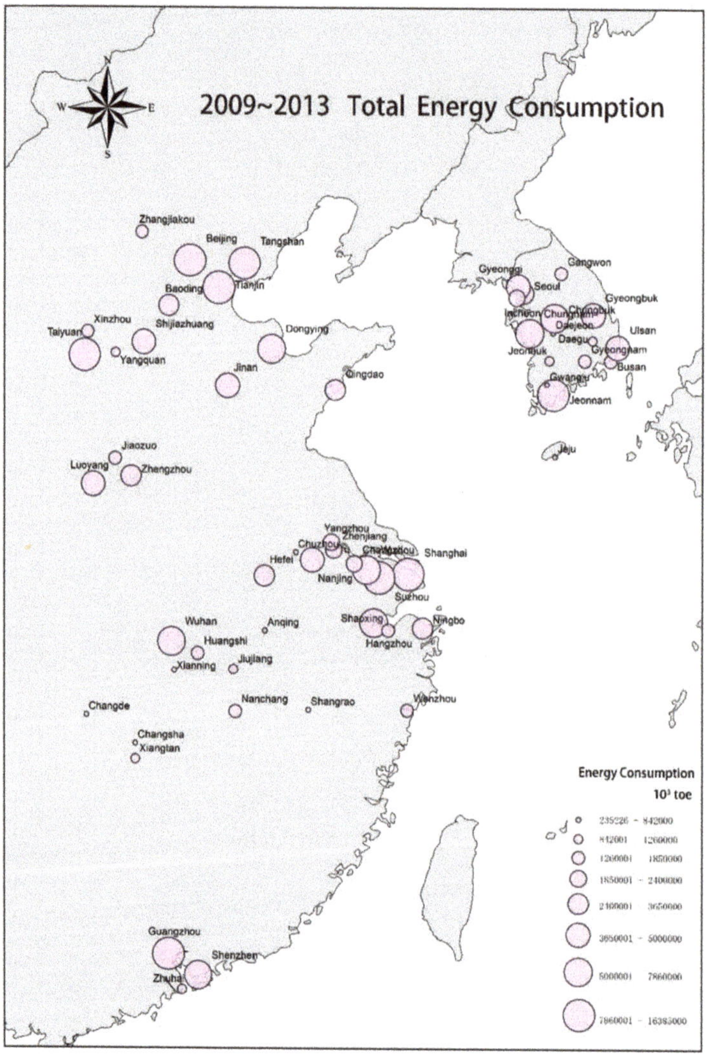

Figure 3. The location of China and Korea Cites energy usage amounts.

The data was sourced from the annual data during 2009 to 2013 for the three groups: Korean Cities, Eastern Cities of China, and Central Cities of China. In Korea, the partly updated GHG emissions data for 2015 is available, but unfortunately, city-level data takes longer to be updated, so 2013 data is the most recent data we could get. In the DEA model, we selected two basic types of production inputs of capital and labor and another new input of energy consumption. Specifically, capital input is based on the perpetual inventory (stock) system used in Zhang et al. (2004) [40]. As capital stock cannot be directly obtained from any statistical yearbooks, following Young (2003) [41] and Zhang et al. (2017) [36], we calculated the capital stock as follows:

$$K_{n,t} = \frac{\Delta K_{n,t+1}}{(\delta_n + g_n)},$$ (17)

where $K_{n,t}$ is the capital stock of city n at time t, $\Delta K_{n,t+1}$ denotes the investment in capital of city n at time $t+1$, δ_n and g_n is depreciation rate and GDP growth rate of city n respectively, the base year is 2009 so the price levels were adjusted to 2009 prices.

We calculated energy input based on energy consumption. For desirable output of GRDP, we take the GRDP of cities in China according to the data in Municipal Yearbook (CNKI (www.cnki.net)). The Korean data for labor, capital, energy, and GRDP was sourced from the Korea National Information Service (KOSIS) (KOSIS (http://kosis.kr/)). The nominal GRDP was adjusted using the 2009 GDP deflator. For undesirable output of CO_2 emissions, the pure CO_2 data for the Korean governments was unavailable. Hence, we followed the methodology used by Choi and Lee (2016) [17] for GHG emissions data for carbon dioxide (CO_2), which accounts for 78% of GHG emissions and other gases like nitrogen (N_2O), methane (CH_4), perfluorinated compounds (PFCs), sulfur hexafluoride (SF_6), and hydrofluorocarbons (HFCs) [42]. According to Xu et al. (2018) [43], the Chinese CO_2 data is cited from the database on http://www.ceads.net/. The basic descriptive statistics of inputs and outputs are shown in Table 2.

Table 2. Descriptive statistics of production inputs and outputs.

Groups	Variable	Obs	Units	Mean	Std. Dev.	Min	Max
Korean cities (k)	Capital	80	10^6 US$	13,268.5	12,137.7	1547.4	61,079.7
	Labor	80	10^3 persons	1516.6	1523.6	283.0	5988.0
	Energy	80	10^3 toe	524,309.7	439,220.6	39,525.3	1,642,058.0
	Desirable outputs GRDP	80	10^6 US$	64,757.5	63,723.1	7959.7	248,914.0
	Undesirable output CO_2	80	10^6 ton	31.5	27.9	2.8	118.3
Eastern cities of China (e)	Capital	115	10^6 US$	34,924.2	20,944.4	5356.4	117,953.8
	Labor	115	10^3 persons	5075.0	2616.8	1015.7	11,410.0
	Energy	115	10^3 toe	1,112,443	805,213.1	176,860.5	3,430,065.0
	Desirable outputs GRDP	115	10^6 US$	43,736.4	32,125.8	6634.7	131,308.6
	Undesirable output CO_2	115	10^6 ton	94.4	62.8	11.1	310.3
Central cities of China (c)	Capital	90	10^6 US$	18,971.2	15,601.6	3163.6	69,947.6
	Labor	90	10^3 persons	2841.5	1481.2	582.8	5380.9
	Energy	90	10^3 toe	448,017.9	471,922.9	79,687.4	2,038,335.0
	Desirable outputs GRDP	90	10^6 US$	14,698.3	12,397.4	2890.7	53,272.8
	Undesirable output CO_2	90	10^6 ton	51.0	47.0	9.5	209.8

Sources: Greenhouse Gas Inventory and Research Center of Korea (http://www.gir.go.kr), KOSIS (http://kosis.kr/), Municipal Yearbook (www.cnki.net), China Emission Accounts Datasets (http://www.ceads.net/).

As Table 2 shows, there are large variations across the three groups in all five variables. We can take Mean as an example. For inputs, the mean of the eastern city group of China (e) is the highest among the three groups. The Korean city group (k) has the lowest mean for capital and labor input, but its mean for energy consumption is more than that of the central city group of China (c). For desirable output of GRDP, the mean for the Korean city group has the highest value with the lowest CO_2 emissions mean.

3.2. Empirical Results

3.2.1. Carbon Emission Performance Indices

Our empirical test begins with the development of carbon emission and energy performance indices under group-frontier technologies to investigate whether there are significant differences in performance indices across groups and across cities. Table 3 presents the energy and CO_2 emission performance indices across groups under group-frontier. Figures 4–6 illustrate the city-level energy and CO_2 emission performance indices for five years.

First, we consider the CO_2 emission performance under group-frontier technologies. Table 3 shows that the average GTCPI value for Korean cities (k) is 0.931 indicating that, on average, Korean cities can reduce their CO_2 intensity by 6.9% if all these cities are on the group frontier. The standard deviation for Korean cities is the lowest implying the smallest technical differences within-group. As for the eastern cities of China (e), the GTCPI value ranges from 0.577 to 1.000, with the average GTCPI value of 0.927. This suggests that the eastern cities of China can improve their CO_2 emission performance by 7.3%, if all within-group cities can operate with group-frontier technologies. In addition, the highest standard deviation of GTCPI for the eastern cities of China shows the greatest variance in within-group technology gaps. Therefore, the eastern cities of China have the greatest potential to improve their CO_2 emissions performance. The GTCPI value ranges from 0.578 to 1.000, with average GTCPI value of 0.912, which implies that the central cities of China can improve their CO_2 emissions performance by 8.8%, if all within-group cities can operate with group-frontier technologies.

Table 3. Energy and CO_2 emission performance under group-frontier technologies.

Variable	GTCPI				GTEPI			
	Mean	Std. Dev.	Min	Max	Mean	Std. Dev.	Min	Max
k	0.931	0.071	0.724	1.000	0.913	0.078	0.645	1.000
e	0.927	0.100	0.577	1.000	0.917	0.110	0.539	1.000
c	0.912	0.092	0.578	1.000	0.824	0.204	0.287	1.000

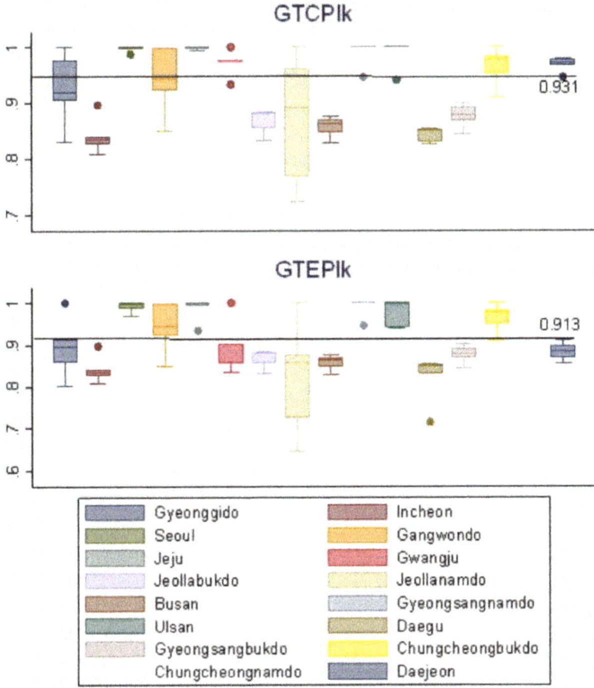

Figure 4. Boxplots of energy and CO_2 emission performance under group-frontier technologies of Korean cities.

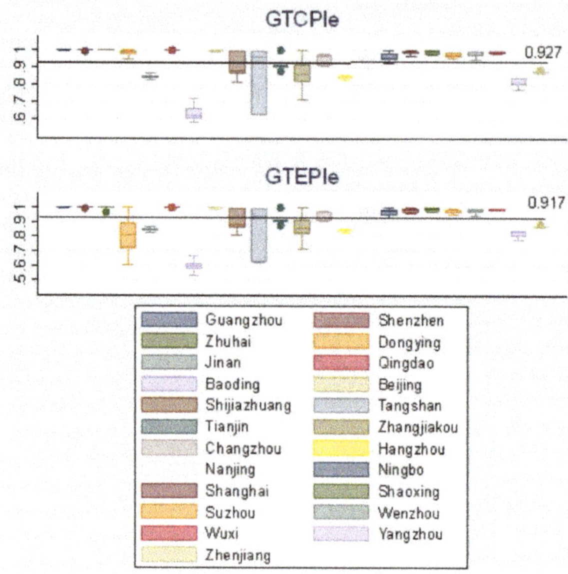

Figure 5. Boxplots of energy and CO2 emission performance under group-frontier technologies of eastern cities of China.

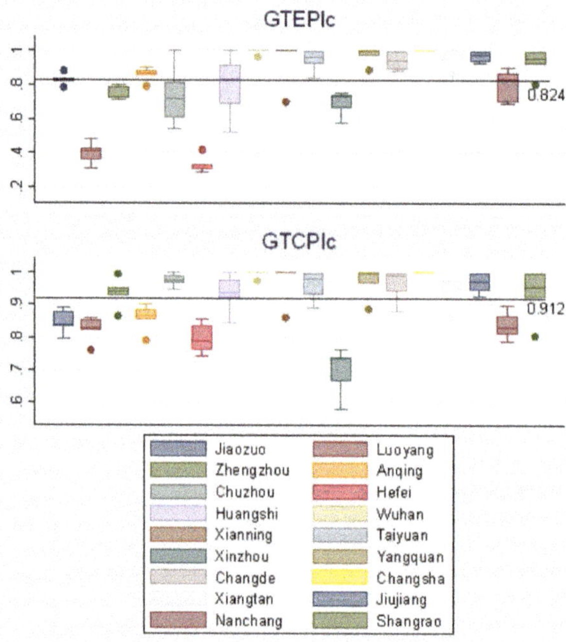

Figure 6. Boxplots of energy and CO2 emission performance under group-frontier technologies of central cities of China.

Next, we consider GTCPI at city levels. For the agglomerated Korean cities, as Figure 4 shows, the value of GTCPIk for five years ranges from 0.724 to 1. Seoul, Jeju, Chungnam, and Ulsan show a higher performance than the other cities. Jeonnam shows the lowest value, implying that Jeonnam has the greatest potential for carbon emission performance improvement. For the eastern agglomerated cities of China, Guangzhou, Shenzhen, Zhuhai, Qingdao, Beijing, and Shanghai attain unity more than three times, indicating that they are at best practice CO_2 emission performance. Baoding has the lowest GTCPIe value, implying that it has the greatest potential for CO_2 emission performance improvements. For the central cities of China, as shown in Figure 6, only Wuhan, Xianning, Yangquan, and Changsha cities attain unity thrice in five years. Xinzhou shows the lowest GTCPI value, implying that it has the greatest potential for CO_2 emission performance improvements.

3.2.2. Energy Performance Indices

We analyzed the energy performance indices under group-frontier technologies. As shown in Table 3, the Group-Frontier Total-Factor Energy Performance Index (GTEPI) value for Korean cities (k) ranges from 0.654 to 1.000 with an average value of 0.913 indicating that, on average, Korean cities can improve their energy performance by approximately 8.7% if all within-group agglomerated cities perform on the best practice group-specific technologies. The standard deviation for Korean cities is the lowest, implying the smallest technology gap within-group. As for the eastern cities of China (e), the GTEPI value ranges from 0.539 to 1.000, with average GTEPI value of 0.917, which suggests that the eastern cities of China can improve energy performance by approximately 8.3%. The GTEPI value for the central cities of China ranges from 0.287 to 1.000, with average GTEPI value of 0.824. These results suggest that the central cities of China can improve energy performance by approximately 17.6%, if all within-group cities operate on group-frontier technologies. Among the three groups, the highest standard deviation of GTEPI is for the central cities of China, implying the greatest within-group technology gaps. Therefore, the central cities of China have the greatest potential to improve energy performance.

Next, we consider GTEPI at city level. As shown in Figures 4–6, there are significant variations in energy performance indices at city level. For the Korean cities, as shown in Figure 4, the value of GTEPI for five years ranges from 0.645 to 1. Seoul, Jeju, Chungnam, and Ulsan show a higher performance than other cities. Jeonnam shows the lowest value, implying Jeonnam has the greatest potential for energy performance improvement. For the eastern cities of China, as shown in Figure 5, Guangzhou, Shenzhen, Zhuhai, Qingdao, Beijing, and Shanghai cities attain unity more than thrice, indicating that they are at best practice energy performance. Baoding shows the lowest GTCPIe value, implying that it has the greatest potential for energy performance improvements. For the central cities of China, as shown in Figure 6, only Wuhan, Xianning, Yangquan, and Changsha cities attain unity more than thrice in five years. Xinzhou shows the lowest GTCPI value implying that it has the greatest potential for energy performance improvements within this group.

Figures 4–6 show that the average GTCPI value of 0.931 is higher than the average GTEPI value (0.913), indicating that the Korean cities show a better CO_2 emission performance than energy performance. In addition, we also found that cities with high CO_2 emission performance also have high energy performance. Further, we found similar results for the eastern and central cities of China. These results are not surprising since we assumed fixed CO_2 emission factors of energy in our research.

From the results, we also found that the average for GTCPI and GTEPI values shows an uptrend between 2009 and 2012, but it shows a mild downtrend in the year 2013, which is consistent with results obtained from other research, such as Lee and Choi (2018) [17]. Thus, we can say that since the target management system (TMS) was implemented in 2009, the strict environmental regulation has increased efficiency and encouraged innovation for a more environmentally-friendly production process. Unfortunately, the successor of President Park Geun-hye changed the paradigm from "green growth" to "creative economy" in 2013. As a result, environmental policies became passive and lax compared to the policies of the former president Lee Myung-bak's government. We find that the average for GTCPI and GTEPI value shows a downtrend from 2009 to 2010, uptrend in 2011, and a downtrend again in 2012 to 2013, which is consistent with other research too [44].

According to the state council regulations, 2010 was the deadline for reporting the targets of the 11th five-year plan in China. If the local government fails to fulfill its task of energy conservation and emission reduction during the 11th five-year plan, relevant leaders are to be held accountable or even removed from their official position.

3.3. Total-Factor and Single-Factor Performance Indices

Next, we compared the total and single factor performance indices, as shown in Table 4. We set the weight vector as (0,0,1/3,1/3, and 1/3) in Equation (8), to measure the single (or pure) factor energy and CO_2 emission performance by fixing the capital and labor inputs. There should be no possibility of substitution between energy and other productive inputs (capital and labor). The average of the GTEPIk (GTCPIk) value is higher than that of the GPEPIk (GPCPIk) value, suggesting that the alternatives between energy and other production inputs contribute to improved energy and carbon performance. However, for the eastern and central cities of China, the GTEPI average is slightly lower than the GPEPI average, suggesting that the substitution between energy and other production inputs reduces energy performance. The results are not surprising as there is a regional gap between the three groups. In general, it requires a large initial capital stock to replace energy in the production process. The capital stock in Korea is quite limited, thus inhibiting the substitution of factors between energy and other production inputs. As a result, substitutability between production inputs is very limited in Korea. The result for the eastern and central cities of China illustrates the Chinese government's erroneous policy of energy saving and power rationing again. Power rationing reduces productivity. The blind pursuit of rapid GDP growth by local policy makers has delayed the elimination of enterprises using outdated production capacity, leading to a reassessment of energy-saving targets and the local government resorting to forced power cuts to meet the targets. First, the mandatory "power rationing" disrupts the normal production of enterprises, destroys the ability of enterprises to use peak and valley regulations and other measures to save electricity. The production process of an enterprise is continuous, and orders received must be delivered on time. Forced switch-offs and power rationing leads to a short-term backlog of production tasks of enterprises, after which enterprises are bound to work overtime, resulting in a growth of power consumption and artificial power supply tension. Secondly, under the forced pressure of power limit, many enterprises are forced to abandon the power supply from the "power grid" and adopt "self-provided diesel power generation" to solve the pressing need to ensure that their production and business activities are not greatly affected. This not only wastes energy, but also causes serious damage to the national energy strategy.

Table 4. Total-factor and single-factor performance indices.

Agglomerated Cities	Total-factor GTEPI	Single-factor GPEPI	Total-factor GTCPI	Single-factor GPCPI
Korean Cities (k)	0.913	0.882	0.931	0.885
Eastern Cities of China (e)	0.917	0.918	0.927	0.919
central Cities of China (c)	0.824	0.831	0.912	0.883

In Sections 3.4 and 3.5, we test for group-heterogeneity across cities, and find whether there are significant differences in CO_2 emission and energy performance across cities. Table 5 compares the meta-frontier energy and CO_2 emission performance. In addition, it also presents the MTREI and MTRCI values across cities. Table 8 gives the result of Pearson and Spearman correlation coefficients and the *p*-values.

Table 5. Meta-frontier energy and CO_2 emission performance and the meta-technology gap.

Variable	Mean	Std. Dev.	Variable	Mean	Std. Dev.	Variable	Mean	Std. Dev.
MTREIk	1.000	0.000	MTREIe	0.294	0.074	MTREIc	0.400	0.187
MTRCIk	1.000	0.000	MTRCIe	0.408	0.112	MTRCIc	0.492	0.201
MTEPIk	0.913	0.078	MTEPIe	0.270	0.079	MTEPIc	0.323	0.173
MTCPIk	0.931	0.071	MTCPIe	0.378	0.113	MTCPIc	0.455	0.211

3.4. Meta-Frontier Energy and CO_2 Emission Performance and the Meta-Technology Gap

The Korean cities have the highest average MTEPIk of 0.913 and average MTCPIk (0.931) in the three groups, signifying that they are relatively efficient in terms of energy performance and CO_2 emissions. In this case, if all the agglomerated cities adopt meta-frontier technology, the energy efficiency of the Korean cities can be improved by 8.7% and the carbon emission performance can be improved by 6.9%. We also found that the average MTCPIk value is higher than the average MTEPIk value, indicating that the Korean cities are relatively more efficient in CO_2 emissions than energy performance. In addition, MTCPIk shows the lowest standard deviation, which means there is the smallest intra-agglomerated city technology gap in CO_2 emissions performance. Since average MTREIk and MTRCIk values are equal to 1, it makes no difference in energy and CO_2 performance indices between meta-frontier and group-frontier. Therefore, the Korean cities play an important role for benchmarking under meta-frontier.

In contrast, the average MTEPIe value of 0.270 and average MTCPI value (0.378) for the eastern cities of China are the lowest, signifying the eastern cities of China are the least efficient in terms of energy or CO_2 emissions performance. Among the three groups, the eastern cities of China have the greatest potential to improve energy efficiency and they can increase their energy efficiency by 73%, if they adopt meta-frontier technologies in all cities. In this context, the CO_2 emission performance of eastern cities of China can also be improved by 62.2%. In addition, MTEPIe and MTCPIe show relatively low standard deviations, signifying a small within-group technical gap in energy and CO_2 performance. In contrast to the Korean cities, there are significant differences between meta-frontier and group-frontier in the eastern cities of China. The average MTREIe value is 0.294 and MTRCIe value is 0.408, implying that the eastern cities of China can improve their energy performance by 70.6% and CO_2 emission performance by 59.2% with meta-frontier technologies. In comparison, the average MTREIe is lower than the average MTRCIe value, indicating that meta-technology gaps are relatively severe for energy efficiency.

Meanwhile, the central cities of China show a medium average MTEPIc value of 0.323 and average MTCPIc value (0.455), signifying the energy and CO_2 performance of the central cities of China are more effective than the eastern cities of China. Given that all cities use cutting-edge technology, the central cities of China could increase their energy performance by 67.7% and carbon performance by 54.5%. The largest within-group efficiency gap is similar for the central and the eastern cities of China. The average MTREIc value of 0.400 and average MTRCIc value of 0.492, indicates that the central cities of China can improve their energy performance by 60% and CO_2 emission performance by 50.8% with meta-frontier technologies.

Next, we compare the efficiency gap from the agglomerated city perspective. For the Korean cities, the average GTEPIk is equal to the average MTEPIk, and the average GTCPIk is equal to the average MTCPIk. These results show there is no technical gap in Korea between the group-frontier and the meta-frontier between GTEPI and MTCPI. For the eastern cities of China, the average value of GTEPI is much higher than of MTEPI, and the average value of GTCPI is significantly higher than of MTCPI. These results imply a considerable technological gap in eastern China between the group-frontier and the meta-frontier technologies. Similar results were found in central cities of China. However, it needs to be mentioned here that because we chose city-level data instead of province-level data as the object of our study, we found that under meta-frontier technology, the central cities of China show a medium average in energy and CO_2 emission performance which is better than the eastern cities of China. These results suggest that group heterogeneity could exist across cities. Without taking heterogeneity into consideration, the energy efficiency index in the central and eastern region might be overestimated. It can also be observed that Korean cities play a more important role in the baseline for efficiency of the meta-frontier technology.

Finally, we compare efficiency gaps at city level. Since it is not easy for cities to catch up with energy conservation and emissions reduction, the central government needs to support the local governments to learn more from the benchmark cities on the meta-frontier. We used the benchmark information for 2009. The cities can learn from the benchmarking cases of effective DMU—each inefficient DMU can learn by matching to an efficient DMU called a "reference set." When similar input and output structures are present, the reference set is assigned to the inefficient DMU. For an inefficient DMU to improve its efficiency, its input target should reach the value derived from Equation (18). Cities for benchmark under the VRS condition can be derived as:

$$\text{Reference set's input} * \lambda = \text{inefficient DMU's input target.} \tag{18}$$

The intercept value can be defined as the degree of influence of DMU on each inefficient DMU. For example, the target value of DMU 1, Gyeonggi = DMU 10 Gyeongnam (2012) * 0.031+ DMU 12 Daegu (2012) * 0.719+ DMU 15 Chungnam (2012) * 0.201 + DMU 16 Daejeon (2010) *0.019 + DMU 16 Daejeon (2012) * 0.030. In the same way, DMU 3 Seoul's target value can come from the input value of DMU 3 Seoul (2009) * 1.000. The result of the benchmark information is shown in Appendix B. In this result, we find DMU 12 Daegu (2009) is the best DMU as it showed up 31 times as a reference set. This is because, among the 57 sample cities for this research, DMU 12 Daegu's (2009) input and output structure would be similar to other inefficient DMUs. Efficient DMUs that are not reported as a benchmark set as much as DMU 12 Daegu (2009), imply that their input and output structure is different from other inefficient DMUs. With the benchmark information the inefficient DMU should learn from the cases of their allocated reference set.

3.5. Test for Group-Heterogeneity Across Agglomerated Cities

Since the cities in the three groups have large differences in their economic environment as well as production technology level, we conducted tests for group-heterogeneity across the three groups. The tests aimed to find out whether there are significant differences between energy and carbon emission performance across groups. Table 6 reports the results of the Wilcoxon-Mann-Whitney U-test. From Table 6, all indexes of MTEPI, MTREI, MTCPI and MTRCI, and all invalid assumptions can be rejected at a significant level of 5% as far as the group heterogeneity is concerned. These results indicate that there is group heterogeneity among the three groups in terms of energy and CO_2 emission performance.

In addition, we used the Kolmogorov–Smirnov tests on MTEPI, MTREI, MTCPI, and MTRCI to determine any significant differences in Kernel density distribution. Table 7 reports the results of the Kolmogorov–Smirnov test. In terms of the performance indexes of MTEPI, MTREI, MTCPI, and MTRCI, it was found that for the Kernel density distribution among three groups, all null hypotheses were rejected at a significant level of 5%. These results indicate that the Kernel density distribution proves the different density distribution for each other, supporting heterogeneity for all three groups (Figure 7).

Next, we tested the Spearman correlation coefficients. We found from Table 8 that most of the performance indicators have a positive correlation. As expected, MTEPI shows a high correlation with MTCPI, while MTREI shows a high correlation with MTRCI, since carbon emissions are calculated based on energy-type carbon emission factors. In contrast, the results do not support a positive correlation between GTEPI (GTCPI) and MTREI, suggesting that there is no significant correlation between the group-frontier performance index and the meta-frontier energy performance technology index, but there is a small one between the group-frontier performance index and the meta-frontier CO_2 emission performance technology index.

Table 6. Results of the Wilcoxon-Mann-Whitney U-test.

Null Hypothesis (H0)	Mann-Whitney U Statistic	*p*-Value
Mean (MTEPIk) = Mean (MTEPIe)	12,440 ***	0.000
Mean (MTEPIk) = Mean (MTEPIc)	10,310 ***	0.000
Mean (MTEPIe) = Mean (MTEPIc)	11,022 *	0.051
Mean (MTREIk) = Mean (MTREIe)	12,440 ***	0.000
Mean (MTREIk) = Mean (MTREIc)	10,440 ***	0.000
Mean (MTREIe) = Mean (MTREIc)	9857 ***	0.000
Mean (MTCPIk) = Mean (MTCPIe)	12,423 ***	0.000
Mean (MTCPIk) = Mean (MTCPIc)	10,112 ***	0.000
Mean (MTCPIe) = Mean (MTCPIc)	11,054 *	0.061
Mean (MTRCIk) = Mean (MTRCIe)	12,440 ***	0.000
Mean (MTRCIk) = Mean (MTRCIc)	10,440 ***	0.000
Mean (MTRCIe) = Mean (MTRCIc)	10,722 **	0.008

*** significant at the level of 0.1%, ** significant at the level of 1%, * significant at the level of 5%.

Table 7. Results of the Kolmogorov–Smirnov test.

Indices	Null Hypothesis (H$_0$)	K-S Statistic	*p*-Value
MTEPI	Distribution (Korean) = Distribution (East)	1.000 ***	0.000
	Distribution (Korean) = Distribution (Central)	0.944 ***	0.000
	Distribution (Central) = Distribution (East)	0.200 *	0.035
MTREI	Distribution (Korean) = Distribution (East)	1.000 ***	0.000
	Distribution (Korean) =Distribution (Central)	1.000 ***	0.000
	Distribution (Central) = Distribution (East)	0.314 ***	0.000
MTCPI	Distribution (Korean) = Distribution (East)	0.979 ***	0.000
	Distribution (Korean) = Distribution (Central)	0.844 ***	0.000
	Distribution (Central) = Distribution (East)	0.239 ***	0.006
MTRCI	Distribution (Korean) = Distribution (East)	1.000 ***	0.000
	Distribution (Korean) = Distribution (Central)	1.000 ***	0.000
	Distribution (Central)= Distribution (East)	0.264 ***	0.002

*** significant at the level of 0.1%, ** significant at the level of 1%, * significant at the level of 5%.

Table 8. Correlations between indices.

	GTCPI	GTEPI	MTCPI	MTEPI	MTRCI	MTREI
GTCPI	1.000					
GTEPI	0.9216 ***	1.000				
	(0.000)					
MTCPI	0.3769 ***	0.3497 ***	1.000			
	(0.000)	(0.000)				
MTEPI	0.3126 ***	0.3350 ***	0.7805 ***	1.000		
	(0.000)	(0.000)	(0.000)			
MTRCI	0.1166 *	0.1213 *	0.9507 ***	0.7278 ***	1.000	
	(0.049)	(0.041)	(0.000)	(0.000)		
MTREI	0.033	0.007	0.7073 ***	0.9200 ***	0.7310 ***	1.000
	(0.585)	(0.905)	(0.000)	(0.000)	(0.000)	

*** significance level is 0.1%, * significance level is 5%.

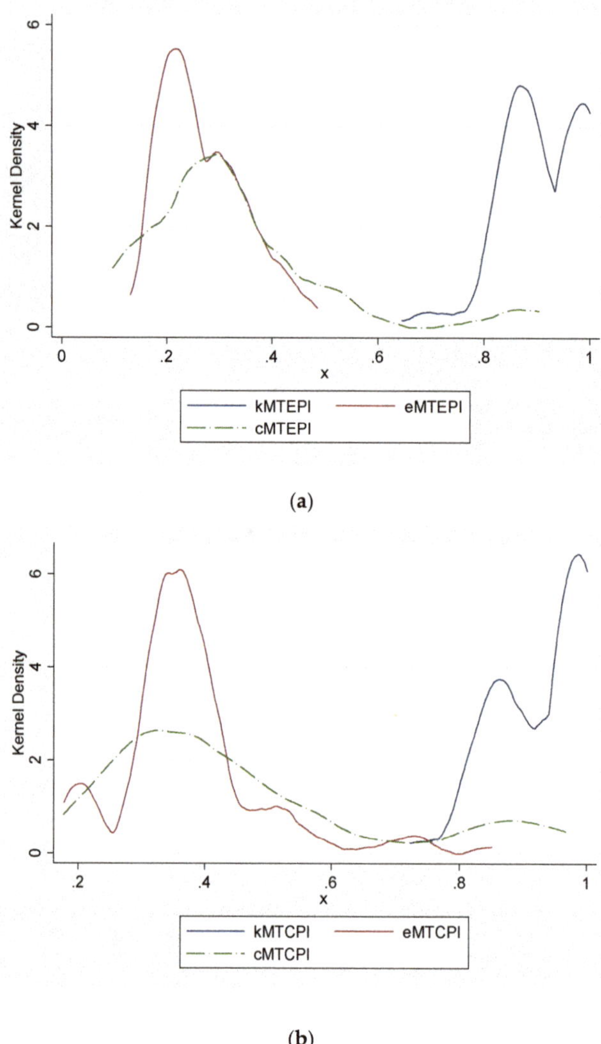

(a)

(b)

Figure 7. Kernel density plots of the Meta-Frontier Total-Factor Energy Performance Index (MTEPI) (a) and the Meta-Frontier Total-Factor Carbon Emission Performance Index (MTCPI) (b) for all three groups.

3.6. Abatement Potential for International Cooperation

Finally, we analyzed the potential of energy and carbon emission abatement for international cooperation between China and Korea. Our purpose was to investigate whether cooperative policies can contribute significantly to energy consumption and carbon emission abatement. Figures 8 and 9 demonstrate the potential of carbon emission abatement across the group-frontier and meta-frontier technologies. We find there will be huge potential CO_2 emission and energy reductions when all agglomerated cities can be on meta-frontier technologies. From 2009 to 2013, the Korean cities, and the eastern and northern Chinese could reduce CO_2 emissions by 0.53, 5.02 and 2.02 billion tons, respectively—a total reduction of 7.58 billion tons of CO_2 emissions for the five years if Korea and China proactively collaborate with each other. The Korean cities and the eastern and central Chinese cities can reduce their energy consumption by 69.86, 1452.74, and 408.01 million toe, respectively—a total reduction of 1930.62 toe energy for the five years.

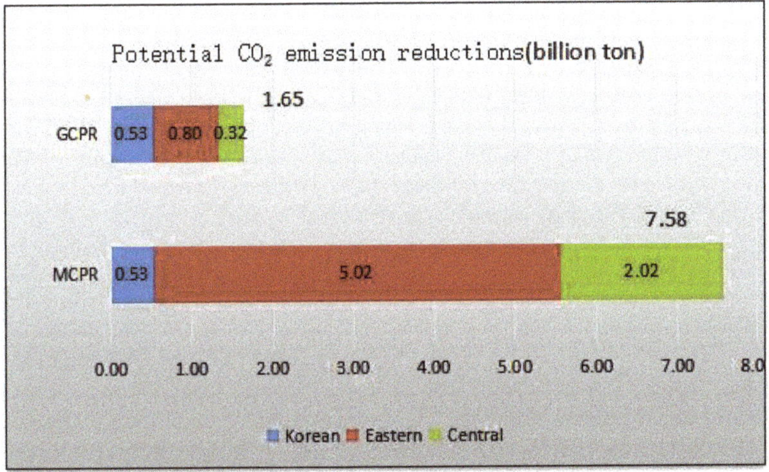

Figure 8. Potential CO_2 emissions reductions under group-frontier and meta-frontier technologies.

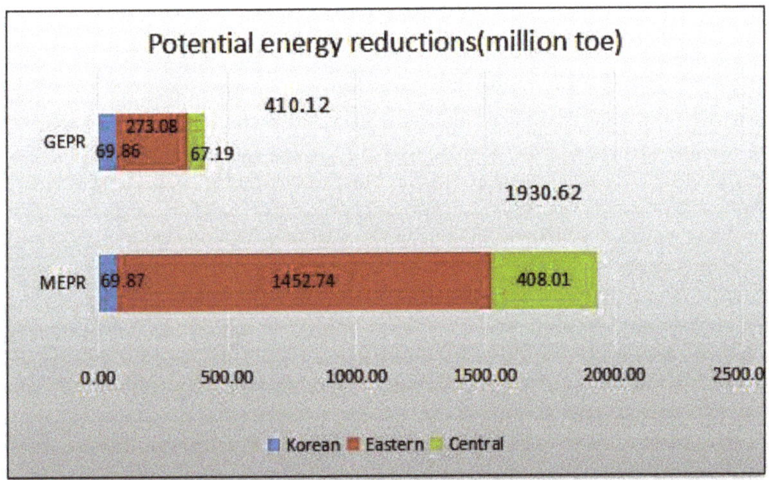

Figure 9. Potential energy reductions under group-frontier and meta-frontier technologies.

4. Conclusions

China, the world's largest energy consumer and carbon emitter, and Korea, another heavy emitter, have set ambitious emissions reduction targets of 60% to 65% per unit of GDP and 37% reduction below the business as usual (BAU) levels till 2030 based on 2005 levels, by China and Korea, respectively. The northeast Asian region is a dynamic region in both its economic activities as well as environmental policies. As the saying goes, "without the lips, the teeth feel the cold." China and South Korea should be intimately interdependent on each other for the effects of environmental policies.

Therefore, as neighboring countries, it is imperative for both governments to develop cooperative policies to curb energy consumption and CO_2 emissions in the region. However, establishing effective policy tools requires more complicated and yet systematic information about energy efficiency, CO_2

emission performance, and energy (CO_2 emission) reduction potential. China is a transitional economy with huge regional disparities and so, the Chinese government needs to pay more attention to their agglomerated cities. In the same way, Seoul has almost 50% of the Korean population, resulting in complex environmental issues from the excessive urbanization. To solve these environmental issues in the metropolitan cities, this study utilizes a GMNDDF of energy efficiency, CO_2 emission performance, and technology gap from a regional perspective. Our main findings are as follows:

First, in terms of energy efficiency and CO_2 emission performance, significant group heterogeneity exists in the three groups. Empirical results show significant differences among groups in energy efficiency, CO_2 emission performance, and meta-technology gaps. Nonetheless, to answer the question whether there is potential for international cooperation, we employed regional heterogeneity in our model. The Korean cities can play a leading role with benchmark efficiency level of the meta-frontier technology, as there are no meta-technology gaps in Korea at least. However, as far as the eastern and central cities of China are concerned, there is a considerable meta-technical gap between the group-frontier and the meta-frontiers. GTCPI and GTEPI values show an uptrend between 2009 and 2012, but a mild downtrend in 2013 in Korea, because the new President Park Geun-hye changed the Korean paradigm from "green growth" to "creative economy" and hence, environmental policies became more lax under the current administration. For the eastern and central cities of China, the average values for GTCPI and GTEPI show a downtrend in 2009 to 2010, increase in 2011, and downtrend again in 2012 to 2013, implying that the power cuts that plague China is the wrong policy, and sudden power outages are not conducive to business operations as it results in reduced production efficiency.

Second, there is no significant difference between total-factor and single-factor performance indices in the Korean cities, since Korea requires extensive labor and land resources to replace energy in the production process, which are quite limited. This inhibits the substitution of factors between energy and other production inputs. As a result, substitutability between production inputs is very limited in Korea. The eastern and central cities of China need technology, while the Korean agglomerated cities need labor and land resources. This illustrates that the Chinese government needs to restructure the production scale in the cities by learning from benchmark cities, and the Korean government needs to increase its production scale. They can also encourage enterprises to set up factories in China.

Third, three groups can make a huge contribution to reduce energy and CO_2 emissions by "catching up." It is important for the local governments to remove barriers and regulations that limit flow of production inputs and encourage investment in advanced meta-frontier technologies, especially in cities with low energy efficiency and low CO_2 performance. In this way, China can significantly reduce its carbon emissions.

Fourth, many cities in China could enhance their environmental performance by international cooperation. There are several benchmarking cases for the possible arrangement of the cooperative activities for CO_2 emissions abatement. For example, Daegu (2009) is the best DMU as it shows up 31 times as a reference set for Chinese cities to emulate. With the benchmark information, the inefficient DMUs should be developed to enhance the learning effect on the cases of their allocated reference sets.

We demonstrate even if the eastern cities of China are much better in their environmental performance as a group, they are still inferior to the central cities of China. Moreover, all central and eastern cities of China lag far behind in environmental performance compared to the Korean cities, implying that the regional cooperation between cities in Korea and China could enhance performance. Many Memoranda of Understanding (MOUs) between cities in China and Korea have been signed for cooperation. This study showed that verbal commitment alone is not helpful for sustainable performance. Instead we show that the comparative city level analysis that could enhance the benchmarking cities among the two countries.

Author Contributions: Conceptualization, N.W. and Y.C.; Methodology, N.W.; Software, N.W.; Validation, Y.C.; Formal Analysis, N.W.; Investigation, N.W.; Resources, Y.C.; Data Curation, N.W.; Writing-Original Draft Preparation, N.W.; Writing-Review & Editing, Y.C.; Visualization, Y.C.; Supervision, Y.C.; Project Administration, Y.C.; Funding Acquisition, Y.C.

Funding: This paper is supported by the National Research Foundation of Korea Grant (NRF-2017K2A9A2A06013582).

Conflicts of Interest: The authors declare no conflict of interest.

Appendix A.

Nomenclature	
GEPRI	Group-Frontier Energy Potential Reduction Index
MEPRI	Meta-Frontier Energy Potential Reduction Index
GTEPI	Group-Frontier Total-Factor Energy Performance Index
MTEPI	Meta-Frontier Total-Factor Energy Performance Index
GCPRI	Group-Frontier Carbon Emission Potential Reduction Index
MCPRI	Meta-Frontier Carbon Emission Potential Reduction Index
GTCPI	Group-Frontier Total-Factor Carbon Emission Performance Index
MTCPI	Meta-Frontier Total-Factor Carbon Emission Performance Index

Appendix B. Cities for Benchmark (2009 Case)

DMU	city	Benchmark DMU-year (Lambda)
1	Gyeonggi	10-12(0.031); 12-12(0.719); 15-12(0.201); 16-10(0.019); 16-12(0.030)
2	Incheon	2-13(0.216); 3-10(0.078); 12-12(0.624); 15-12(0.041); 16-10(0.041)
3	Seoul	3-09(1.000)
4	Gangwon	10-11(0.007); 12-09(0.847); 15-12(0.101); 16-12(0.045)
5	Jeju	7-09(0.624); 12-09(0.345); 15-13(0.031)
6	Gwangju	3-09(0.144); 10-11(0.002); 10-12(0.140); 12-10(0.669); 16-12(0.045)
7	Jeolbuk	7-09(1.000)
8	Jeolnam	3-13(0.152); 8-13(0.121); 15-13(0.727)
9	Busan	2-13(0.153); 3-10(0.180); 12-12(0.206); 15-12(0.119); 16-10(0.342)
10	Gyeongnam	3-10(0.076); 10-12(0.205); 10-13(0.142); 12-12(0.444); 15-13(0.133)
11	Ulsan	10-11(0.183); 10-12(0.349); 12-10(0.347); 15-12(0.008); 16-12(0.113)
12	Daegu	12-09(1.000)
13	Gyeongbuk	2-13(0.121); 3-10(0.009); 12-12(0.753); 15-12(0.067); 16-10(0.048)
14	Chungbuk	3-09(0.069); 6-11(0.111); 14-11(0.116); 16-12(0.704)
15	Chungnam	15-09(1.000)
16	Daejeon	3-11(0.005); 12-12(0.219); 16-10(0.325); 16-12(0.451)
17	Jiaozuo	7-09(0.663); 12-09(0.337)
18	Luoyang	12-09(0.932); 15-12(0.065); 16-12(0.002)
19	Zhengzhou	10-12(0.035); 12-09(0.321); 15-13(0.025)
20	Anqing	7-09(0.034); 12-09(0.966)
21	Chuzhou	7-09(0.873); 15-09(0.127)
22	Hefei	12-09(0.963); 15-12(0.014); 16-12(0.023)

DMU	city	Benchmark DMU-year (Lambda)
23	Huangshi	12-09(0.911); 15-12(0.070); 16-12(0.019)
24	Wuhan	12-12(0.938); 15-13(0.016); 16-10(0.011); 16-11(0.036)
25	Xianning	2-13(0.117); 3-11(0.061); 12-12(0.488); 16-10(0.334)
26	Taiyuan	10-12(0.019); 12-09(0.698); 15-13(0.036)
27	Xinzhou	7-09(0.249); 12-09(0.741); 15-13(0.010)
28	Yangquan	6-11(0.894); 12-09(0.018); 16-12(0.087)
29	Changde	2-13(0.373); 3-10(0.011); 12-12(0.131); 15-12(0.313); 16-10(0.173)
30	Changsha	10-11(0.034); 12-09(0.913); 15-12(0.016); 16-12(0.037)
31	Xiangtan	12-09(1.000)
32	Jiujiang	6-11(0.043); 12-09(0.957)
33	Nanchang	6-11(0.104); 12-09(0.896)
34	Shangrao	7-09(0.197); 12-09(0.536); 15-13(0.267)
35	Guangzhou	7-09(0.907); 15-09(0.093)
36	Shenzhen	15-09(0.450); 15-13(0.550)
37	Zhuhai	7-09(0.246); 12-09(0.569); 15-13(0.185)
38	Dongying	2-13(0.084); 3-10(0.035); 12-12(0.321); 15-12(0.028); 16-10(0.532)
39	Jinan	12-09(0.212); 15-12(0.677); 16-12(0.111)
40	Qingdao	12-09(0.526); 15-12(0.414); 16-12(0.059)
41	Baoding	7-09(0.417); 12-09(0.332); 15-12(0.251)
42	Beijing	12-12(0.492); 15-13(0.309); 16-11(0.198)
43	Shijiazhuang	12-09(0.474); 15-12(0.307); 16-12(0.219)
44	Tangshan	7-09(0.306); 12-09(0.335); 15-13(0.359)
45	Tianjin	8-13(0.202); 15-12(0.266)
46	Zhangjiakou	10-11(0.032); 12-09(0.788); 15-12(0.163); 16-12(0.017)
47	Changzhou	12-09(0.367); 15-12(0.525); 16-12(0.108)
48	Hangzhou	10-11(0.164); 12-09(0.591); 15-12(0.214); 16-12(0.7-09)
49	Nanjing	03-09(0.020); 10-11(0.065); 15-12(0.536); 16-12(0.378)
50	Ningbo	10-11(0.046); 12-09(0.428); 15-12(0.129); 16-12(0.397)
51	Shanghai	12-12(0.234); 15-13(0.605); 16-11(0.161)
52	Shaoxing	10-11(0.098); 12-09(0.747); 15-12(0.138); 16-12(0.018)
53	Suzhou	12-12(0.527); 15-13(0.361); 16-11(0.112)
54	Wenzhou	7-09(0.904); 15-09(0.034); 15-12(0.062)
55	Wuxi	12-09(0.947); 15-12(0.042); 16-12(0.03-09)
56	Yangzhou	12-09(0.829); 12-12(0.056); 15-13(0.114)
57	Zhenjiang	7-09(0.712); 12-09(0.288)

References

1. BP. BP Statistical Review of World Energy. 13 June 2018. Available online: http://www.bp.com/statisticalreview (accessed on 30 June 2018).
2. Sorrell, S. Reducing energy demand: A review of issues, challenges and approaches. *Renew. Sustain. Energy Rev.* **2015**, *47*, 74–82. [CrossRef]
3. Ning, Y.; Miao, L.; Ding, T.; Zhang, B. Carbon emission spillover and feedback effects in China based on a multiregional input-output model. *Resour. Conserv. Recycl.* **2019**, *141*, 211–218. [CrossRef]
4. Park, H.; Hong, W. Korea's emission trading scheme and policy design issues to achieve market-efficiency and abatement targets. *Energy Policy* **2014**, *75*, 73–83. [CrossRef]
5. Chavez, A.; Ramaswami, A. Progress toward low carbon cities: Approaches for transboundary GHG emissions' foot printing. *Carbon Manag.* **2014**, *2*, 471–482. [CrossRef]
6. Hoornweg, D.; Sugar, L.; Gomez, C.L.T. Cities and greenhouse gas emissions: Moving forward. *Environ. Urban.* **2011**, *23*, 207–227. [CrossRef]
7. Kennedy, C.; Steinberger, J.; Gasson, B.; Hansen, Y.; Hillman, T.; Havranek, M.; Pataki, D.; Phdungsilp, A.; Ramaswami, A.; Mendez, G.V. Methodology for inventorying greenhouse gas emissions from global cities. *Energy Policy* **2010**, *38*, 4828–4837. [CrossRef]
8. Kennedy, C.; Demoullin, S.; Mohareb, E. Cities reducing their greenhouse gas emissions. *Energy Policy* **2012**, *49*, 774–777. [CrossRef]
9. Wang, H.; Zhang, R.; Liu, M.; Bi, J. The carbon emissions of Chinese cities. *Atmos. Chem. Phys.* **2012**, *12*, 6197–6206. [CrossRef]
10. Li, A.; Hu, M.; Wang, M.; Cao, Y. Energy consumption and CO_2 emissions in Eastern and Central China: A temporal and a cross-regional decomposition analysis. *Technol. Forecast. Soc. Chang.* **2016**, *103*, 284–297. [CrossRef]
11. Kim, K.; Kim, Y. International comparison of industrial CO_2 emission trends and the energy efficiency paradox utilizing production-based decomposition. *Energy Econ.* **2012**, *34*, 1724–1741. [CrossRef]
12. Lin, B.Q.; Du, K.R. Technology gap and China's regional energy efficiency: A parametric meta-frontier approach. *Energy Econ.* **2013**, *40*, 529–536. [CrossRef]
13. Bian, Y.W.; He, P.; Xu, H. Estimation of potential energy saving and carbon dioxide emission reduction in China based on an extended non-radial DEA approach. *Energy Policy* **2013**, *63*, 962–971. [CrossRef]
14. Wang, K.; Wei, Y.M. China's regional industrial energy efficiency and carbon emissions abatement costs. *Appl. Energy* **2014**, *130*, 617–631. [CrossRef]
15. Wang, Z.H.; Feng, C.; Zhang, B. An empirical analysis of China's energy efficiency from both static and dynamic perspectives. *Energy* **2014**, *74*, 322–330. [CrossRef]
16. Yao, X.; Zhou, H.C.; Zhang, A.Z.; Li, A.J. Regional energy efficiency, carbon emission performance and technology gaps in China: A meta-frontier non-radial directional distance function analysis. *Energy Policy* **2015**, *84*, 142–154. [CrossRef]
17. Lee, H.; Choi, Y. Greenhouse gas performance of Korean local governments based on non-radial DDF. *Technol. Forecast. Soc. Chang.* **2018**, *135*, 13–21. [CrossRef]
18. Farrel, M.J. The measurement productive efficiency. *J. R. Stat. Soc. Ser. A Gen.* **1957**, *120*, 253–290. [CrossRef]
19. Charnes, A.; Cooper, W.W.; Rhodes, E. Measuring the efficiency of decision making units. *Eur. J. Oper. Res.* **1978**, *2*, 429–444. [CrossRef]
20. Banker, R.D.; Charnes, A.; Cooper, W.W. Some Models for Estimating Technical and Scale Inefficiencies in Data Envelopment Analysis. *Manag. Sci.* **1984**, *30*, 9. [CrossRef]
21. Tone, K. A Slacks-based Measure of Efficiency in Data Envelopment Analysis. *Eur. J. Oper. Res.* **2001**, *130*. [CrossRef]
22. Färe, R.; Grosskopf, S.; Norris, M.; Zhang, Z. Productivity growth, technical progress, and efficiency change in industrialized countries. *Am. Econ. Rev.* **1994**, *84*, 66–83.
23. Chung, Y. Directional Distance Functions and Undesirable Outputs. Ph.D. Dissertation, Southern Illinois University at Carbondale, Carbondale, IL, USA, 1996.
24. Färe, R.; Grosskopf, S.; Pasurka, C.A. Environmental production functions and environmental directional distance functions. *Energy* **2007**, *32*, 1055–1066. [CrossRef]

25. Oggioni, G.; Riccardi, R.; Toninelli, R. Eco-efficiency of the world cement industry: A data envelopment analysis. *Energy Policy* **2011**, *39*, 2842–2854. [CrossRef]
26. Riccardi, R.; Oggioni, G.; Toninelli, R. Efficiency analysis of world cement industry in presence of undesirable output: Application of data envelopment analysis and directional distance function. *Energy Policy* **2012**, *44*, 140–152. [CrossRef]
27. Fukuyama, H.; Weber, W.L. A directional slacks-based measure of technical inefficiency. *Socio-Econ. Plan. Sci.* **2009**, *43*, 274–287. [CrossRef]
28. Barros, C.P.; Managi, S.; Matousek, R. The technical efficiency of the Japanese banks: Non-radial directional performance measurement with undesirable outpu. *Omega* **2012**, *40*, 1–8. [CrossRef]
29. Tian, Y.H.; He, S.B.; Hu, S.Q. Re-estimation of total factor productivity growth in a region under environmental constraints:1998-2008. *China's Ind. Econ.* **2011**, *1*. [CrossRef]
30. Wang, B.; Wu, Y.R.; Yan, P.F. China's regional environmental efficiency and total factor productivity growth. *Econ. Res.* **2010**, *5*. Available online: http://en.cnki.com.cn/Article_en/CJFDTOTAL-JJYJ201005008.htm (accessed on 30 June 2018).
31. Pastor, J.T.; Knox Lovell, C.A. A Global Malmquist Productivity index. *Econ. Lett.* **2005**, *88*, 2. [CrossRef]
32. Zhou, P.; Ang, B.W.; Wang, H. Energy and CO_2 emission performance in electricity generation: A non-radial directional distance function approach. *Eur. J. Oper. Res.* **2012**, *221*, 625–635. [CrossRef]
33. Choi, Y.; Zhang, N.; Zhou, P. Efficiency and abatement costs of energy-related CO_2 emissions in China: A slacks-based efficiency measure. *Appl. Energy* **2013**, *98*, 198–208. [CrossRef]
34. Färe, R.; Grosskopf, S. *New Directions: Efficiency and Productivity*; Springer: New York, NY, USA, 2005.
35. Färe, R.; Grosskopf, S.; Lovell, C.A.K.; Pasurka, C. Multilateral productivity comparisons when some outputs are undesirable: A nonparametric approach. *Rev. Econ. Stat.* **1989**, *71*, 90–98. [CrossRef]
36. Zhang, N.; Chen, Z.F. Sustainability characteristics of China's Poyang Lake Eco-Economics Zone in the big data environment. *J. Clean. Prod.* **2017**, *142*, 642–653. [CrossRef]
37. Chiu, C.R.; Liou, J.L.; Wu, P.I.; Fang, C.L. Decomposition of the environmental inefficiency of the meta-frontier with undesirable output. *Energy Econ.* **2012**, *34*, 1392–1399. [CrossRef]
38. Battese, G.E.; Rao, D.S.P.; O'Donnell, C.J. A meta-frontier production function for estimation of technical efficiencies and technology gaps for firms operating under different technologies. *J. Prod. Anal.* **2004**, *21*, 91–103. [CrossRef]
39. O'Donnell, C.J.; Rao, D.S.P.; Battese, G.E. Meta-frontier frameworks for the study of firm-level efficiencies and technology ratios. *Empir. Econ.* **2008**, *34*, 231–255. [CrossRef]
40. Zhang, J.; Wu, G.Y.; Zhang, J.P. The Estimation of China's provincial capital stock: 1952–2000. *Econ. Res. J.* **2004**, *10*, 35–44. [CrossRef]
41. Young, A. Gold into base metals: Productivity growth in the People's Republic of China during the reform period. *J. Polit. Econ.* **2003**, *111*, 1220–1261. [CrossRef]
42. Olivier, J.G.J.; Schure, K.M.; Peters, J.A.H.W. Trends in global CO_2 and total greenhouse gas emissions. PBL Netherlands Environmental Assessment Agency, 2017; p. 5. Available online: https://www.pbl.nl/sites/default/files/cms/publicaties/pbl-2017-trends-in-global-co2-and-total-greenhouse-gas-emissons-2017-report_2674.pdf (accessed on 28 October 2018).
43. Xu, X.; Huo, H.; Liu, J.; Shan, Y.; Li, Y.; Zheng, H.; Guan, D.; Ouyang, Z. Patterns of CO_2 emissions in 18 central Chinese cities from 2000 to 2014. *J. Clean. Prod.* **2018**, *172*, 529–540. [CrossRef]
44. Wang, B.; Lai, P.H.; Yang, Y.S.; Yu, L.J. Study on the potential of energy efficiency and energy conservation and emission reduction in China considering the differences of natural environment. *Rev. Ind.* **2016**. (In Chinese) Available online: http://www.cnki.com.cn/Article/CJFDTotal-TQYG201601008.htm (accessed on 28 October 2018).

© 2019 by the authors. Licensee MDPI, Basel, Switzerland. This article is an open access article distributed under the terms and conditions of the Creative Commons Attribution (CC BY) license (http://creativecommons.org/licenses/by/4.0/).

Article

System and Cost Analysis of Stand-Alone Solar Home System Applied to a Developing Country

Chowdhury Akram Hossain [1], Nusrat Chowdhury [2], Michela Longo [3] and Wahiba Yaïci [4,*]

[1] Department of Electrical and Electronic Engineering, American International University-Bangladesh, Dhaka 1229, Bangladesh; chowdhury.akram@aiub.edu

[2] Department of Electrical and Electronic Engineering, Daffodil International University, Dhaka 1207, Bangladesh; nusrat.eee@diu.edu.bd

[3] Department of Energy, Politecnico di Milano, 34–20156 Milano, Italy; michela.longo@polimi.it

[4] CanmetENERGY Research Centre, Natural Resources Canada, Ottawa, ON K1A 1M1, Canada

* Correspondence: wahiba.yaici@canada.ca; Tel.: +1-613-996-3734

Received: 17 February 2019; Accepted: 1 March 2019; Published: 6 March 2019

Abstract: Power is one of the key requirements for the development of economies and upgrading of standards of living of developing countries. Countries such as Bangladesh depend largely on fossil fuels such as diesel fuel and natural gas to produce the main proportion of their electricity. However, this country's combination of limited natural gas reserves high fuel prices and escalating costs of transmission and distribution lines has greatly increased the unit cost of electricity generation and it is becoming difficult for customers to pay for electricity. On the other hand, burning fuel causes environmental pollution that leads to global warming which is ultimately responsible for climate change and its devastating consequences. In this study, we have recommended a stand-alone system for the traditional consumption of domestic electric use at residential units in Bangladesh. We have shown a comparison of using the stand-alone photovoltaic (PV) system with the traditional grid connection. Although the initial set-up cost is high, it becomes profitable as people are supplied with electricity, which is being generated from PV as a result minimizing the energy cost from the grid, and in addition, they can later make savings from this system. This paper, therefore, aims at determining the optimum size of the rooftop solar home system that will fulfil all the criteria for powering up electrical appliances at an affordable price. Comparative analysis of both energy systems based on the cost calculation has been performed by means of the Hybrid Optimization of Multiple Energy Renewables (HOMER) software. The validity of this proposal and its usefulness is also analysed.

Keywords: solar home systems (SHS); levelized cost of energy (LCOE); photovoltaic system; HOMER

---

1. Introduction

Bangladesh's economy is mostly dependent on agriculture and its industrial sector. To enhance employment opportunities, policies and incentives were put in place to boost the growth of both these sectors. However, electricity generation and supply in the country is much lower than is required for the growing demand, a shortfall that has impeded sustainability of the country's economic growth. Despite achievements, a large proportion of the country's inhabitants even now does not gain access to power. Per the study carried out by the International Energy Agency (IEA), 1.1 billion persons totally did not get access to electricity in 2016. This represents a decrease of 35% in the total people compared to 2000 [1]. In Bangladesh only 60% of the population get access to power and it would require an estimated 15 years to extend the service to every one [2]. Currently, 99% of energy consumed in rural households is based on fossil fuels. Relentlessly, energy demands in the country continue to rise daily and a power crisis has become imminent.

Natural gas plays a crucial role in enabling the country to meet its energy needs. It accounts for 72% of the total commercial energy use and 82% of the total electricity produced [3,4]. Nonetheless, limited availability of the resource in the country translates to a deficit of 4.02×10^6 m^3/day in 2011. It is estimated that this will rise to 48.54×10^6 m^3/day by 2019–2020. As a result, Bangladesh will require to complement 19,000 MW of extra power, even if its national GDP increase stays low at 5.5% until 2025. This factor will likely cause the gas demand to escalate up to 129.32×10^6 m^3/day by 2019–20 [5]. In order to resolve the power problem, the national grid's capacity was recently enhanced substantially by the addition of several power-generating units although sustained high demand and increasing pervasive needs for power have minimised these accomplishments and created a serious challenge for power stations. To solve this energy challenge, this paper analyses that renewable energy, is a preferable alternative power source. Since the year of 2010 the reduced cost of electricity from utility-scale solar photovoltaic (PV) projects has been remarkable. The worldwide weighted average levelized cost of power (LCOE) of utility-scale solar PV has fallen 73% since 2010, to USD 0.10/kWh for new ventures dispatched in 2017, though the non-renewable energy source terminated electricity cost extend in 2017 was assessed to range from USD 0.05 to USD 0.17/kWh, contingent upon the fuel and nation [6]. While investment expenditures of renewable energy are usually higher compared to fossil fuels, this option becomes economically viable when all other factors such as health hazards, environmental problems and lower operating costs are taken into consideration. Increasing the renewable energy usage in the European Union (EU), many directives and agreements have been signed. As the crucial year 2020 is coming back nearer, the European Commission revealed a proposal for a revised Renewable Energy Directive to create the EU a worldwide leader in renewable energy and make sure that the target of a minimum of 27% renewables within the final energy consumption in the EU by 2030 is met [1]. Bangladesh's Renewable Energy Policy (REP) set targets for exploiting renewable energy resources to meet 5% of the total power demand by 2015 and 10% by 2020 [7]. During 2016, a minimum of 75 GW of solar PV capability was added worldwide and at the top of 2016 international solar PV capacity in total had reached 303 GW [1].

The Italian government has effectively implemented the concept of rooftop PV in a project. The project clearly described the calculation for the reimbursement time along with the economic value. Similar type of investigations were also executed in India and Indonesia [8,9]. Many researches have also been performed considering the technical and economic aspects of solar PV system for Bangladesh specially focusing the urban and remote areas [10,11]. A basic solar PV system consists of a small solar panel, a battery, and a charge controller [12]. The system costs approximately $350, which can be sourced from a financial institution (preferably a microfinance bank) at a modest rate [13]. Through discussions with local banks and microfinance companies, convenient payment plans can be negotiated, which would ultimately help to finance the large initial investment cost involved in purchasing the new system. It is notable that the government of Bangladesh is currently making efforts to popularize renewable energy and also make it available to ordinary people. Its efforts are complemented by those of private financing companies including the Rural Electrification Board (REB) and the Infrastructure Development Company Limited (IDCOL), which are similarly investing in solar systems.

For different household appliances and lighting purpose, by 2017 a total number of 5.2 million standalone PV systems, which are also known as Solar Home Systems (SHSs) with a capacity of 218 MW, were installed. The total capacity of the installed SHSs was 218 MW. This information was published in "Renewables 2018-Global Status Report", which is released annually by the Paris-based energy think-tank REN21 in June. In the report it was also mentioned that with a total number of 17 million users, Bangladesh stood the second highest users of SHSs after India with a total of 148 million users. This statistic supports that approximately 13% of the total population uses electricity provided by SHSs [14,15].

The world's most abundant and permanent energy source is solar radiation. In every year per minute solar energy received by the surface of the Earth is much greater than the energy utilisation by the entire population. For the time being, solar power, being accessible all over, is enticing for complete systems notably within the rural parts of all the developing nations [8]. Solar energy has a significant potential for success in Bangladesh. The average solar radiation during the year for Bangladesh is 4.57 kWh/m^2/day. The system is endearing because once payment for initial set up is completed, there are no further additional cost for fuel incurred throughout its lifetime. Savings can later be generated through this innovative scheme. Savings, which result from discontinuing the use of candles, kerosene fuel, can be channeled into making the monthly payments whilst at the same time promoting a greener environment.

The software Hybrid Optimization Model for Electric Renewables (HOMER 2.68 beta) [16,17], was utilised as a modelling tool, which in recent years has been employed for optimal sizing and simulations of micro-grids by many researchers in several counties [18–22]. A literature review reveals that many studies were undertaken in various counties on the advantages of rooftop solar off grid system [23–25]. The authors' previous work has contributed to the proposal of a new solar energy system for an electric vehicle system design and for its optimal size using HOMER [26]. In this study, a recommendation that the renewable energy-based rooftop stand-alone solar home system is a viable alternative energy source. According to our knowledge, it appears that no study has been accomplished on such residential system scale for use in developing country, and it assures the novelty of this publication. The objective of the current investigation is to examine the technical and economic feasibility of a rooftop solar home system to evaluation the advantages of users swapping over from grid to solar PV as an alternative source of electric energy. HOMER tool is utilised to specify the optimal size of the equipment taking into account the geographical and meteorological data of the location under study. HOMER applies the net present cost method for ranking the system's suitability.

Therefore, the reminder of the paper is structured as follows. Section 2 gives full details of the rooftop solar home system utilised to generate heat and power. Section 3 provides basic model data such as input data and component specifications for sizing the various system components. Simulation results are discussed in Section 4, including the optimal sizing of the system and the cost analysis. Finally, conclusions are drawn in Section 5.

2. Rooftop Solar Home System Description

A rooftop solar home system (Figure 1) may be a possible alternative supply of electrical energy. This method is changing into tremendously popular and is taken into account as a response to the energy drawback within the country's rural areas. Bangladesh is placed between 88.04 and 92.44 degrees east and 20.30 and 26.38 degrees north, a novel location that provides the country prepared access to solar power for trappings and use [2,27]. Indeed, it experiences a minimum of 10 hour of sunlight daily throughout the year. The average solar irradiance of Bangladesh is around 5 kWh/m^2/day [27,28]. Since the solar modules solely produce electrical energy throughout the day, it is necessary to store this energy to be used at night time and through cloudy days. Storage systems utilized typically employ rechargeable lead-based batteries, because of their ability to optimally receive energy at each low- and high-input voltages. A battery regulator is installed in to stop overcharging and deep discharges. The solar photovoltaic system can offer battery output voltage of 12- or 24-Volts electricity (DC) in most cases. To supply devices, that are solely obtainable for AC voltage, a power inverter is used. it is important to arrange such a complete PV system which will match the potential energy consumption with the native average solar irradiation, the ensuing energy production and therefore the needed storage capability [29].

The rooftop stand-alone SHS is now very popular in Bangladesh due to its accessibility and relative affordability as discussed earlier. Owing to the high rate of adoption of SHSs in rural areas, researchers generated a list of probable factors that determine an individual's likelihood of purchasing or adopting it. Easy access to loans from several financing companies now makes it possible for the middle- and lower-income groups to afford the SHS.

Figure 1. Rooftop solar home system.

The government of Bangladesh has pledged to use 5% of their electricity using renewable energy sources in 2008 Washington International Renewable Energy Conference. Following that in 2009, a fund of $ 29 M was set up by Bangladesh Bank to create public awareness in the solar sector. The central bank also had several agreements with state-owned and private sector banks to provide funds to different financing schemes related with renewable sources with a very low interest rate of about 5% [30].

Therefore, it appears that the solar home system is systematically becoming more entrenched and established in the country as a viable means of eliminating the collective problems associated with fossil fuels [31].

Promoters of this system perform feasibility and techno-economic analysis to enhance affordability of the SHS. Some innovative financing structures have been developed and are discussed later in this paper. Studies have also been conducted on the cost of energy for the stand-alone SHS system. As mentioned in the previous section, an optimization tool known as HOMER has been used for all the simulation and analysis of the data. The HOMER software, NREL's (National Renewable Energy laboratory), and the micro-power optimization model, can evaluate a range of equipment options over varying constraints and sensitivities to optimise small power systems. HOMER's flexibility is suitable in the assessment of design, planning and decision-making areas for ensuring feasible projects [16,17]. A detailed description of the available mathematical models implemented in HOMER goes beyond the scope of this study. For a comprehensive analysis, the interested readers can refer to the work of Bahramara et al. [18]. After the input data are fed into HOMER, optimal sizes of SHS equipment are established in three stages comprising simulation, optimisation, and sensitivity analysis as illustrated in Figure 2 [16–18].

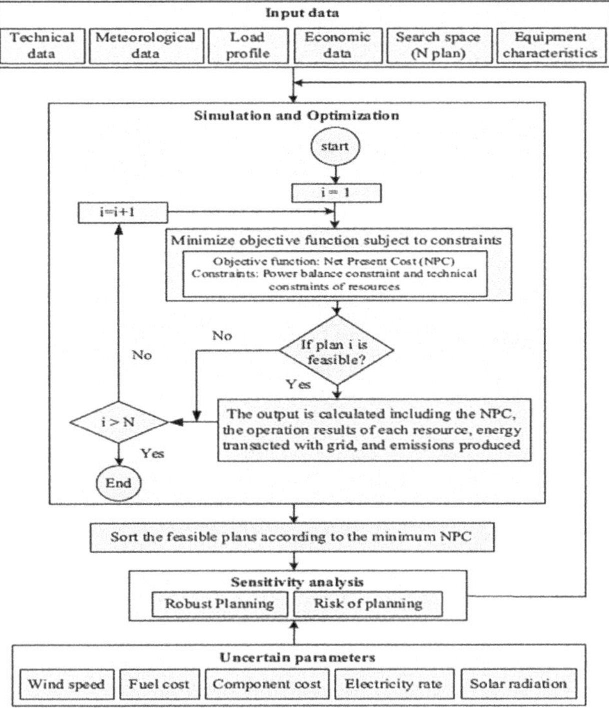

Figure 2. The comprehensive structure of Hybrid Optimization of Multiple Energy Renewables (HOMER) simulation, optimization and sensitivity procedures [18].

3. Data Collection and Component Specifications

3.1. Load Definition

Consider a typical household, which has 4 CFL (compact fluorescence lamps), 3 alternating current (AC) fans and 1 television (TV). Operating time for the CFL is assumed to be 8 h/day, fan is 12 h/day and TV is 4 h/day. The required power for every CFL is 23 W, AC fan is 80 W, and the TV is 100 W. Therefore, one household requires an approximate load of 432 W. Table 1 describes the average load consumption. It should be mentioned that the consideration of CFL was taken into account based on the typical condition of the proposed research area. The energy efficiency can be increased considering LED lamps and other energy-efficient devices.

Table 1. Average load consumption.

Loads	Power Rating [W]	No. of Pieces	Operating [h/day]
CFL	23	4	8
Fan	80	3	12
TV	100	1	4

The energy demand per day (E_d) can be calculated as shown in Equation (1):

$$E_d = \frac{\sum_{k=1}^{N} A_k P_k Q_k}{1000} \ (kWh) \tag{1}$$

where:

k = Index of each type of load such as fan, lights, TV etc.;

A_k = number of hours k^{th} device type used per day;

P_k = power rating of kth device type;

Q_k = number of devices of kth type.

According to the information provided in Table 1, the approximate energy demand per day is 4.02 kWh for each household.

3.2. Daily Load Profile

In HOMER according to the load consumption, there is an assumption that off-peak time runs from 12 a.m. to 6 a.m where we consider only fans as a load as it is night time, and therefore the load consumption is almost constant and low during these hours, while the peak time is from 5 p.m. to 11 p.m. The load variation for a day for different months are shown in Figure 3.

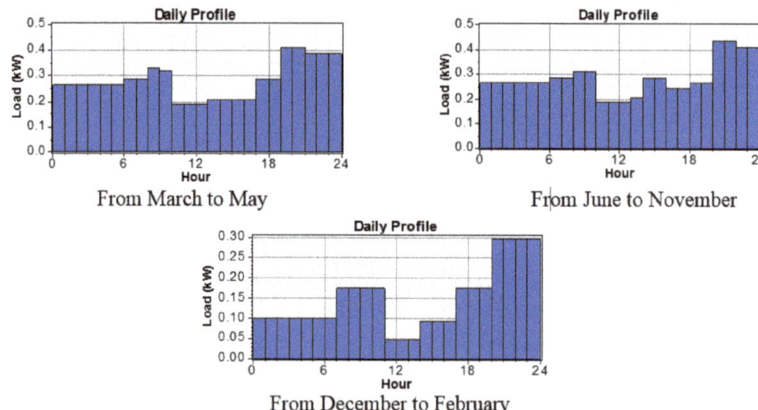

Figure 3. Typical daily load profiles.

The seasonal profile for the household load has been represented in Figure 4. It is observed that the maximum value achievable is in June with the value equal to 0.6 kW. Overall, the average value is above 0.4 kW. For January, February and December, the value is low and is around 0.45 kW.

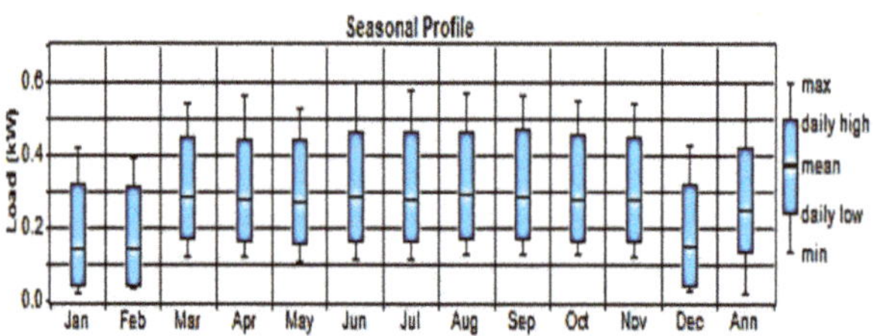

Figure 4. Seasonal load profile.

3.3. Size Consideration of Photovoltaic Array

To calculate the PV size, different combinations along with other parameters for panel was given as input in the software to get optimized performance for the proposed system. Here we have considered a lifetime of 25 years for the PV with no tracking system. The cost of the PV has been considered as $0.36/ W_P [32]. For the other costs such as installation we have considered $360, and for operation/maintenance $5/year considering the standard current market value. Eq. (2) has been used to find the PV generation capacity.

$$PV_{capcity} = \frac{E_d}{I_d \times d} \ (kW) \tag{2}$$

where:

I_d = average number of hours the sun shines;
d = de-rating factor, which takes into account the effects of efficiency and changes in the solar generation during the day.
The typical values of I_d and d are 5 h and 80%, respectively in Bangladesh.

3.4. Battery Size

Electricity is supplied from battery at night and also used in emergency situations such as bad weather. This study considered a battery which has nominal voltage of 12 V, and a nominal capacity of 200 Ah (2.4 kWh). The lifetime of battery is assumed to be five years. The installation, replacement, operating and maintenance costs of one battery are taken as $180, $180 and $7/year, respectively [32]. The number of batteries can be determined using the following mathematical equation:

$$N_{Batteries} = \frac{E_d \times n_d}{V_{battery} \times AH \times DOD} \tag{3}$$

where:

n_d = the number of days of backup power required;
$V_{battery}$ = the battery voltage rating;
AH = the ampere-hour rating;
and DOD is the depth of discharge of the battery system.

3.5. Charge Controller

A charge controller is similar to the voltage regulator. It regulates the voltage and current that is emitted from the solar panels and going into the battery. Mainly "12 Volt" panels are set to about 16 to 20 volts, so if there is no voltage regulation the batteries face the risks of being damaged from overcharging. Thus, the charge controller protects against overly high voltage, over current, short circuit, polarity reverse and lighting. The LCD display indicates levels of voltage, current and short circuit. This study considered a charge controller, which is priced at $70 [32]. This is an electronic device which guarantees longer lifespan if properly used in the system and thereby helps reduce the cost of the system.

3.6. Inverter

An inverter is utilised to convert the power from DC to AC. For 1-kw inverter the installation and replacement costs are considered as $310 and $310, respectively [32]. Various sizes of inverters are applied in the system. The lifetime of an inverter is taken to be 15 years with an efficiency of 90%. As this is power electronics, a higher lifespan can be utilised.

3.7. Solar Radiation and Clearness Index

Bangladesh's geographical location in terms of latitude and longitude are 23°59′ N and 90°27′ E respectively. The scaled annual average radiation (Figure 5) for Bangladesh is five kWh/m^2/day.

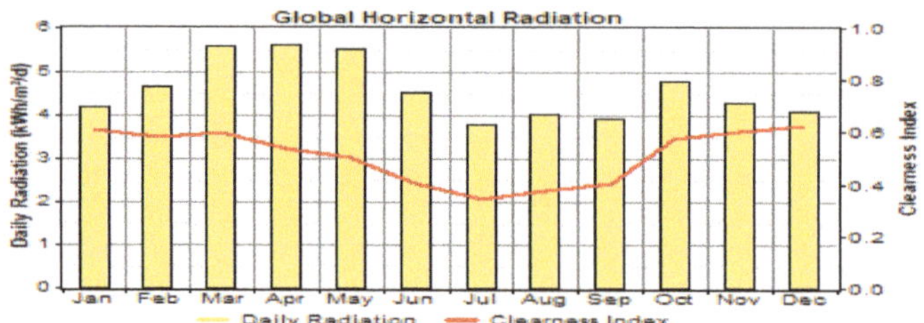

Figure 5. Clearness index and average daily solar radiation.

It should be well noted that using the Solar Resource window in HOMER we can enter the value for either the average radiation or the average clearness index. If any of the value is provided then the other one is automatically calculated. On the other hand, the clearness index is given in a scale of 0 and 1 which give provides the fraction of the solar radiating striking the top of the atmosphere which can reach the Earth's surface. Equation (4) defines the monthly average clearness index:

$$K_T = \frac{H_{ave}}{H_{o,ave}} \tag{4}$$

where:

H_{ave} = monthly average radiation on the horizontal surface of the earth [kWh/m^2/day];
$H_{o,ave}$ is extra-terrestrial horizontal radiation, meaning the radiation on a horizontal surface at the top of the Earth's atmosphere [kWh/m^2/day].

4. Results and Discussion

4.1. System Analysis

The system is fed by solar PV arrays and batteries and it will be installed in the rooftop of households as depicted in Figure 6. There is no grid connection in the system. In this system, during the day, the PV array supplies power to the load directly and also charges the battery. At any time, day or night, the load is able to draw power from the battery. The charge controller prevents excessive overcharging and deep discharging of the battery.

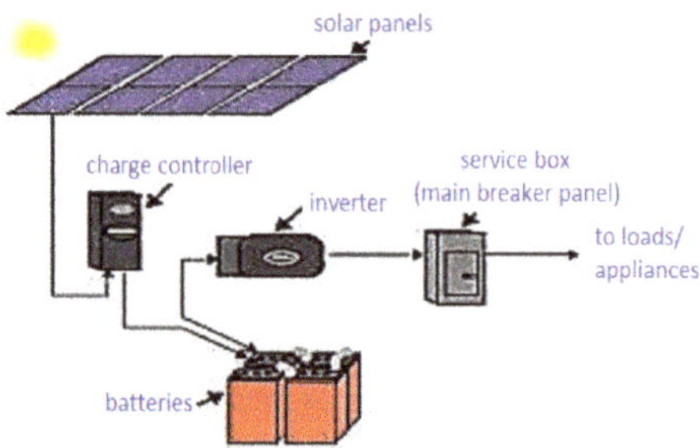

Figure 6. Block diagram of stand-alone SHS.

The total solar home system is designed in HOMER. In HOMER, we have considered 10% day–to-day load variation. Considering the noise, system loss and random variability of the load, the system is designed for 602 W peak load, which is much higher than the actual load consumption mentioned in Section 3.1. The daily energy consumption is considered to be 6 kWh for the proposed case.

The system under study is based on AC load, which is supplied by PV and at night or emergency backup battery is used. For AC supply, an inverter is used which converts the DC power that is obtained from PV and battery into AC power. For system analysis this model is ran by HOMER and after simulation with best possible combinations, we get the following optimum result shown in Figure 7 for this system.

			PV (kW)	6FM200D	Conv. (kW)	Initial Capital	Operating Cost ($/yr)	Total NPC	COE ($/kWh)	Ren. Frac.
			2.0	8	1.0	$ 2,770	302	$ 10,331	0.211	1.00
			3.0	4	2.0	$ 2,720	305	$ 10,347	0.212	1.00
			2.0	6	2.0	$ 2,720	308	$ 10,426	0.217	1.00

Figure 7. Optimised result from HOMER.

From the HOMER analysis, the optimal result is attained for the arrangement of 2 kW PV panels, 1 kW inverter and 8 batteries (each has 200 Ah, 12 V). This result is based on the lowest LCOE and operating cost for SHS.

To calculate the LCOE, which is expressed as the average cost per kWh of useful electrical energy generated by the system, HOMER divides the annualised cost of generating electricity (the cost of serving the load (Cs) subtracted from the total annualised cost (*AnC*) by the total useful electric energy production (WT). In HOMER, a 6% annual real interest rate is considered. Hence, it calculates the LCOE using Equation (5):

$$LCOE = \frac{(AnC - Cs)}{WT} \tag{5}$$

where LCOE = 0.211 $/kWh (16.88 Tk/kWh).

The total system's power is supplied by 2 kW$_p$ PV, 24 V system with 4 panels in parallel to maintain the PV voltage 24 V and 4 string of batteries (each string has two 12 V, 200 Ah batteries connected in series) to meet the load demand with one charge controller and an inverter. Figure 8 presents average monthly production of SHS throughout the year. It is clear that production of the solar power varies with respect to different seasons, but it always produces enough necessary power. From the monthly electrical production, it can be seen that the production is high from March to June.

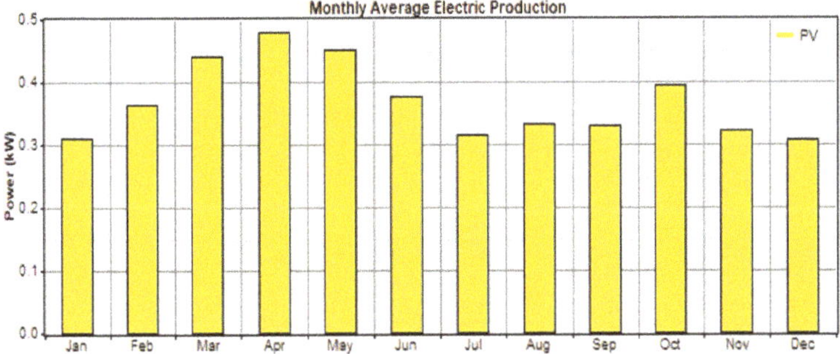

Figure 8. Monthly average electrical production SHS.

It can be calculated from the input data that this system will be producing a total of 3232 kWh/yr from the PV system and the consumption of the proposed system will be around 1959 kWh/yr, which allows us to conclude that the proposed system will be producing enough power for additional electrical equipment to use. The total power surplus from the system is equal to 1273 kWh/yr, which can be sold to or shared with neighbours under certain conditions.

4.2. Cost Analysis

In the existing grid connection system if the user paid the monthly electricity for 180 units (6 units per day) then the total cost of the utility power from the grid (C_{Grid}) can be calculated using Equation (6):

$$C_{Grid} = 365 \times E_d \times C_{unit} \times L \tag{6}$$

where:

C_{unit} = the cost of per unit energy and L is the project lifetime.

Therefore, the total electricity bill for 25 years for grid-connected supply without using solar connection is \$3901 (312,075 Tk) considering per unit cost of utility is \$0.0713 (5.7 Tk). On the other hand, the total cost of the rooftop solar system without taking into account any subsidy (C_{pv}) can be evaluated using Equation (7):

$$C_{pv}(without\ subsidy) = C_{PV\ panel} + C_{battery} + C_{inverter} + C_{Charge\ controller} + C_{others} \tag{7}$$

where:

$C_{PV\ panel}, C_{battery}, C_{inverter}, C_{Charge\ controller}$ are the capital costs of PV, battery, inverter and charge controller, respectively. The other system capital cost is denoted by C_{others}.

The initial cost of this proposed rooftop PV system is \$2770 (221,600 Tk). Thus, the overall profit during the full project life realised by switching from utility grid connection to rooftop solar PV system is \$1131 (90,480 Tk). Accordingly, the payback period can be calculated using Equation (8):

$$Payback\ Period = \frac{C_{pv}}{E_d \times C_{unit} \times 365}\ (yr) \tag{8}$$

If users replace their existing system with rooftop solar home system, then within 18 years they can recover their initial cost; that means the payback period is 18 years. Here in the rooftop PV system no subsidy has been considered for that reason the payback period is much higher. If we consider 40% government subsidy as per process, then the payback period will be much lower than the present value. Therefore, by adopting this proposed system users can save their money and improve their economic conditions. In a SHS, generally a large portion of cost around 80% of total cost has been added because of the battery and its short lifespan. The cost summary of the proposed project is shown in Figure 9.

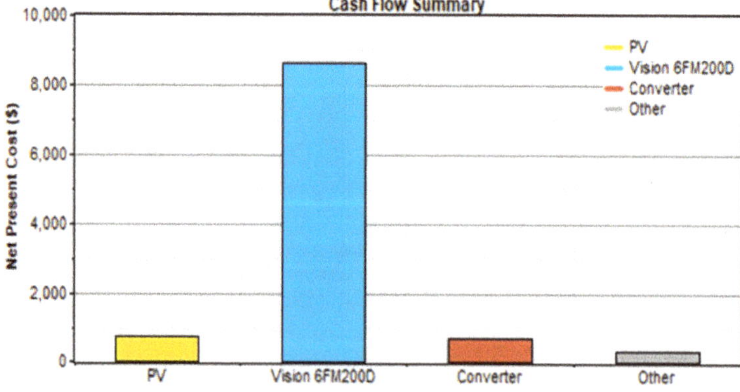

Figure 9. Cash flow summary of SHS over lifespan.

From the cost summary displayed in Figure 9, it can be seen that the bulk of expenses is due to the specific battery model 6FM200D. The long lifespan of PV cells and converter contribute to the system's feasibility for use. Figure 10 presents the yearly cash flow of each component of SHS for the lifespan.

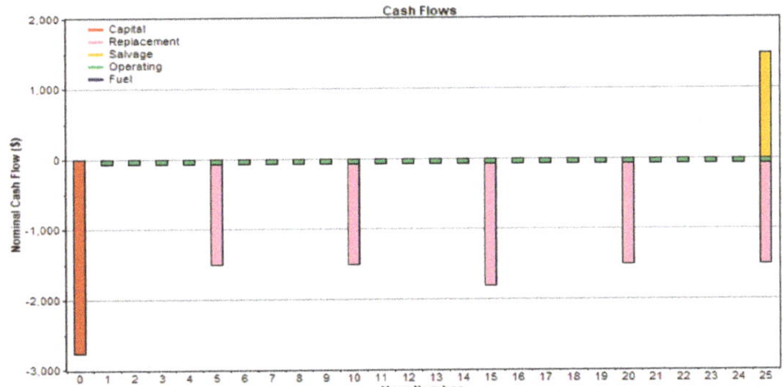

Figure 10. Cash flow of SHS for entire life span.

The cash flow diagram provides us a clear idea of the typical expenses of different components of SHS. For our system, a user will initially need to spend a fairly amount of money and then spend smaller amounts periodically. It is important to note that every five years users will need to replenish the battery because of its limited lifespan. The inverter and charge controller cost will add once 15 year is passed and at the end of the life span of the system salvage value will return or the system will be replaced completely. The nominal cost of the system is also mentioned for every year, which includes periodic servicing of the inverters and cleaning the panels.

4.3. Environmental Impacts

According to the CO_2 baseline database [33] prepared by the Department of Environment, Government of People's Republic of Bangladesh, the weighted average grid emission factor (GEF) for power plants is 0.67 Tonnes of CO_2 per MWh of energy generated. Using this factor, we can determine the reduction of CO_2 emissions because of shifting to rooftop solar for a lifetime of 25 years as disclosed in Equation (9).

$$CO_2(Emission) = \frac{E_d \times 0.67 \times 365 \times 25}{1000} \; (t/Mwh) \tag{9}$$

In the proposed rooftop PV system, we can reduce 36.68 tonnes CO_2 emissions during the 25 years of project life. Therefore, we can reveal that beside the monetary benefit it should also be mentioned that the current grid supply that is generated from the fossil fuel as a result of the CO_2 emissions is much higher. Whereas using the proposed stand-alone system can reduce the demand of energy from the grid as a result reducing the overall CO_2 emissions.

5. Conclusions

This paper highlighted the benefits of using a rooftop stand-alone Solar Home System both to minimise energy costs, and as a viable alternative to fossil fuels. The study demonstrated that by using this system, the total cost of energy will be decreased as the per unit cost is around $0.211 (16.88 Tk) and that there would gradually be no more additional fuel charges for consuming energy. In fact, within a few years, income can be earned by installing this rooftop stand-alone Solar Home System. The system is further endeared by availability of financing options that payment and installation of this system. From our cost analysis, users can approach micro finance companies or obtain loans from commercial banks. The analysis also illustrated that Bangladesh has surplus energy, which can be shared. The system can be improved by the installation of PV in between open neighbouring places for more daylight. Through this method, many off-grid areas will gain light sources. Further, total load pressure on the electricity grid network will also be reduced. Bangladesh is a developing country and conventional energy sources may no longer be feasible in the future in view of increasing fuel costs and environmental concerns. The rooftop SHS system can minimise the load pressure on the fuel system, which ultimately reduces greenhouse gas emissions and can contribute to a greener environment. Finally, in undertaking this scheme, people can be enlightened about all its benefits including the potential for income savings, which would ultimately enhance the economic activities of the people without any adverse effects on the environment.

Author Contributions: C.A.H. and N.C. proposed the core idea, developed the models. They performed the simulations, exported the results and analysed the data. W.Y. and M.L. revised the paper. C.A.H., N.C., M.L. and W.Y. contributed to the design of the models and the writing of this manuscript.

Funding: This research received no external funding. The APC was funded by Natural Resources Canada and by W. Yaïci.

Conflicts of Interest: The authors declare no conflict of interest.

References

1. Golub, V.; Raff, R.; Topic, D. Optimization of an off-Grid PV System in Respect to the Capacity Shortage Value. In Proceedings of the 2018 IEEE PES Innovative Smart Grid Technologies Conference Europe (ISGT-Europe), Sarajevo, Bosnia and Herzegovina, 21–25 October 2018; pp. 1–6.

2. Islam, M.Z.; Mashsharat, A.; Mim, M.S.; Rafy, M.F.; Pervei, M.S.; Rahman Ahad, M.A. A Study of Solar Home System in Bangladesh: Current Status, Future Prospect and Constraints. In Proceedings of the 2nd International Conference on Green Energy and Technology, Dhaka, Bangladesh, 5–6 September 2014; pp. 110–115.

3. International Energy Agency, World Energy Outlook, Executive Summary. 2008. Available online: https://www.eia.gov/outlooks/ieo/pdf/exec_summ.pdf (accessed on 10 February 2019).

4. Flavian, C.; Aeck, M.H. The potential role of renewable energy in meeting the millennium development goals. In *REN21 Network*; The Worldwatch Institute: Washington, DC, USA, 2005.

5. Khan, S. The search for alternatives. *Star Weekend Magazine*, Volume 8, 5 April 2009.

6. Renewable Power Generation Costs in 2017, IRENA, International Renewable Energy Agency. Executive Summary. Available online: https://www.irena.org/-/media/Files/IRENA/Agency/Publication/2018/Jan/IRENA_2017_Power_Costs_2018_summary.pdf?la=en&hash=6A74B8D3F7931DEF00AB88BD3B339CAE180D11C3 (accessed on 27 December 2018).

7. Bangladesh Power Development Board. Available online: http://www.bpdb.gov.bd/bpdb/index.php?option=com_content&view=article&id=26 (accessed on 10 February 2019).

8. Kumar, S.; Taneja, L.; Kaur, R. Design and control of residential off-grid connected PV systems. In Proceedings of the 2015 International Conference on Recent Developments in Control, Automation and Power Engineering (RDCAPE), Noida, India, 12–13 March 2015; pp. 204–208.

9. Hiendro, A.; Kurnianto, R.; Rajagukguk, M.; Simanjuntak, Y.M.; Junaidi. Techno-economic analysis of photovoltaic/wind hybrid system for on shore/remote area in Indonesia. *Energy* **2013**, *59*, 652–657. [CrossRef]

10. Sharma, P.; Bojja, H.; Yemula, P. Techno-economic analysis of off-grid rooftop solar PV system. In Proceedings of the 2016 IEEE 6th International Conference on Power Systems (ICPS), New Delhi, India, 4–5 March 2016; pp. 1–5.

11. Kumar Nandi, S.; Ranjan Ghosh, H. Techno-economical analysis of off-grid hybrid systems at Kutubdia Island, Bangladesh. *Energy Policy* **2010**, *38*, 976–980. [CrossRef]

12. Longo, M.; Roscia, M.; Lazaroiu, G.C.; Pagano, M. Analysis of sustainable and competitive energy system. In Proceedings of the 3rd International Conference on Renewable Energy Research and Applications, ICRERA 2014, Milwaukee, WI, USA, 19–22 October 2014; pp. 80–86.

13. Make Your Home Energy Efficient. Available online: http://www.makeyourhomeenergyefficient.com/solar-home-appliances.html (accessed on 10 February 2019).

14. Renewables 2018-Global Status Report. Available online: www.ren21.net/wp-content/.../2018/06/17-8652_GSR2018_FullReport_web_final_.pdf (accessed on 3 February 2019).

15. Renewable Energy Projects, IDCOL Solar Energy Program. August 2013. Available online: http://www.idcol.org/prjshsm2004.php (accessed on 10 February 2019).

16. Getting Started Guide for HOMER Legacy (Version 2.68), January 2011. Available online: http://www.science.smith.edu/~{}jcardell/Courses/EGR325/Readings/HOMERGettingStartedGuide.pdf (accessed on 6 December 2018).

17. Givler, T.; Lilienthal, P. *Using HOMER®Software, NREL's Micropower Optimization Model, to Explore the Role of Gen-sets in Small Solar Power Systems*; Technical Report NREL/TP-710-36774; National Renewable Energy Laboratory: Golden, CO, USA, May 2005.

18. Bahramara, S.; Moghaddam, M.P.; Haghifam, M.R. Optimal planning of hybrid renewable energy systems using HOMER: A review. *Renew. Sustain. Energy Rev.* **2016**, *62*, 609–620. [CrossRef]

19. Nandi, S.K.; Ghosh, H.R. Prospect of wind–PV-battery hybrid power system as an alternative to grid extension in Bangladesh. *Energy* **2010**, *35*, 2047–3040. [CrossRef]

20. Bekele, G.; Palm, B. Feasibility study for a stand-alone solar–wind-based hybrid energy system for application in Ethiopia. *Appl. Energy* **2010**, *87*, 487–495. [CrossRef]

21. Sen, R.; Bhattacharyya, S.C. Off-grid electricity generation with renewable energy technologies in India: An application of HOMER. *Renew. Energy* **2014**, *62*, 388–398. [CrossRef]

22. Adaramola, M.S.; Paul, S.S.; Oyewola, O.M. Assessment of decentralized hybrid PV solar-diesel power system for applications in Northern part of Nigeria. *Energy Sustain Dev* **2014**, *19*, 72–82. [CrossRef]

23. Ramli, M.A.M.; Hiendro, A.; Sedraoui, K.; Twaha, S. Optimal sizing of grid-connected photovoltaic energy system in Saudi Arabia. *Renew. Energy* **2015**, *75*, 489–495. [CrossRef]

24. Roy, P.; Arafat, Y.; Upama, M.B.; Hoque, A. Technical and financial aspects of solar PV system for city dwellers of Bangladesh where green energy installation is mandatory to get utility power supply. In Proceedings of the 2012 7th International Conference on Electrical and Computer Engineering (ICECE), Dhaka, Bangladesh, 20–22 December 2012; pp. 916–919.

25. Yodkhuanga, A.; Ngaopitakkul, A. Performance Evaluation of Solar Rooftop System. In Proceedings of the 3rd IIAE International Conference on Intelligent Systems and Image Processing, Fukuoka, Japan, 2–5 September 2015; pp. 288–293. [CrossRef]

26. Chowdhury, N.; Hossain, C.A.; Longo, M.; Yaïci, W. Optimization of Solar Energy System for the Electric Vehicle at University Campus in Dhaka, Bangladesh. *Energies* **2018**, *11*, 2433. [CrossRef]

27. Amin, A.; Sultana, A.; Hasan, J.; Islam, M.T.; Khan, F. Solar home system in Bangladesh: Prospects, challenges and constraints. In Proceedings of the 2014 3rd International Conference on the Developments in Renewable Energy Technology (ICDRET), Dhaka, Bangladesh, 29–31 May 2014; pp. 1–5.

28. Hasan, F.; Hossain, Z.; Rahman, M.; Ar Rahman, S. Design and Development of a Cost Effective Urban Residential Solar PV System. In Proceedings of the 3rd International Conference on the Developments in Renewable Energy Technology, Dhaka, Bangladesh, 29–31 May 2014.

29. Shariar, K.F.; Ovy, E.G.; Hossainy, K.T.A. Closed Environment Design of Solar Collector Trough using lenss and reflectors. In Proceedings of the World Renewable Energy Congress 2011, Linkoping, Sweden, 8–13 May 2011.

30. The Basics of Solar Power for Producing Electricity [online]. Available online: http://www.sunforceproducts.com/Support%20Section/Solar%20Panel%20&%20Charge%20Controllers/The%20Basics%20of%20Solar%20Power%20for%20Producing%20Electricity.pdf (accessed on 6 December 2018).

31. Jahan, N.; Hasan, M.A.; Hossain, M.T.; Subayer, N. Present Status of Installed Solar Energy for Generation of Electricity in Bangladesh. *Int. J. Sci. Eng. Res.* **2013**, *4*, 604–608.

32. Alibaba.com. Available online: https://www.alibaba.com (accessed on 16 February 2019).

33. Department of Environment, Government of People's Republic of Bangladesh. Available online: http://www.doe.gov.bd/site/notices/059ddf35-53d3-49a7-8ce6-175320cd59f1/Grid-Emission-FactorGEF-of-bd (accessed on 10 February 2019).

© 2019 by the authors. Licensee MDPI, Basel, Switzerland. This article is an open access article distributed under the terms and conditions of the Creative Commons Attribution (CC BY) license (http://creativecommons.org/licenses/by/4.0/).

Article

Comparisons of Acid and Water Solubilities of Rice Straw Ash Together with Its Major Ash-Forming Elements at Different Ashing Temperatures: An Experimental Study

Yi Zhang [1], Guanmin Zhang [1,*], Min Wei [1], Zhenqiang Gao [2,3], Maocheng Tian [1] and Fang He [2,*]

[1] School of Energy and Power Engineering, Shandong University, Jinan 250061, Shandong, China; zhy20360@163.com (Y.Z.); weimin@sdu.edu.cn (M.W.); tianmc65@sdu.edu.cn (M.T.)

[2] Department of Energy and Power Engineering, Shandong University of Technology, Zibo 255049, Shandong, China; gaozq@sdut.edu.cn

[3] School of Energy and Power Engineering, Xi'an Jiaotong University, Xi'an 710049, Shanxi, China

* Correspondence: zhgm@sdu.edu.cn (G.Z.); hf@sdut.edu.cn (F.H.)

Received: 6 March 2019; Accepted: 31 March 2019; Published: 3 April 2019

Abstract: Recycling utilization of straw ash as a fertilizer in farmland is expected to play an important role in the sustainable development of both agriculture and biomass energy. However, the ashing temperature and the aqueous solution characteristics may affect the recycling properties of the nutrients contained in the ash. The solubilities of both the ash and its elements can represent the above recycling properties. This paper presents a systematic experimental investigation on the acid solubilities of both rice straw ash and its major elements produced from combustion at 400–800 °C, and these findings are compared with the corresponding water solubilities obtained from the authors' previous work. Meanwhile, the correlations of two solubilities with the ashing temperature were given based on the experimental data. Results show that the acid solubility of rice straw ash decreases linearly by approximately 76% as the ashing temperature increases from 400 to 800 °C, while it is significantly higher than the corresponding water solubility at different temperatures. The acid solubilities of K, P, Ca, Mg, and Na are higher than their water solubilities, whereas two solubilities of S and Cl have almost no dependence on the temperature and the acidity of solution. This study also reveals a strong negative linear relationship between the solubility of K and the temperature. The solubilities of other elements (P, S, Na, Ca, Mg, and Cl) with the temperature have quadratic curve or cubic curve relationships. Furthermore, it is recommended that the ashing temperature should be lower than 600 °C to avoid the loss of some nutrients in the straw ash.

Keywords: rice straw ash; ash-forming elements; solubility; sustainable development of both agriculture and biomass energy; recycling property; ashing temperature

---

1. Introduction

Applications of biomass ash have been attracting a lot of attention recently in various fields, such as farmland fertilizer [1–3], construction material [4,5], adsorbent material [6,7], production material of ceramics, and raw material of metal recovery [7,8]. In all these applications, the soil application as a green, cyclic, and sustainable way has a huger potential and better prospect [9,10]. According to the statistics, approximately 85–95% of nutrients taken up from soil by plants during growth—including potassium (K), phosphorus (P), calcium (Ca), and magnesium (Mg)—are contained in biomass ash [11]. The recirculation of these nutrients to farmland soil is important in agricultural sustainable development [12,13]. Furthermore, using biomass ash as an alternative to chemical fertilizer [14,15] would not only balance the fertility of agricultural soil, but also reduce the amount of both the ash disposal and the chemical fertilizer of consumption, and thus alleviate loss of soil nutrients and environment pollution [16] and reduce the investment cost of agricultural production process. Comparing to direct straw returning, ash recirculation could prevent plants from being infected by pathogens and pests contained in the straw [17], and thus reduce the utilization of chemical pesticide. Therefore, the recycling of biomass ash as fertilizer to farmland soil plays a great role in the sustainable development of both agriculture and biomass energy.

In farmland soil, most of the nutrients are accessible to plants only in their free ionic form. One of the fundamental questions in the soil application of straw ash is transforming biomass ash to accessible (soluble or bioavailable) nutrients [18]. Therefore, the solubilities of biomass ash together with its major ash-forming elements are vital properties for the recirculation of nutrients, as shown in Figure 1. These solubilities are greatly determined by the pH values or acidity of solution and soil [11]. It is common knowledge that soil acidification is a serious global problem [19]. The main reasons for the soil acidification can be explained as follows: (1) the osmosis effect of the natural or artificial acidic rain resulting from dissolving of CO_2, SO_2, and NO_x in air [20] and (2) the extensive application of chemical nitrogen fertilizers [7]. This seriously limits the sustainable development of eco-agriculture in the new era. According to the reference [19], 40–50% of the potentially arable lands were acidic in the word. In China, the area of acid land (about 200 million ha) accounted for 23% of total land area [19]. Biomass ash contains various alkali metal compounds, which can improve acidic soil. As a result, the acid-soluble characteristics of both biomass ash and its ash-forming elements can partially represent the recycling property of the nutrients in the ash.

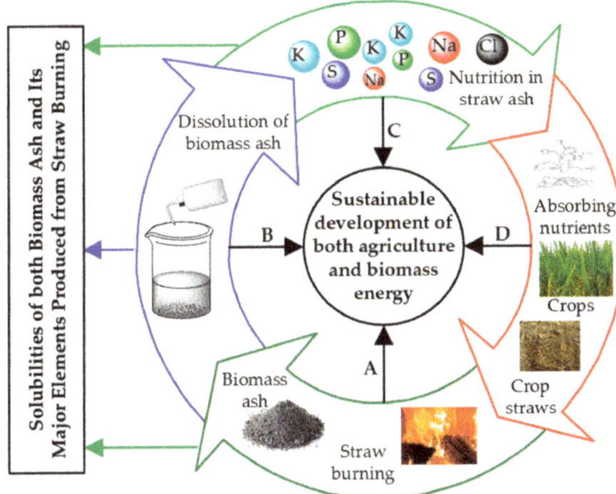

Figure 1. Schematic diagram of recycling application of biomass ash as a fertilizer in farmland.

The solubilities of both biomass ash and its ash-forming elements have been extensively reported in the published literatures. The total dissolved mass content of wood ash increased by 500% as the pH value of aqueous solution decreased from 13 to 5 [21]. In the bottom ash of municipal solid waste incineration, the solubilities of aluminum (Al), calcium (Ca), magnesium (Mg), sulfur (S), sodium (Na), and iron (Fe) in nitric acid solution (pH = 4, 6) increased significantly with the increase of the leaching time [22]. At the same time, the elements solubilities differ significantly with each other. Most of Ca and Fe, 30% of K, 32% of Na, 48% of Al, 25% of S, and 48% of P in the solid residue of cane bagasse gasification were acid-soluble [23]. Moreover, in wood combustion ash, 81% of Ca, 57% of Mg, 34% of K, and 20% of P were soluble in the ammonium acetate aqueous solution of pH = 4.2 [24]. In addition, based on their water solubilities, these elements were classified into the following three categories: (1) easily soluble nutrients (K, boron (B), Na, chlorine (Cl), and S); (2) slightly soluble nutrients (Ca, Mg, silicon (Si), Fe, and Al); and (3) highly insoluble nutrients (P) [22,25].

Rice is the third-largest agricultural product in the world. In China, rice accounts for about 40% of the total agricultural products according to the statistical data of Food and Agriculture Organization of the United States, meaning that a large amount of biomass ashes are produced from the burning of rice straw every year. Biomass ash properties are significantly affected by the combustion temperature or ashing temperature [26–32]. Effects of the temperature on the water solubilities of rice residue ashes or ash-forming elements have been investigated by some scholars. As reported by the references [17,33], after rice straw was pretreated at 300–1000 °C for 2 h, the water-soluble Si, P, and K in the treated residues firstly increased and gradually decreased with increasing the heating temperature. According to our previous study [22], the water solubilities of both rice straw ash and rice husk ash decreased obviously with the increase of ashing temperature from 400 °C to 800 °C. The water solubilities of K and Na contained in rice straw ash decreased with increasing the temperature, while those of S and Cl did not depend on the ashing temperature. In addition, Si, Ca, Mg, and P contained in rice residue ashes prepared at different temperatures were almost insoluble in water [22].

As an important follow-up work of our previous study, this paper will present an investigation on the solubilities of rice straw ash together with major ash-forming elements in 20% acetic acid aqueous solution of pH = 2.11. Meanwhile, through the least square method, the functional relationship between the solubilities of rice straw ash together with its major elements and the ashing temperature will be established, aiming at providing a reference for the sustainable development of biomass energy and the developing of low-temperature biomass combustion equipment. In addition, based on the practical background of agricultural residues ashes in soil application, this study will attempt to clarify the effects of the ashing temperature on the recycling properties of the nutrients contained in rice straw ash in different solutions.

In the rest of this paper, we will firstly introduce the material and methods used in this study. Then, the characteristics of rice straw ash, and the acid solubilities of the ash together with its major ash-forming elements will be given and compared with the corresponding water solubilities obtained from the authors' previous work [22]. Finally, some main conclusions will be summarized and the future studies related to the utilization of biomass ash also will be proposed.

2. Materials and Methods

2.1. Experimental Material

The rice straw raw material used in this study was collected from farmland of rural area of Gaoqing County, Zibo City, Shandong Province, China. Then it was pulverized to the rice straw powder sample with the particle size <1 mm using a blade pulverizer (DXF-20C) in Key Laboratory of Low-grade Energy and Waste Heat Utilization, Shandong University of Technology. The properties of the prepared rice straw powder are listed in Table 1. To facilitate the pulverized process, the rice straw material was dried in a constant-temperature drying oven at 105 °C for 12 h. This also causes the moisture of the rice straw powder to be less than 4% as shown in Table 1. It should be noted that in order to compare with the water solubility, the straw material used in this paper is the same as that in the authors' previous research.

Table 1. Properties of experimental rice straw material.

Experimental Material	Proximate Analysis (%)				Ultimate Analysis (%)		
	Moisture	Volatiles	Ash	Fixed Carbon	C	H	O
Rice straw	3.6	68.9	13.2	14.3	42.4	6.9	48.8

2.2. Experimental Scheme

2.2.1. Selection of Ashing Temperature

According to the published references [34,35], the temperatures of the reactors in various biomass thermo-chemical conversion technologies are generally controlled within the range of 250–1200 °C. The results of our pre-experimental study showed that (1) when the temperature was below 400 °C, the colors of rice straw combustion residues were obviously black, indicating that the rice straw may not be completely combusted; and (2) when the temperature was above 800 °C, almost all of the rice straw ash was sintered and adhered to the crucible bottom. The main components of sintered ash are silicate or aluminosilicate and it is difficult to be taken out from the crucible for further analysis and utilization [36,37]. In addition, most of the other components contained in straw are volatilized in the form of gas phase during the straw ashing at high temperature [38]. Therefore, we consider the temperature range of 400–800 °C with the interval of 50 °C in this study.

In addition, the scanning electron microscopy (SEM) experiments of the rice straw ashes at 400 °C and 800 °C are employed to confirm whether the sintering phenomenon of rice straw ash happens.

2.2.2. Selection of Residence Time

Residence times of biomass fuels in various thermo-chemical conversion devices are different according to the literature. For example, the residence time varies from a few seconds to a few minutes in a fluidized bed [39], and from several minutes to a few hours in a grate-fired furnace [35]. During the proximate analysis of solid biomass fuel in a muffle furnace, the fuel usually can be kept for several hours [40].

According to the pre-experimental data in our previous work, when the residence time was more than 4 h, the changes of the water solubilities of rice straw ashes were not significant at different ashing temperatures. Thus, in the presented work, the residence time is selected for 4 h.

2.3. Experimental Method

2.3.1. Ash Sample Preparation Process

The rice straw ash sample was prepared in a muffle furnace at different ashing temperatures (400–800 °C). The ash preparation process is similar to that in our previous study [22], which can be summarized in the following five steps:

- Step 1: Approximately 5 g samples of the prepared rice straw powder sample were put into a crucible (90 × 60 × 15 mm).
- Step 2: The crucible being put the rice straw powder sample was heated in the muffle furnace from the room temperature to the experimental temperature with the heating rate of 10 °C per minute.
- Step 3: After staying 4 h at the experimental temperature, the muffle furnace was closed and cooled naturally to below 200 °C.
- Step 4: The residual ash in the crucible was taken out to place in a desiccator with silica gel.
- Step 5: After being further cooled to the room temperature, the rice straw ash was sealed in a sample bag for analysis.

2.3.2. Ash Dissolution Process

The sample preparation process of the acid solubilities measurements of rice straw ash and its elements is shown in Figure 2. From Figure 2, as we can see that the dissolving process of the rice straw ash sample is mainly divided into four steps:

- Step 1: Approximately 0.5 g samples of the original rice straw ash were accurately weighed into the beaker. 25 mL of 20% acetic acid aqueous solution of pH = 2.11 was added to the sample and stirred in the beaker of 50 mL for 20 min.

- Step 2: After being stored overnight, the mixture consisting of ash and acetic acid aqueous solution was boiled for 2 min, and then filtered using a quantitative filter paper with pore size of 30–50 microns.
- Step 3: The acid-insoluble rice straw ash was washed using deionized water (25 mL each time) five times. Subsequently, the acid-insoluble rice straw ash wrapped in the filter paper was dried in the constant-temperature drying oven at 105 °C for 30 min and then it was burned at 550 °C for 2 h (according to the Chinese standard GB/T 8307-2002).
- Step 4: The acid-insoluble rice straw ash was weighted and kept in another sample bag for the measurement of element contents.

It should be noted that the dissolution experiments of the rice straw ashes at different ash-forming temperatures were repeated three times.

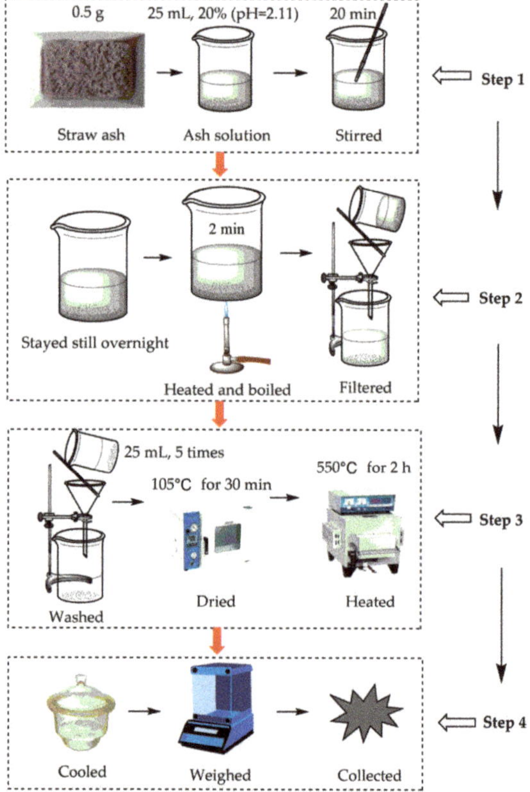

Figure 2. The dissolving process of rice straw ashes at different ashing temperatures.

2.3.3. Element Content Measurement in Ashes

A semi-quantitative X-ray fluorescence spectrometer (XRF) is widely used to analyze inorganic non-metallic materials. The XRF can measure various kinds of elements contents in the sample of biomass ash at one time [22,26,36,41]. Thus, to obtain the solubilities of various main nutrient elements in the rice straw ash, these elements contents in the original rice straw ash and the acid-insoluble rice straw ash were measured using the semi-quantitative ZSX-100e XRF analyzer.

In our pre-test, tests were carried out in triplicate to determine the accuracy of XRF for measurement of elements in straw ash [42]. The results showed that the standard deviations for all elements were less than 1%, illustrating that the results of each test are reliable. As a result, in the presented work, the XRF measurement of elements contents in the original ash and the acid-insoluble ash is only carried out once at different ashing temperatures.

2.4. Experimental Data Processing

According to the mass conservation of rice straw ash and its ash-forming elements before and after dissolution, the acid solubilities of rice straw ash together with its elements can be obtained using the above measurement data. It should be noted that the rice straw ashes from different ashing temperatures are taken as the research object in this paper. Thus, the volatilization of the elements is not taken into account during combustion when calculating the solubilities of ash and its major elements.

The acid solubility of the ash is the ratio of the mass of acid-soluble rice straw ash to the mass of original rice straw ash. It was calculated as

$$AS_A = (1 - m_r/m_0) \times 100\% \tag{1}$$

where AS_A is the acid solubility of rice straw ash; m_0 and m_r are the mass of original rice straw ash and acid-insoluble rice straw ash, respectively. It should be noted that the acid solubilities of the ashes at different temperatures were the average values of experiments repeated in triplicate.

The acid solubility of element i in rice straw ash is the ratio of the acid-soluble mass of i to the total mass of i in the original rice straw ash. It was calculated using Equation (2).

$$AS_{A,i} = [1 - (m_r \times Y_{r,i})/(m_o \times Y_{o,i})] \times 100\% \tag{2}$$

where $AS_{A,i}$ is the acid solubility of element i in rice straw ash; $Y_{r,i}$ and $Y_{o,i}$ are the mass fractions of element i in the acid-insoluble rice straw ash and the original rice straw ash, respectively.

In addition, in order to further clarify the relationships between the solubilities of rice straw ash together with its elements and the ashing temperature, the correlation functions between both of them are given based on experimental data using least squares method.

Of note, in the authors' previous work [22], the dissolution process of rice straw ash in water, the methods of calculation of the water solubility of both rice straw ash and its elements have been given in detail. Therefore, in order to avoid the identical content, they were not given again in the materials and methods section in the presented work.

3. Results and Discussion

3.1. Ash Characteristics

3.1.1. SEM Analysis

The SEM images of the rice straw ashes prepared at 400 °C and 800 °C are shown in Figure 3a–f. As can be seen, the particles size of the rice straw ash prepared at 800 °C are larger and more compact than that prepared at 400 °C, suggesting that ash sintering has occurred after the rice straw is kept at 800 °C for 4 h. The ash sintering is referred to the bonding or welding of adjacent particles under the influence of the excess surface tension as shown Figure 3g [43,44]. According to some previous works, it is also a phenomenon of the combination of particles caused by the synthetic effect of partial melting, viscous flow, and gas–solid chemical reactions [22,45]. Obviously, the rice straw ash with high sintering tendency is expected to have greater ash particle sizes.

In addition, the high alkali metals contents in rice straw ash can also accelerate the straw ash sintering [44], and the surface of sintered ash particle is smoother in comparison to unsintered ash. This resulted from the molten ash forming a coating and blocking up the holes on the surface of the ash particles [17]. As a result, some inorganic nutrients may be trapped inside rice straw ash particle at high combustion temperatures.

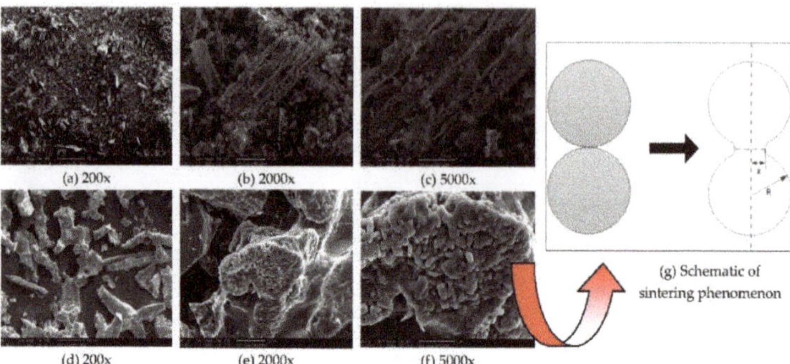

Figure 3. SEM images of rice straw ash at 400 °C (**a–c**) and 800 °C (**d–f**) and schematic of sintering phenomenon (**g**).

3.1.2. Ash-Forming Elements

The XRF measurement results show that there are O, Si, K, Na, Cl, Ca, Mg, S, P, Fe, Al, Mn, Ni, As, Br, Sr, Pb, Rb, and Ba in both original rice straw ash and two insoluble rice straw ashes (the acid-insoluble ash and the water-insoluble ash [22]) at different ashing temperatures. Mass fractions of major ash-forming elements (more than 0.2%) in these ashes, including O, Si, K, Na, Ca, Mg, Cl, P and S, are listed in Table 2.

Table 2. Mass fractions of major ash-forming elements in original and insoluble rice straw ashes at different ashing temperatures.

Mass Fractions of Major Ash-Forming Elements in Original Rice Straw Ashes (>0.2%)									
T (°C)	O	Si	K	Na	Ca	Mg	Cl	P	S
400	41.20	22.50	10.80	6.33	4.63	2.60	6.15	1.29	2.64
450	40.50	23.00	11.30	6.15	4.50	2.69	6.11	1.32	2.63
500	42.30	22.70	10.90	5.74	4.26	2.70	5.96	1.26	2.44
550	41.50	23.00	10.90	5.91	4.51	2.76	5.55	1.33	2.22
600	42.30	24.30	10.20	5.20	4.00	2.10	4.90	1.03	1.94
650	40.80	23.80	11.30	5.04	4.95	1.90	4.90	1.09	1.72
700	39.10	21.70	11.80	5.82	5.64	2.44	6.36	1.53	2.74
750	44.30	26.00	9.07	6.04	5.35	2.26	0.55	1.14	2.89
800	44.30	25.10	8.98	6.05	6.33	2.22	0.25	1.24	3.04

Mass Fractions of Major Ash-Forming Elements in Water-Insoluble Rice Straw Ashes (>0.2%)									
T (°C)	O	Si	K	Na	Ca	Mg	Cl	P	S
400	46.90	32.90	2.53	1.31	7.28	3.85	0.34	1.67	0.45
450	51.50	29.80	3.19	1.97	5.62	3.39	0.47	1.43	0.53
500	41.70	34.40	5.08	3.50	6.27	3.68	0.71	1.77	0.54
550	42.50	31.80	6.03	3.97	6.12	3.51	1.17	1.69	0.64
600	46.30	29.50	6.52	3.82	5.45	3.02	1.08	1.64	0.48
650	46.70	29.00	7.06	3.85	4.98	3.15	1.31	1.45	0.46
700	44.10	30.60	7.33	4.16	5.55	2.93	0.98	1.43	0.42
750	44.00	31.30	7.43	4.14	5.32	2.90	0.76	1.40	0.41
800	42.30	31.30	8.18	4.39	6.14	2.80	0.34	1.63	0.34

Mass Fractions of Major Ash-Forming Elements in Acid-Insoluble Rice Straw Ashes (>0.2%)									
T (°C)	O	Si	K	Na	Ca	Mg	Cl	P	S
400	49.80	45.10	0.86	0.41	0.37	0.30	0.15	0.20	0.16
450	46.10	41.40	3.07	2.93	1.78	1.01	0.62	0.26	0.29
500	44.60	39.00	4.91	2.62	2.98	1.43	0.87	0.41	0.40
550	46.20	34.20	5.68	3.39	3.65	1.95	0.98	0.81	0.46
600	44.10	35.50	6.62	3.53	3.64	2.33	1.01	0.38	0.47
650	47.50	32.20	6.95	3.40	3.57	2.41	0.97	0.47	0.43
700	44.30	32.50	7.72	4.17	4.36	2.66	0.89	0.52	0.38
750	48.60	31.50	6.70	2.97	4.04	2.81	0.45	0.37	0.41
800	43.00	33.60	8.19	3.68	4.83	2.96	0.25	0.49	0.37

From Table 2, it can be seen that the mass fractions of major ash-forming elements in the original rice straw ash at different ashing temperatures vary basically in the following decreasing order of quantity: O > Si > K > Na > Cl > Ca > Mg > S > P. Also, the elements contents in both the acid-insoluble ashes and water-insoluble ashes vary in the following decreasing order of quantity: O > Si > K > Ca > Na > Mg > Cl > P > S and O > Si > K > Ca > Na > Mg > P > Cl > S, respectively. It should be noted that the contents of the individual elements (such as Na, Ca, Mg, Cl, and P) in the original and insoluble rice straw ashes at different temperatures do not strictly follow the order mentioned above. The reason is that the contents of these elements present different changing trends with the increase of the ashing temperature.

As shown in Table 2, the major ingredients of both original ashes and insoluble ashes are O and Si (the sum of mass fractions of them >60%) at different ashing temperatures. The contents of K, Na, Ca, Mg, Cl, P, and S in the water-insoluble ashes are basically higher than these in the acid-insoluble ashes, indicating that some compounds consisting of these elements are only soluble in acid solution but not insoluble in deionized water [11].

The XRF measurement data of two insoluble ashes in Table 2 also show that the mass fractions of K increase with increasing the ashing temperature. This is due to the following two reasons [7]: (1) the high temperature makes soluble salts transform into different insoluble silicate compounds, and some key reactions can see the reference [46]; and (2) the sintered materials or insoluble components on the surface of rice straw ash particles prevent internal substances from dissolving in the deionized water.

3.2. Comparisons of Acid and Water Solubilities of Rice Straw Ashes

Two solubilities of rice straw ashes, created at different ashing temperatures, in the acetic acid solution and the deionized water are shown in Figure 4. It can be seen that they are significantly lower than the water solubility (61%) of fly ash from straw combustion [47]. This can be attributed to the release of many soluble substances. It can be also found from the Figure 4 that the acid solubilities (45.3–11.1%) of the ashes in the acetic acid solution are approximately 5–15% higher than the corresponding water solubilities (37.5–6.3%) of those in the deionized water in the temperature range of 400–800 °C. However, their values are low and more than 50% of rice straw ash is insoluble in the acetic acid solution and the deionized water. According to the literature [21], for the same biomass ash sample, the total dissolved mass content increased by 500% with the increased acidity of aqueous solutions (the pH value decreased from 13 to 5). This means that most of the ash-forming elements are acid-soluble, Zevenhoven-Onderwater et al. [48] also confirmed this result. For example, some compounds (e.g., $CaCO_3$, $MgCO_3$) are acid-soluble but water-insoluble. In our experiments, we found that there is gas escaping from the solution while the acetic acid solution is added into the water-insoluble ash. The possible reactions of this phenomenon are R1.

$$CaCO_3 \text{ (s)} + 2CH_3COOH \text{ (aq)} \rightarrow 2CH_3COO^- + Ca^{2+} + H_2O \text{ (l)} + CO_2 \text{ (g)}$$
$$MgCO_3 \text{ (s)} + 2CH_3COOH \text{ (aq)} \rightarrow 2CH_3COO^- + Mg^{2+} + H_2O \text{ (l)} + CO_2 \text{ (g)}$$

(R1)

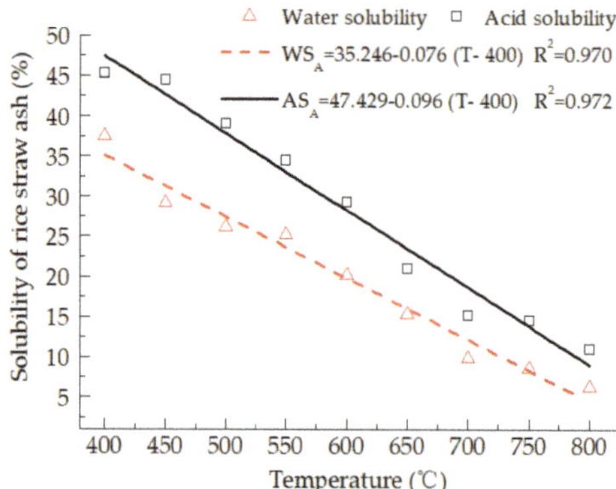

Figure 4. Solubilities of rice straw ash in acetic acid and water at different ashing temperatures. WS_A is the water solubility of the ash; AS_A is the acid solubility of the ash; T is temperature; R^2 is determination coefficient.

Additionally, the Figure 4 also shows that as the ashing temperature increases from 400 °C to 800 °C, two solubilities of rice straw ash decrease by about 70–80%. To further clarify the relationship between the ash solubility and the ashing temperature, the functional dependencies and determination coefficients between the solubilities of rice straw ash and the ashing temperature are given in Figure 4. It is found that a close negative linear relationship exists between the solubility (including acid solubility and water solubility) of rice straw ash and the ashing temperature. This suggests that a low temperature combustion is more beneficial to the resource utilization of straw ash as a fertilizer.

One of the main reasons for the decline is the severe sintering of straw ash at high temperature [49,50]. The ash sintering leads to the transformation of Ca and Mg in the ashes from oxides or sulphates to acid-insoluble Ca-silicate, Ca-Mg-silicate [7], and the melted insoluble inorganic components covering the soluble nutrients [22]. Some important reactions are seen in R2 [7,51].

$$2CaO + SiO_2 \rightarrow Ca_2SiO_4$$
$$2CaSiO_3 + MgO \rightarrow Ca_2MgSi_2O_7$$
$$14CaSiO_3 + 2MgO \rightarrow 2Ca_7Mg(SiO_4)_4 + 6SiO_2 \tag{R2}$$
$$3CaSO_4 + Ca_3Al_2O_6 + 32H_2O \rightarrow Ca_6Al_2(SO_4)_3(OH)_{12}\cdot26H_2O$$

Besides ash sintering, the decreasing of both solubilities are also attributed to the release [7,38] or the transformation [52] of partially soluble compounds (e.g., HCl(g), SO₃(g), MCl(g) and MOH(g), where M represents K and Na, similarly hereinafter). According to the reference [2], approximately 80–98% of Cl is released between 700 °C and 800 °C, and 40% of the total K is released at 800 °C. For S, 25–35% is released to the gas phase at 500 °C, and up to 40–50% when the combustion temperature increases from 500 °C to 800 °C. The potential reactions are listed below as R3 [7].

$$2MCl\ (g) + nSiO_2\ (s,l) + H_2O\ (g) \rightarrow M_2O{\cdot}nSiO_2\ (s,l) + 2HCl\ (g)$$
$$2MCl\ (g) + nSiO_2\ (s,l) + Al_2O_3\ (s,l) + H_2O\ (g) \rightarrow 2MAlSi_{n/2}O_{2+n}\ (s,l) + 2HCl\ (g)$$
$$M_2CO_3\ (s,l) + H_2O\ (g) \rightarrow 2MOH\ (g) + CO_2\ (g) \tag{R3}$$
$$M_2SO_4\ (g) + nSiO_2\ (s,l) \rightarrow M_2O{\cdot}nSiO_2\ (s,l) + SO_3\ (g)$$
$$M_2SO_4\ (g) + nSiO_2\ (s,l) + Al_2O_3\ (s,l) \rightarrow 2MAlSi_{n/2}O_{2+n}\ (s,l) + SO_3\ (g)$$

3.3. Comparisons of Acid and Water Solubilities of Major Elements

It is well known that crop straw ash contains some essential nutrients and beneficial elements (e.g., Na, Si), which can contribute to plant growth. The essential nutrients can be classified into three types: (1) macronutrients (K, P); (2) medium nutrients (Ca, Mg, S); and (3) micronutrients (Cl). In this section, the acid and water solubilities of K, P, Ca, Mg, S, Cl, and Na are analyzed. Of note, according to our experimental data, Si is almost insoluble in both the acetic acid solution and the deionized water, so it will not be discussed in this section.

3.3.1. Solubility of Macronutrients

K and P are indispensable as essential nutrients for the growth of organisms in the ecosystem. In general, K in biomass ash is mainly in the form of KCl, K_2SO_4 and K_2CO_3 [53] as well as potassium silicate, aluminosilicate, and sulphate ($K_2Ca(SO_4)_2$, $K_3Na(SO_4)_2$) [54]. P is mainly in the form of phosphates.

The acid and water solubilities of macronutrients K in rice straw ashes prepared at different ashing temperatures are shown in Figure 5. As shown in Figure 5, the acid solubility of K is about 6% higher than its water solubility in the temperature range from 400 °C to 800 °C. This can be explained as follows: (1) some sintered compounds in the ashes, such as the mixture of potassium-rich silicate and phosphate melts [55,56], are only acid-soluble; (2) the transport channels of some internal K may be occluded by some of acid-soluble salts in the surface of ash particles formed during the rice straw powder combustion [17]. Moreover, Liu et al. [57] indicated that 93% of K in rice straw is mainly enriched by acid-soluble salts, and 73% of K is water-soluble in rice straw. This trend is similar to the results of K solubilities contained in rice straw ash in this paper.

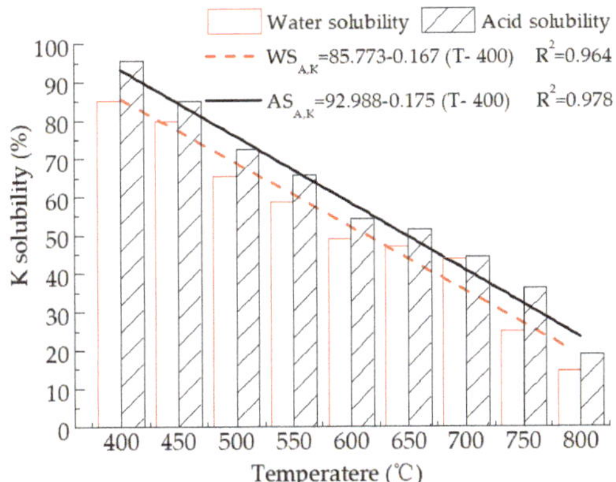

Figure 5. Acid and water solubilities of K in rice straw ashes at different ashing temperatures. $WS_{A,i}$ is the water solubility of element *i* in the ash; $AS_{A,i}$ is the acid solubility of *i* in the ash.

From the figure, we can also see there is a strong negative linear relationship between the solubility of K and the ashing temperature. As the temperature increases from 400 °C to 800 °C, the acid and water solubility of K decrease linearly by approximately 80% (from 95.6% to 18.9%) and 83% (from 85.4% to 14.7%), respectively. Vassilev et al. [11] also reported similar trends in K solubility. Of note, when the ashing temperature is lower than 600 °C, both solubilities are more than 50%, indicating that a lower combustion temperature is beneficial to the recycling application of K contained in ash in farmland. The main reason of this decrease with temperature, as pointed out in previous works, is the transformation of K from K-sulphate, K-chloride, and K-hydroxide to K-aluminosilicates via the reactions R3 and R4 [7,37,50,53,58].

$$Al_2O_3 \cdot 2SiO_2 + 2KCl + H_2O \rightarrow 2KAlSiO_4 + 2HCl$$
$$Al_2O_3 \cdot 2SiO_2 + 2KCl + 2SiO_2 + H_2O \rightarrow 2KAlSi_2O_6 + 2HCl$$
$$Al_2O_3 \cdot 2SiO_2 + 2KOH \rightarrow 2KAlSiO_4 + H_2O \tag{R4}$$
$$Al_2O_3 \cdot 2SiO_2 + 2KOH + 2SiO_2 \rightarrow 2KAlSi_2O_6 + H_2O$$

Furthermore, as reported in [17,59], K solubility is determined by the dissolution of Si, and this dissolution rate increases with increasing the pH value of solution. However, according to the experimental data in both the present study and our previous study [22], Si in rice straw ash is almost insoluble in the acetic acid solution and the deionized water, but K is easily soluble. It suggests that K solubility does not depend on Si solubility. This significant difference of Si solubility in the references and our studies is due to the differences of both the experimental conditions and the characteristics of extraction solutions. Therefore, the correlations between the solubilities of other elements and Si solubility are not considered and analyzed in this paper.

The acid and water solubilities of macronutrients P in rice straw ashes prepared at different ashing temperatures are shown in Figure 6. As we can see, when the ashing temperature is less than 600 °C, the acid solubility of P decreases slowly with the increase of the temperature. When the temperature is higher than 600 °C, the effect of the ashing temperature on the acid solubility of P is not significant. In addition, Figure 6 indicates that in the whole experimental temperature range (400–800 °C), the acid solubility of P contained in rice straw ashes is tremendously higher than its water solubility. More than 60% of P is acid-soluble at different temperatures, while its water solubility is less than 25%. This indicates that most of phosphates (e.g., 72–80% of $Ca_3(PO_4)_2$, 8–10% of $Na_6(PO_4)_2$) [11,60] are dissolved in the acetic acid solution through the neutralization reaction of acetic acid and hydroxide ion (OH^-) produced from the hydrolysis of phosphate radical (PO_4^{3-}) (R5).

$$PO_4^{3-} + H_2O \leftrightarrow HPO_4^{2-} + OH_2CH_3COOH + OH^- \rightarrow CH_3COO^- + H_2O \tag{R5}$$

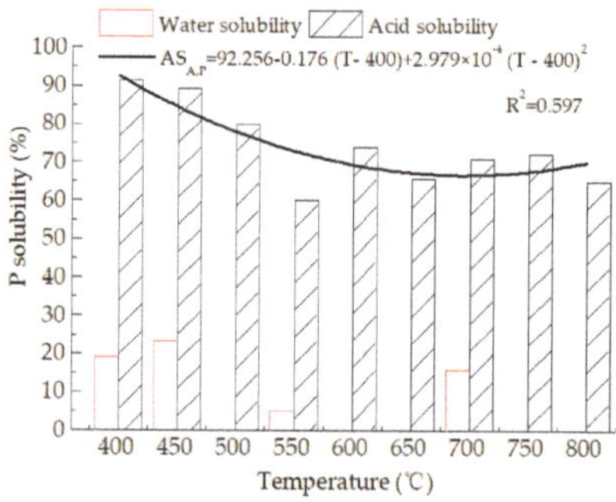

Figure 6. Acid and water solubilities of P in rice straw ashes at different ashing temperatures.

As Figure 6 shows, the determination coefficient (R^2) of the correlation of the acid solubility with the ashing temperature is equal to 0.597, which is obviously less than 1.0. Thus, there is not a significant quadratic curve relationship between the acid solubility of P and the temperature.

3.3.2. Solubility of Medium and Micronutrients

The acid and water solubilities of medium nutrients Ca, Mg, and S contained in rice straw ashes at different ashing temperatures are presented in Figures 7–9, respectively. As shown in Figures 7 and 8, the acid solubilities of both Ca and Mg are significantly higher than their water solubilities. Both Ca and Mg are almost water-insoluble because most of them are in the form of carbonates in the whole experimental temperature range (400–800 °C) [53]. These carbonates are insoluble in water but soluble in acetic acid solution as shown in reaction R1. Especially, when the temperature is less than 600 °C, more than 50% of both Ca and Mg contained in rice straw ash is acid-soluble.

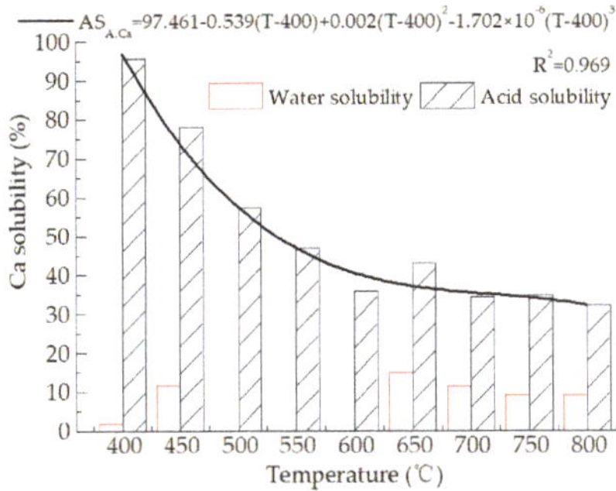

Figure 7. Acid and water solubilities of medium nutrient Ca at different ashing temperatures.

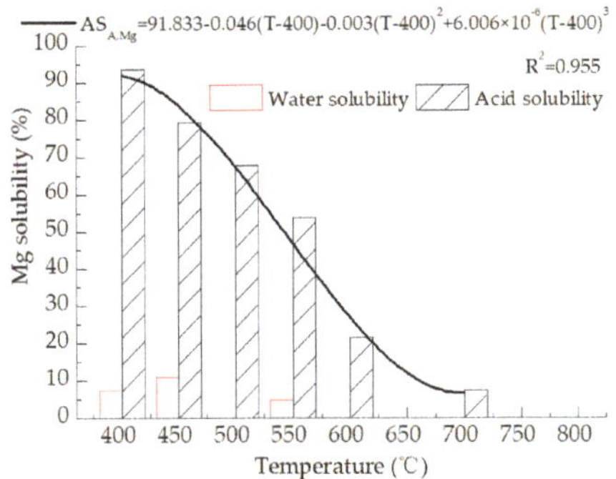

Figure 8. Acid and water solubilities of medium nutrient Mg at different ashing temperatures.

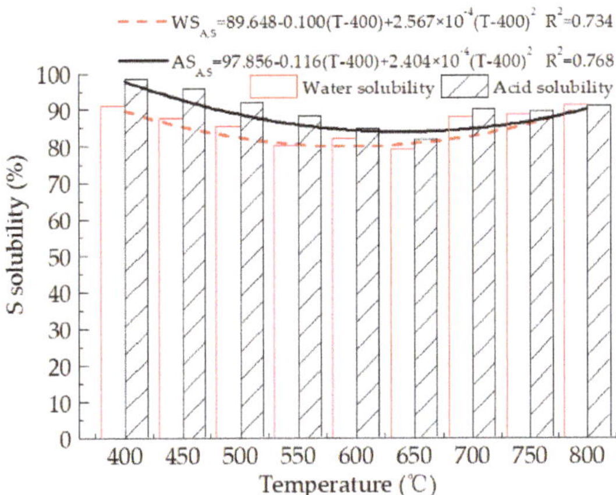

Figure 9. Acid and water solubilities of medium nutrient S at different ashing temperatures.

It is clearly shown in Figures 7 and 8 that a strong cubic curve relationship exists between the acid solubility of Ca/Mg and the ashing temperature. When the temperature increases from 400 °C to 600 °C, the acid solubilities of Ca and Mg decrease significantly from 95.68% to 35.73% (decrease by about 63%) and from 93.69% to 21.64% (decrease by about 77%), respectively. This change trend is similar to that of P. The partial reason of this decrease, as pointed out in previous works [7], may be that alkaline-earth metals transform into molten silicates (via the equation R4) which coat the surface of straw ash particles to prevent the dissolution of nutrients at high temperature. In addition, the change of the acid solubility of Ca is small as the temperature increases from 600 °C to 800 °C, indicating that the calcium compounds in the ashes may approach phase equilibrium at the temperatures above 600 °C.

For medium nutrient S, it is mainly concentrated in straw ash in the form of the sulphates of K, Na, etc. [11]. It is well known that most of sulphates are soluble in both the acetic acid solution and the deionized water. As shown in Figure 9, the acid solubility of S contained in the rice straw ashes is slightly higher than the corresponding water solubility at 400–750 °C. The acid and water solubilities of S are more than 75% at different temperatures. With the increase of the ashing temperature, the changes of both solubilities are firstly decreased and then slightly increased. Consequently, the effect of the ashing temperature on the solubility of S contained in rice straw ash is slight in different solutions.

Micronutrient Cl is an important component of rice straw ash. It can form a variety of chlorides with Ca, Na, K, Zn, etc. [11] and they are the main forms of Cl in rice straw ashes at different ashing temperatures. Like the sulphates, the chlorides are also soluble in the acetic acid solution and the deionized water.

Figure 10 describes the acid and water solubilities of micronutrient Cl contained in rice straw ashes prepared at different ashing temperatures. As we can see, both the acid and water solubilities of Cl are higher than 75% in the ashing temperature range of 400–700 °C. In this temperature range, two solubilities of Cl are slightly decreased with the increase of the temperature. However, when the temperature is higher than 700 °C, two solubilities of Cl contained in the rice straw ashes decrease rapidly. The reason for that is that most of Cl in the ashes are released, as shown in Table 2.

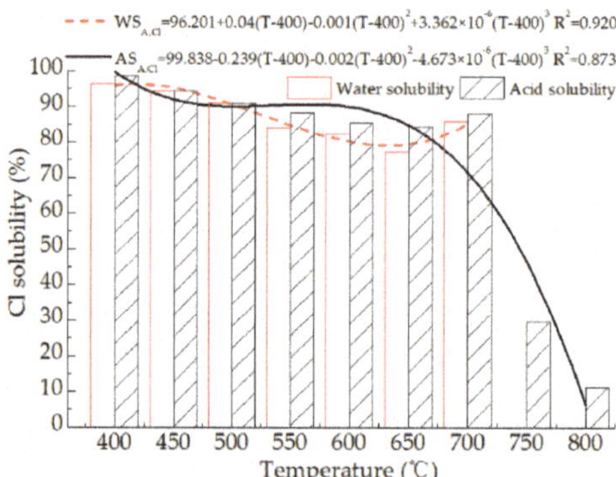

Figure 10. Acid and water solubilities of micronutrient Cl at different ashing temperatures.

3.3.3. Solubility of Beneficial Element

The acid and water solubilities of beneficial element Na contained in rice straw ashes prepared at different ashing temperatures are shown in Figure 11. It can be seen that the acid solubilities of Na contained in the ashes are about 4–7% higher than the corresponding water solubilities at different temperatures.

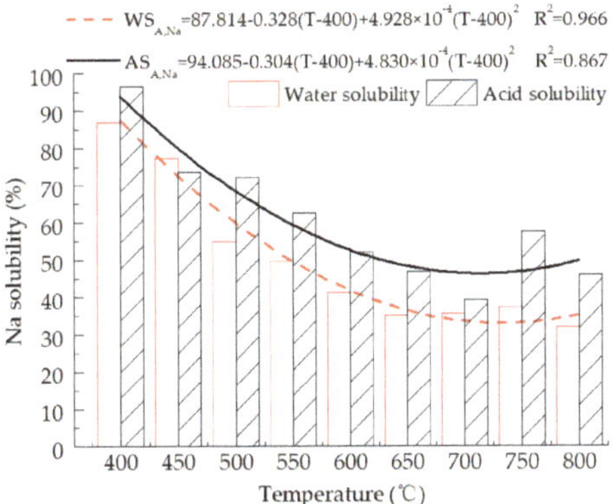

Figure 11. Acid and water solubility of beneficial element Na at different ashing temperatures.

Figure 11 also shows that there is a close quadratic curve relationship between the solubility of Na and the temperature. The acid and water solubility of Na decrease gradually from about 96% to 39% and from about 87% to 32% with increasing the temperature, respectively. The reasons for these decreases are similar to those of K.

From the comparison results of two solubilities of different nutrient elements mentioned above, it is not difficult to find that the acid solubilities of the major ash-forming elements are always higher than the water solubilities of those. When the ashing temperature is less than 600 °C, their solubilities are basically higher than 50%. Especially, most of P, Ca, and Mg are acid-soluble but water-insoluble. This conclusion is also confirmed by some previous work [23,24]. Hence, to better realize the closed cycle of nutrients in rice straw, the following two principles should be followed: (1) the ashing temperature should not be higher than 600 °C, (2) the rice straw ash should be applied to acidic soil.

4. Conclusions

The acid solubilities of rice straw ash together with its major ash-forming elements in the ashing temperature range of 400–800 °C have been experimentally studied and compared with the corresponding water solubilities of them in this paper. The main results are as follows:

- The acid solubility of rice straw ash is approximately 5–15% higher than the water solubility of that in the experimental temperature range. Two solubilities of the ash decrease linearly with the increase of the ashing temperature, and the acid solubility of the ash decreases by about 76% when the temperature increases from 400 °C to 800 °C.
- The acid and water solubility of K decreases linearly by approximately 80% and 83% as the ashing temperature increases, respectively. The acid solubility of K is about 6% higher than its water solubility. P, S, and Na solubilities with the temperature have quadratic curve relationships. The solubilities of Ca, Mg, and Cl have significant cubic curve correlations with the temperature.
- P, Ca, and Mg are soluble in the acetic acid solution but almost insoluble in the deionized water. The solubilities of K, P, Ca, Mg, and Na vary obviously with increasing the temperature, and the acid solubilities of them are higher than the corresponding water solubilities. Whereas two solubilities of S and Cl are slightly affected by the ashing temperature.
- When the ashing temperature of rice straw is lower than 600 °C, the rice straw ash will have a more accessible (soluble or bioavailable) nutrient content. Otherwise, most nutrients contained in the ash will be sintered or released with the flue gas.

It can be concluded that the recycling property of the nutrients in rice straw ash is significantly affected by the combustion temperature and the acidity of solution. A lower combustion temperature and a lower pH value can lead to a better recycling property of the nutrients in straw ash. In other words, the application of straw ash produced from low-temperature in acid soil will have a better prospect. This can not only contribute to the sustainable application of biomass energy, but also reduce the amount of utilization of chemical fertilizer, the area of acid soil and the cost of agricultural production, thus achieving the sustainable development of modern agriculture. Therefore, low temperature combustion technology for biomass should be considered and developed, and some fundamental studies on recycling of straw ash as a fertilizer to agricultural land should be also strengthened in the future.

5. Patents

According to the above conclusion that the low temperature combustion technology of biomass can be beneficial to the recycling utilization of biomass ash as a fertilizer in farmland, a patent for invention of a method and device for low-temperature solid phase, high-temperature gas phase combustion of biomass briquette fuel has been applied by the authors, and it has been authorized by the State Intellectual Property Office of the P.R.C (Patent no. CN 201710174853.0).

Author Contributions: Conception of the idea, Y.Z., G.Z., and F.H.; Methodology of the study, Y.Z. and G.Z.; Material collection and treatment, Y.Z.; Data curation, Y.Z.; Result analysis, Y.Z. and F.H.; Writing—original draft preparation, Y.Z.; Writing—review and editing, Y.Z., M.W., M.T., Z.G., G.Z., and F.H.; Funding acquisition, G.Z. and F.H.; Contributed equally, G.Z. and F.H.

Funding: This research was funded by two National Natural Science Foundations of China, grant numbers 51576115 and 51676115.

Acknowledgments: The authors gratefully acknowledge financial support from the National Natural Science Foundation of China (51576115) and the National Natural Science Foundation of China (51676115).

Conflicts of Interest: The authors declare no conflict of interest.

References

1. Brannvall, E.; Wolters, M.; Sjoblom, R.; Kumpiene, J. Elements availability in soil fertilized with pelletized fly ash and biosolids. *J. Environ. Manag.* **2015**, *159*, 27–36. [CrossRef]
2. Väätäinen, K.; Sirparanta, E.; Räisänen, M.; Tahvanainen, T. The costs and profitability of using granulated wood ash as a forest fertilizer in drained peatland forests. *Biomass Bioenergy* **2011**, *35*, 3335–3341. [CrossRef]
3. Zhang, Z.; He, F.; Zhang, Y.; Yu, R.; Li, Y.; Zheng, Z.; Gao, Z. Experiments and modelling of potassium release behavior from tablet biomass ash for better recycling of ash as eco-friendly fertilizer. *J. Clean. Prod.* **2018**, *170*, 379–387. [CrossRef]
4. Memon, S.A.; Wahid, I.; Khan, M.K.; Tanoli, M.A.; Bimaganbetova, M. Environmentally Friendly Utilization of Wheat Straw Ash in Cement-Based Composites. *Sustainability* **2018**, *10*, 1322. [CrossRef]
5. Payá, J.; Roselló, J.; Monzó, J.M.; Escalera, A.; Santamarina, M.P.; Borrachero, M.V.; Soriano, L. An Approach to a New Supplementary Cementing Material: Arundo donax Straw Ash. *Sustainability* **2018**, *10*, 4273. [CrossRef]
6. Novais, R.M.; Ascensão, G.; Tobaldi, D.M.; Seabra, M.P.; Labrincha, J.A. Biomass fly ash geopolymer monoliths for effective methylene blue removal from wastewaters. *J. Clean. Prod.* **2018**, *171*, 783–794. [CrossRef]
7. Niu, Y.; Tan, H.; Hui, S. Ash-related issues during biomass combustion: Alkali-induced slagging, silicate melt-induced slagging (ash fusion), agglomeration, corrosion, ash utilization, and related countermeasures. *Prog. Energy Combust. Sci.* **2016**, *52*, 1–61. [CrossRef]
8. Rosales, J.; Cabrera, M.; Beltrán, M.G.; López, M.; Agrela, F. Effects of treatments on biomass bottom ash applied to the manufacture of cement mortars. *J. Clean. Prod.* **2017**, *154*, 424–435. [CrossRef]
9. Prasara-A, J.; Gheewala, S.H. Sustainable utilization of rice husk ash from power plants: A review. *J. Clean. Prod.* **2017**, *167*, 1020–1028. [CrossRef]
10. Nunes, L.J.R.; Matias, J.C.O.; Catalão, J.P.S. Biomass combustion systems: A review on the physical and chemical properties of the ashes. *Renew. Sustain. Energy Rev.* **2016**, *53*, 235–242. [CrossRef]

11. Vassilev, S.V.; Baxter, D.; Andersen, L.K.; Vassileva, C.G. An overview of the composition and application of biomass ash. Part 1. Phase-mineral and chemical composition and classification. *Fuel* **2013**, *105*, 40–76. [CrossRef]
12. Ekvall, H.; Löfgren, S.; Bostedt, G. Ash recycling—A method to improve forest production or to restore acidified surface waters? *For. Policy Econ.* **2014**, *45*, 42–50. [CrossRef]
13. Bonanno, G.; Cirelli, G.L.; Toscano, A.; Giudice, R.L.; Pavone, P. Heavy metal content in ash of energy crops growing in sewage-contaminated natural wetlands: Potential applications in agriculture and forestry? *Sci. Total Environ.* **2013**, *452–453*, 349–354. [CrossRef]
14. Nurmesniemi, H.; Mäkelä, M.; Pöykiö, R.; Manskinen, K.; Dahl, O. Comparison of the forest fertilizer properties of ash fractions from two power plants of pulp and paper mills incinerating biomass-based fuels. *Fuel Process. Technol.* **2012**, *104*, 1–6. [CrossRef]
15. Freire, M.; Lopes, H.; Tarelho, L.A.C. Critical aspects of biomass ashes utilization in soils: Composition, leachability, PAH and PCDD/F. *Waste Manag.* **2015**, *46*, 304–315. [CrossRef]
16. Kalembkiewicz, J.; Chmielarz, U. Ashes from co-combustion of coal and biomass: New industrial wastes. *Resour. Conserv. Recycl.* **2012**, *69*, 109–121. [CrossRef]
17. Nguyen, M.N.; Dultz, S.; Picardal, F.; Bui, A.T.; Van, P.Q.; Schieber, J. Release of potassium accompanying the dissolution of rice straw phytolith. *Chemosphere* **2015**, *119*, 371–376. [CrossRef]
18. Vassilev, S.V.; Baxter, D.; Andersen, L.K.; Vassileva, C.G. An overview of the composition and application of biomass ash.: Part 2. Potential utilisation, technological and ecological advantages and challenges. *Fuel* **2013**, *105*, 19–39. [CrossRef]
19. Li, S.; Liu, Y.; Wang, J.; Yang, L.; Zhang, S.; Xu, C.; Ding, W. Soil Acidification Aggravates the Occurrence of Bacterial Wilt in South China. *Front. Microbiol.* **2017**, *8*, 1–12. [CrossRef] [PubMed]
20. Blake, L. Acid Rain and Soil Acidification. In *Encyclopedia of Soils in the Environment*; Hillel, D., Ed.; Elsevier: Oxford, UK, 2005; pp. 1–11.
21. Etiégni, L.; Campbell, A.G. Physical and chemical characteristics of wood ash. *Bioresour. Technol.* **1991**, *37*, 173–178. [CrossRef]
22. Zhang, Y.; He, F.; Gao, Z.; You, Y.; Sun, P. Effects of ash-forming temperature on recycling property of bottom ashes from rice residues. *Fuel* **2015**, *162*, 251–257. [CrossRef]
23. Andrea Jordan, C.; Akay, G. Speciation and distribution of alkali, alkali earth metals and major ash-forming elements during gasification of fuel cane bagasse. *Fuel* **2012**, *91*, 253–263. [CrossRef]
24. Werkelin, J.; Skrifvars, B.-J.; Zevenhoven, M.; Holmbom, B.; Hupa, M. Chemical forms of ash-forming elements in woody biomass fuels. *Fuel* **2010**, *89*, 481–493. [CrossRef]
25. Khanna, P.; Raison, R.; Falkiner, R. Chemical properties of ash derived from Eucalyptus litter and its effects on forest soils. *For. Ecol. Manag.* **1994**, *66*, 107–125. [CrossRef]
26. Xing, P.; Mason, P.E.; Chilton, S.; Lloyd, S.; Jones, J.M.; Williams, A.; Nimmo, W.; Pourkashanian, M. A comparative assessment of biomass ash preparation methods using X-ray fluorescence and wet chemical analysis. *Fuel* **2016**, *182*, 161–165. [CrossRef]
27. Vassilev, S.V.; Vassileva, C.G.; Song, Y.-C.; Li, W.-Y.; Feng, J. Ash contents and ash-forming elements of biomass and their significance for solid biofuel combustion. *Fuel* **2017**, *208*, 377–409. [CrossRef]
28. Katare, V.D.; Madurwar, M.V. Experimental characterization of sugarcane biomass ash—A review. *Constr. Build. Mater.* **2017**, *152*, 1–15. [CrossRef]
29. Maresca, A.; Hyks, J.; Astrup, T.F. Recirculation of biomass ashes onto forest soils: Ash composition, mineralogy and leaching properties. *Waste Manag.* **2017**, *70*, 127–138. [CrossRef]
30. Zhang, Z.; He, F.; Zhang, Y.; Li, X.; Gao, Z. Simulation of combustion process of a single biomass pellet based on heterogeneous-dimension discretization. *J. Energy Inst.* **2018**. [CrossRef]
31. Ziemiański, L.; Zak, G.; Duda, A.; Wojtasik, M. Modification of the characteristic melting temperatures of sawdust ashes. *J. Renew. Sustain. Energy* **2017**, *9*, 043102. [CrossRef]
32. Wang, S.; Hu, Y.; He, Z.; Wang, Q.; Xu, S. Study of pyrolytic mechanisms of seaweed based on different components (soluble polysaccharides, proteins, and ash). *J. Renew. Sustain. Energy* **2017**, *9*, 023102. [CrossRef]
33. Trinh, T.K.; Nguyen, T.T.H.; Tu, N.N.; Wu, T.Y.; Meharg, A.A.; Nguyen, M.N. Characterization and dissolution properties of phytolith occluded phosphorus in rice straw. *Soil Tillage Res.* **2017**, *171*, 19–24. [CrossRef]

34. He, F.; Yi, W.; Li, Y.; Zha, J.; Luo, B. Effects of fuel properties on the natural downward smoldering of piled biomass powder: Experimental investigation. *Biomass Bioenergy* **2014**, *67*, 288–296. [CrossRef]

35. Van Loo, S.; Koppejan, J. *Handbook of Biomass Combustion and Co-Firing*; Earthscan: London, UK, 2008.

36. Wang, L.; Skreiberg, Ø.; Becidan, M. Investigation of additives for preventing ash fouling and sintering during barley straw combustion. *Appl. Ther. Eng.* **2014**, *70*, 1262–1269. [CrossRef]

37. Chen, C.; Yu, C.; Zhang, H.; Zhai, X.; Luo, Z. Investigation on K and Cl release and migration in micro-spatial distribution during rice straw pyrolysis. *Fuel* **2016**, *167*, 180–187. [CrossRef]

38. Johansen, J.M.; Aho, M.; Paakkinen, K.; Taipale, R.; Egsgaard, H.; Jakobsen, J.G.; Frandsen, F.J.; Glarborg, P. Release of K, Cl, and S during combustion and co-combustion with wood of high-chlorine biomass in bench and pilot scale fuel beds. *Proc. Combust. Inst.* **2013**, *34*, 363–2372. [CrossRef]

39. Werther, J.; Saenger, M.; Hartge, E.U.; Ogada, T.; Siagi, Z. Combustion of agricultural residues. *Prog. Energy Combust. Sci.* **2000**, *26*, 1–27. [CrossRef]

40. Sluiter, A.; Hames, B.; Ruiz, R.; Scarlata, C.; Sluiter, J.; Templeton, D. Determination of ash in biomass. *Nat. Renew. Energy Lab.* **2008**, *1*, 1–5.

41. Febrero, L.; Granada, E.; Patiño, D.; Eguía, P.; Regueiro, A. A Comparative Study of Fouling and Bottom Ash from Woody Biomass Combustion in a Fixed-Bed Small-Scale Boiler and Evaluation of the Analytical Techniques Used. *Sustainability* **2015**, *7*, 5819. [CrossRef]

42. Zhang, Y.; Yu, R.; Zhang, Z.; Gao, Z.; Yang, B.; He, F. A comparison on three methods for measuring of water-soluble potassium content in straw ash. *Renew. Energy Resour.* **2017**, *35*, 1588–1594.

43. Yu, Y.; Xu, M.; Yao, H.; Yu, D.; Qiao, Y.; Sui, J.; Liu, X.; Cao, Q. Char characteristics and particulate matter formation during Chinese bituminous coal combustion. *Proc. Combust. Inst.* **2007**, *31*, 1947–1954. [CrossRef]

44. Luan, C.; You, C.; Zhang, D. Composition and sintering characteristics of ashes from co-firing of coal and biomass in a laboratory-scale drop tube furnace. *Energy* **2014**, *69*, 562–570. [CrossRef]

45. Skrifvars, B.-J.; Hupa, M.; Backman, R.; Hiltunen, M. Sintering mechanisms of FBC ashes. *Fuel* **1994**, *73*, 171–176. [CrossRef]

46. Niu, Y.; Tan, H.; Wang, X.; Liu, Z.; Liu, H.; Liu, Y.; Xu, T. Study on fusion characteristics of biomass ash. *Bioresour. Technol.* **2010**, *101*, 9373–9381. [CrossRef] [PubMed]

47. Lima, A.T.; Ottosen, L.M.; Pedersen, A.J.; Ribeiro, A.B. Characterization of fly ash from bio and municipal waste. *Biomass Bioenergy* **2008**, *32*, 277–282. [CrossRef]

48. Zevenhoven-Onderwater, M.; Blomquist, J.P.; Skrifvars, B.J.; Backman, R.; Hupa, M. The prediction of behaviour of ashes from five different solid fuels in fluidised bed combustion. *Fuel* **2000**, *79*, 1353–1361. [CrossRef]

49. Boström, D.; Skoglund, N.; Grimm, A.; Boman, C.; Öhman, M.; Broström, M.; Backman, R. Ash Transformation Chemistry during Combustion of Biomass. *Energy Fuels* **2011**, *26*, 85–93. [CrossRef]

50. Wang, L.; Hustad, J.E.; Grønli, M. Sintering Characteristics and Mineral Transformation Behaviors of Corn Cob Ashes. *Energy Fuels* **2012**, *26*, 5905–5916. [CrossRef]

51. Steenari, B.M.; Karlsson, L.G.; Lindqvist, O. Evaluation of the leaching characteristics of wood ash and the influence of ash agglomeration. *Biomass Bioenergy* **1999**, *16*, 119–136. [CrossRef]

52. Du, S.; Yang, H.; Qian, K.; Wang, X.; Chen, H. Fusion and transformation properties of the inorganic components in biomass ash. *Fuel* **2014**, *117*, 1281–1287. [CrossRef]

53. Olanders, B.; Steenari, B.-M. Characterization of ashes from wood and straw. *Biomass Bioenergy* **1995**, *8*, 105–115. [CrossRef]

54. Magdziarz, A.; Gajek, M.; Nowak-Woźny, D.; Wilk, M. Mineral phase transformation of biomass ashes—Experimental and thermochemical calculations. *Renew. Energy* **2018**, *128*, 446–459. [CrossRef]

55. Dan, B.; Skoglund, N.; Grimm, A.; Boman, C.; Öhman, M.; Broström, M.; Backman, R. Ash Transformation Chemistry during Combustion of Biomass. *Energy Fuels* **2012**, *26*, 85–93.

56. Hupa, M. Ash-Related Issues in Fluidized-Bed Combustion of Biomasses: Recent Research Highlights. *Energy Fuels* **2012**, *26*, 4–14. [CrossRef]

57. Liu, H.; Zhang, L.; Han, Z.; Xie, B.; Wu, S. The effects of leaching methods on the combustion characteristics of rice straw. *Biomass Bioenergy* **2013**, *49*, 22–27. [CrossRef]

58. Vassilev, S.V.; Baxter, D.; Vassileva, C.G. An overview of the behaviour of biomass during combustion: Part II. Ash fusion and ash formation mechanisms of biomass types. *Fuel* **2014**, *117*, 152–183. [CrossRef]

59. Nguyen, M.N.; Dultz, S.; Guggenberger, G. Effects of pretreatment and solution chemistry on solubility of rice-straw phytoliths. *J. Plant Nutr. Soil Sci.* **2014**, *177*, 349–359. [CrossRef]
60. Thy, P.; Jenkins, B.M.; Lesher, C.E. High-Temperature Melting Behavior of Urban Wood Fuel Ash. *Energy Fuels* **1999**, *13*, 839–850. [CrossRef]

© 2019 by the authors. Licensee MDPI, Basel, Switzerland. This article is an open access article distributed under the terms and conditions of the Creative Commons Attribution (CC BY) license (http://creativecommons.org/licenses/by/4.0/).

 *sustainability*

Article

Local Economic Impact of Wind Energy Development: Analysis of the Regulatory Framework, Taxation, and Income for Galician Municipalities

Damián Copena [1,*], David Pérez-Neira [2] and Xavier Simón [1]

[1]   Department of Applied Economy, FCEE, Vigo University, 36310 Vigo, Spain; xsimon@uvigo.gal
[2]   Department of Economy and Statistics, Leon University, 24071 León, Spain; dpern@unileon.es
*   Correspondence: decopena@uvigo.gal; Tel.: +34-986-812506

Received: 13 March 2019; Accepted: 17 April 2019; Published: 23 April 2019

Abstract: Wind energy has rapidly developed in the last decades, generating economic impacts at different territorial scales and contributing to rural development. However, few research works have analysed its economic impact at a local scale, especially in rural areas. Galicia is a Spanish region in which 3300 MW of wind energy have been installed in rural municipalities with low levels of socioeconomic activity and important socio-environmental problems. In this sense, the objective of this work is to analyse the local revenues directly derived from wind power activity in relation to changes in the regulatory framework (1995–2017), as well as to quantify those revenues for the year 2017. For this purpose, information has been systematically collected from secondary sources and complemented with 10 years of field and monitoring work on site at the wind farms. This article reveals the relationship between the regulatory framework and the main sources of income associated with wind power generation (conventional and specific taxes, municipal ownership, and other revenues). In 2017, these revenues amounted to 17.8 million euros. This work discusses how the public policies implemented during the analysed time period limited the direct economic impacts of the installation of wind farms on Galician rural municipalities, and consequently hindered rural development.

Keywords: municipalities; public policies; rural development; wind farms; renewable energy

---

1. Introduction

The development of the different forms of renewable energy has, in the last decades, become a priority at both a global level [1] and at the European scale [2]. Among those forms, wind power is one of the most prominent renewable technologies, as has been widely stated in the literature [3]. Thus, this type of energy has rapidly expanded across the world [4,5], facilitating the supply of energy from local and renewable resources and contributing to the fight against climate change [6]. Besides, the installation of wind farms has generated socioeconomic impacts at different territorial levels (national, regional, local), fostering rural development to a greater or lesser extent [7,8]. For instance, Varela and Sánchez [9] have analysed the importance of wind energy impacts on employment generation and the gross domestic product of peripheral regions. Slattery et al. [10] have shown how the installation of wind farms has had socioeconomic effects at a regional and, most especially, at a local level in rural areas of Texas (United States). In addition, recent works have underlined how renewable energy can be an opportunity to create more dynamic local communities [11], to achieve sustainable development at the local government level [12], and to encourage rural development through the sustainable exploitation of local resources [13] (see Section 2).

Galicia is a region situated in the northwest of Spain. The number of wind farms in the region has increased exponentially in the last few decades. In 20 years, around 4000 wind turbines have been installed in more than 150 wind farms across Galician rural areas. These wind farms have over 3300 MW distributed to 107 out of 315 municipalities in Galicia (Figure 1). In 2016, Galicia was, after Germany and Denmark, the third largest territory of the European Union in terms of wind power per unit area. On the other hand, it is important to point out that almost all of the wind turbines are located in rural municipalities with serious structural socioeconomic problems. To the low population density (98% of wind turbines are found in municipalities with less than 150 inhabitants per km^2), it is possible to add the high ageing index (Galicia is the second autonomous region in Spain with the highest index) [14], the loss of agricultural employment (which has reduced by 65% in the last 15 years) and the increase in the number of abandoned towns and villages (from 1064 to 1726 in the 2000–2017 period) [15]. All of these factors have led to the progressive abandonment of cropland and forest areas [16], causing important environmental problems such as forest fires [17].

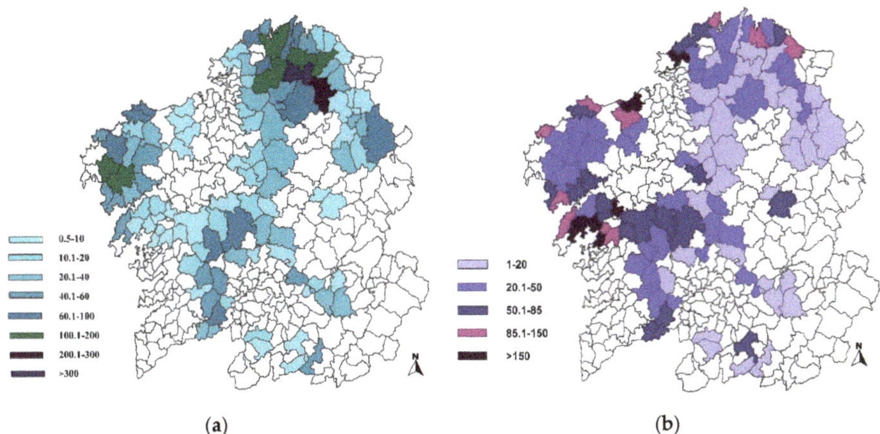

Figure 1. Territorial distribution of: (**a**) number of wind turbines (2018), and (**b**) population density (inhabitants per km^2) in municipalities with wind farms in Galicia (2016). As can be seen, most municipalities with wind farms in their territories have low population densities, often below 50 inhabitants per square kilometre, which makes this new economic activity even more relevant for rural areas. Municipalities with more than 100 wind turbines have an average population density of 20.6 inhabitants per square kilometre, while the density of those with less than 10 wind turbines is 113.5 inhabitants per square kilometre. Source: Own elaboration from: (a) Socioeconomic Information Database of Wind Energy in Galicia (SIDWEG) [18]; (b) Instituto Galego de Estatística (IGE) [19].

In this context, wind energy emerges as a great opportunity to reverse negative trends and revitalise rural municipalities in the region. In this sense, the new local revenues associated with wind power development and the corresponding public policies are two key aspects of the analysis of potential opportunities for the development of rural territories [20]. The objective of this work is three-fold: (a) to identify, characterise, and analyse the dynamics of the main direct revenues derived from wind power activity and earned by rural municipalities between 1995–2017; (b) to make a quantitative estimation of the local revenues in the last year for which information was available (2017); and (c) to discuss the influence of the Galician and Spanish regulatory framework on the results. This article reflects on the economic importance of the main sources of income identified: the revenues obtained from singular wind farms, conventional taxes, the canon eólico (wind power tax), and the Environmental Compensation Fund, and other possible revenues such as those derived from collaboration agreements and the renting of public property (see Section 4). In order to conduct this research, a systematic collection of information from secondary sources has been carried out and completed with field and monitoring work at the wind farms over 10 years (2008–2017). As a result of this research process, a database of socioeconomic information on wind energy in Galicia was created (Socioeconomic Information Database of Wind Energy in Galicia, SIDWEG) (see [18]); this database has been the main source of information for the present work (see Section 3). Thus, this article contributes new and relevant socioeconomic information to the political debate on the relevance of wind energy and its regulatory framework for rural development processes in relation to the endogenous potentialities of the different territories.

2. Local Economic Impacts, Regulatory Framework and Rural Development Associated with Wind Energy

The local socioeconomic impacts of wind farms in rural areas have been poorly studied [10,21]. In spite of this, it is possible to identify four different research lines among the works published. First of all, there are studies focused on the direct economic impacts on the local territory. Taxes are one of the local sources of revenue that are most often analysed in this literature [22,23]. Thus, in some countries, local administrations receive, via taxes, direct contributions from the turnover of wind farms located in their territories. For instance, this is the case of Greek municipalities, which receive annual payments representing 3% of the turnover of wind farms built within their territories [24]; in Portugal, the annual amount corresponds to 2.5% of the turnover [25]. For the United States, Slattery et al. [10] have estimated that the payment of taxes associated with wind power activity to local governments can range between 4000–12,000 US dollars per MW. On the other hand, the establishment of locally owned wind farms also has direct economic impacts on the area, as seen in England [26], the United States [27], Germany, or Denmark, where in 2010, 15% of the wind turbines were owned by cooperatives controlled by the local population [28].

Second, there are research works that discuss the possibilities of local development and revitalisation associated with wind power activity. In the literature, there is a strong consensus on the potential of renewable energy as a driver for the sustainable development of territories [11,12,29,30], which is a fact that is also recognised by the European regulations [2]. Thus, Munday et al. [31] have analysed the benefits that wind energy has for local communities in Scotland, while Copena and Simón [20] have shown how payments received by landowners may foster initiatives to revive local areas in Galicia. Delicado et al. [21] have studied how, in the case of Portugal, the benefits and positive impacts of wind farms are perceived as indirect and quite modest. From another perspective, Castleberry and Greene [32] have examined the specific case of public schools in Oklahoma, finding significant differences in revenues among school districts with and without wind farms in their territories. Okkonen and Lehtonen [33] have highlighted the importance of wind revenues in the provision and maintenance of basic services for the population of northern Scotland.

Third, there are research works focused on studying the role of economic impacts on the local acceptance of wind farms. Economic benefits in particular are presented as strongly linked to the local acceptance of wind farm projects [22]. Thus, Liebe et al. [34], studying the case of Germany, revealed a positive relationship between the local ownership of wind farms and the local acceptance of renewable energy plants, while Slattery et al. [35] pointed to the increase in local revenues via taxes as one of the elements explaining the greater support for this type of project in the United States. Other authors such as Jami and Walsh [36] have also described local ownership and the increase in revenues via taxes as key for the acceptance of wind farms in Ontario (Canada), while Mundaca et al. [37] connected local ownership with distributive justice and the consent for wind farms in Denmark. In the same line, the increment of local revenues through taxes was a strong incentive for Portuguese local authorities to approve the construction of wind farms [21], which is a dynamic that has also been specifically pointed out in relation to rural contexts with low economic activity [38].

Finally, the fourth research line includes works that analyse the importance of the institutional and regulatory frameworks of the wind industry and their influence on the three previous aspects. Certainly, the scope of the local economic impacts of wind energy is limited by the public policies regulating wind power activity [39]. For instance, in Denmark, 20% of the ownership of a wind farm is offered to the public in the form of shares at a cost price, which is a norm that fosters the increase of local revenues [40,41]. In France, a specific wind tax has been created for the purpose of granting municipalities the possibility of participating in the income generated by wind farms [42]. In the United States, local administrations have had a crucial role in the planning and development of regulations; they are even entitled to decide the location of wind farms [43]. Something similar occurs in Sweden, where municipalities hold veto power over the projects [44], or in the United Kingdom, where governance processes at a local level are common [45]. On the contrary, in contexts where local ownership of wind farms is rare and the local authorities have a limited negotiation capacity, the structural interests of wind power companies (whether national or international) weaken the possible benefits and positive impacts of this activity [21].

3. Materials and Methods

3.1. Gathering and Selection of Information

In order to carry out this research, a systematic gathering of information from secondary sources has been performed and completed with field and monitoring work at wind farms over a 10-year period (2008–2017). Among the secondary sources, which are mostly public databases, regulatory documents and official journals were systematised for the 1995–2017 period (see Table 1). Complementarily, the field work consisted of the collection of socioeconomic information on the Galician municipalities affected by wind power projects. For this purpose, participatory research techniques were used [46,47], and a total of 106 semi-structured interviews were conducted on the wind farms in operation. The main source of information for this research paper came from the systematisation of data gathered from secondary sources; the field work was a complementary source that was used mainly to have access to the renting contracts signed by the tenants of municipally-owned land and to compile the few collaboration agreements drawn up by companies and municipalities. The interviews were directly conducted with key actors in the municipalities with wind farms in their territories (71% of them were local inhabitants, 19% were political representatives at the local government, and 10% were promoters and managers of companies providing auxiliary and other services to the wind farms). The semi-structured interviews aimed at gathering information on wind power projects that could not be obtained from the above-mentioned secondary sources. This information included: the type of local participation in the approval procedure for the installation of wind farms; local revenues from wind farms and instruments used for the transfer of those revenues; role of the municipalities in the development of wind farms; and the ultimate destination of wind revenues.

As a result of this process, a Socioeconomic Information Database of Wind Energy in Galicia was created [18], containing wind-related administrative, economic, social, and institutional data of the 1995–2017 period for all the operating wind farms and the municipalities affected by them in the Galician territory. The SIDWEG is an innovative database that includes both quantitative and qualitative information at four different scales: wind farm, municipality, province, and the whole region of Galicia. The SIDWEG contains systematised public information on technological aspects (unit capacity, total capacity, number of wind turbines, technology used, etc.), economic aspects (investments made, annual production, premiums received, local public revenue, etc.), and territorial aspects (areas affected, land occupied, number of wind turbines per municipality, etc.). Finally, the SIDWEG comprises social information (actors identified, the role played by each of them, etc.) and economic information for the above-mentioned four scales regarding the income obtained by the owners of wind farms and the main destination of that income. For this work, we have specifically used the sub-databases included in Table 1 with disaggregated local information. The four sub-databases were the ones containing: (1) socio-administrative data, (2) economic data, (3) public income data, and (4) landowner income data. The first one allows characterising wind farms in operation and those awaiting administrative processing, facilitating a constant monitoring of the administrative situation of every wind farm. The second database synthesises information about the production and economic turnover of wind farms, including those owned by the local administrations. The third database systematises the flows of wind revenues earned by the municipalities, which were classified as follows: (i) revenues derived from wind taxation (the *canon eólico* and the Environmental Compensation Fund, hereinafter ECF); (ii) inflowing revenues from conventional taxes; and (iii) revenues obtained from the local ownership of wind farms (municipal singular wind farms). This database includes as well, to the extent that it is possible, information on the final destination of the local revenues associated with wind power activity. Finally, it reflects other possible wind-related incomes, more specifically, those linked to the leasing contracts and collaboration agreements drawn up between companies and municipalities.

Table 1. Summary of the main sources of information and fields of the SIDWEG used for the analysis of the local economic impacts of wind energy in Galicia.

Database	Sources of Information	Content	Relevance for This Article
1. Socio-administrative	- Official Journal of Galicia - Official State Gazette - Register of special regime installations - Register of pre-allocation of remuneration - Administrative record of electric energy production - Regulatory documents, especially those issued by the Galician regional administration	- Temporary dynamics concerning the installation of wind farms and MW - Information on the developers - Information classified according to the second administrative phase	- To understand the dynamics of wind energy development in relation to the legal framework - To understand the future prospects for the installation of new wind farms and the creation of new economic opportunities at the local level
2. Economic	- Statistical information on the electric energy industry - Statistical information on special-regime energy sales - Specialised reports published by public and private entities	- Wind farm turnover	- To measure the importance of this new industrial activity in rural areas - To estimate the turnover of wind farms in order to compare it with local revenues
3. Public income	- Public statistical databases - Official Journal of Galicia - Official Journal of each Galician province - Specialised scientific literature on the object of study - Public registers related to renewable energy activities - Geographic Information System tools - Websites of the different municipalities - Semi-structured interviews	- Revenues obtained at a local level, disaggregated by categories (*canon eólico* and ECF, conventional taxes, and municipal ownership of wind farms) - Destination of public wind revenues (only available in certain cases)	- Identification, characterisation, and estimation of the new income flows earned by municipalities with wind farms in their territories for each of the categories identified (*canon eólico* and the Environmental Compensation Fund (ECF), conventional taxes, and municipal ownership of wind farms) - To discuss wind revenues for the municipalities and their destination, as well as their impact on local public income
4. Landowner income	- Statistical information on the electric energy industry - Statistical information on special-regime energy sales - Geographic Information System tools - Semi-structured interviews	- Renting contracts and collaboration agreements between companies and municipalities	- To have access to the renting contracts of municipally-owned land signed by wind power companies - To have access to the collaboration agreements drawn up between municipalities and wind power firms

Source: Own elaboration from the SIDWEG.

3.2. Systematisation of the Information and Categories of Analysis

From the information collected (Table 1) and the analysis of the regulatory framework, it was possible to identify the main sources of revenue, which were classified into two categories (Table 2): (i) income from conventional and specific taxes; and (ii) other incomes, which were obtained from the leasing of municipal land for the construction of wind farms and from the collaboration agreements drawn up between wind power firms and municipalities. Among these categories, there are two sources of revenue associated with the specific regulatory framework of Galicia [18]. The first one is a wind power tax (*canon eólico*) that partially benefits municipalities with wind farms in their territories. This wind power tax is a levy on the number of wind turbines, and it feeds the ECF [48]. Its taxable events are the negative visual, environmental, and other impacts of wind farm installations on the territory. The second specific source of income, i.e., singular wind farms, allows municipalities to promote wind power plants. The income flows for 2017, the last year for which there is information available, were estimated as follows. (i) Conventional taxes were calculated from the data of operating power collected in the SIDWEG and the updated average values obtained by Saladié [49] for Catalonia; (ii) ECF revenues were estimated from the systematisation of the municipal information published by the Official Journal of Galicia and reflected in the SIDWEG; (iii) The income derived from singular wind farms was calculated from information gathered from wind farm tenders and during the field work. This estimation allowed calculating the average annual income per MW, which were used to quantify the total revenues derived from singular wind farms in operation (SIDWEG); (iv) Finally, the data collected during the field work and systematised in the SIDWEG were also used to calculate the revenue obtained from the leasing of municipal land and from collaboration agreements.

Table 2. Comparative analysis of the categories of wind revenues in Galician municipalities.

Category	Subcategory	Tax or Taxing Figure	Year of First Possible Revenues	Specific of Galician Legislation	Limitations for the Use of Revenues	Characteristics of the Income
i. Taxes	Conventional taxes	IAE (Tax on economic activities)	1995	No	No	Annual income
		IBI (Property tax)	1995	No	No	Annual income
		ICIO (Tax on Constructions, Installations, and Works)	1995	No	No	One-time income
	b. Wind taxes	*Canon eólico* and ECF	2010	Yes	Yes	Income depending on the number of wind turbines and the number of meters of power evacuation lines
ii. Municipal ownership of wind farms		Singular wind farm	2001	Yes	No	Percentage of the turnover and/or annual fix amount
iii. Other incomes		Renting of municipal land	1995	No	No	Annual income/One-time income
		Collaboration agreements	1995	No	No	Annual income/One-time income

Source: Own elaboration from the SIDWEG.

4. Results

Figure 2 shows the temporal dynamics of the possibility of income flows for the municipalities during the period analysed. Thus, in 1997, with the installation of the first wind farms, conventional taxes started having a local economic impact. In 2001, the mechanism that allows developing small wind farms owned by the municipalities was established. The most recent regulatory change is connected to the creation of the wind power tax and the ECF in 2009. That was also the year in which the installed power levelled off; since then and until 2017, it has only increased by 3%. In 2017, the total revenue of municipalities with wind farms in their territories amounted to 17.8 million euros (Table 3). The most important category of local income was that of taxes (92.7%), followed by the municipal ownership of wind farms (7.0%). The category of other incomes was the less significant one (0.2%). In the following sections, each revenue category and subcategory will be analysed in detail. As shown in Figure 2, the rapid expansion of wind farms in Galicia has been favoured by the existence of incentives, which were mainly associated with feed-in tariff mechanisms [50]. However, this support model has greatly changed in the last few years. Thus, in 2013, the feed-in tariffs incentive system was eliminated, and new support mechanisms have been created since. On the one hand, there is an auction system in which the compensation-for-operation amounts are determined, and, on the other, there is a compensation-for-investment mechanism [51]. The halt in the expansion of wind farms observed since 2008 was not a consequence of the establishment of specific taxes such as the *canon eólico*, but rather of other factors mainly including legal uncertainty, elimination of the compensation schemes, and decline of the demand because of the crisis.

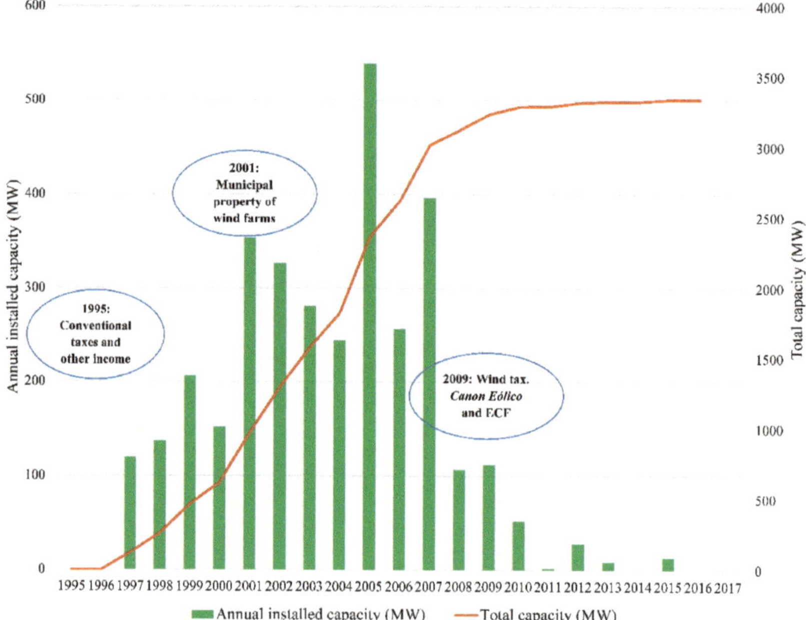

Figure 2. Temporal dynamics of the possibility of income flows for municipalities in the process of developing wind farms in Galicia (MW). Source: Own elaboration from the SIDWEG.

Table 3. Characterisation and estimation of revenues for Galician municipalities with wind farms according to the income flow (2017).

Category	Subcategory	Tax/Mechanism	Number of Municipalities	Income (1000 €)	Income (%)
i. Taxes	a. Conventional taxes (*)	IBI	107	6719	37.7
		IAE	103	3499	19.6
	b. Wind taxes	*Canon eólico* and ECF	117 (**)	6307	35.4
ii. Municipal ownership of wind farms		Singular wind farm	13	1255	7.0
iii. Other incomes		Renting of municipal land	5	<20	0.1
		Collaboration agreements	3	<20	0.1
Total income estimation				17,820	100

Source: Own elaboration from the SIDWEG. Notes: (*) The ICIO is not taken into consideration because it is only paid once, with the construction of the wind farm (no new farms were installed in 2017); (**) This tax includes municipalities with power evacuation lines, in addition to the 107 municipalities with wind turbines in their territories.

4.1. Municipal Tax Revenues

4.1.1. Conventional Taxes

The wind energy model established in Galicia is based almost exclusively on the installation of wind farms promoted by private companies. The development of this business activity requires the payment of taxes to the local administration that become a source of income for the territories where the wind farms are located. In particular, there are three conventional taxes positively affecting those municipalities: the tax on economic activities (IAE), the property tax (IBI), and the tax on constructions, installations, and works (ICIO). The first two taxes are annually paid, while the third one is paid only once, with the construction of the wind farm. Among these three taxes, the IBI is the most important one (37.7% of the total tax revenue in 2017). This tax changed substantially after 2007 as a result of a judicial process initiated by the Galician Federation of Municipalities and Provinces. This process determined that wind farms belong to a specific category of taxable goods (BICES), on which higher tax rates are imposed. Tax rates for BICES are set by municipalities through a tax ordinance and amount to between 0.4–1.3% of the cadastral value of the property.

The IAE is another annual tax managed by the municipalities (19.6% of the total revenue in 2017). It is a levy on the wind power activity that imposes a rate per unit of power. Wind farms with a turnover of less than 1 million euros per year are exempt from this tax. The taxable base is linked to the net annual turnover, and the tax rate is around 1.5% [52]. Finally, the ICIO has also generated, within the wind industry, some legal disputes regarding the rates to be paid [53]. The debate revolved around whether the cost of wind turbines ought to be included in the taxable base of the ICIO or not. Various court sentences have been clear on the matter, and have included the cost in the taxable base. As a result, there has been an increase of income flow for the municipalities, given that wind turbines account for 70% of the cost of a wind farm [54]. The municipality establishes the ICIO rate, which ranges between 2–4% of the cost of executing the works, and is only paid once.

4.1.2. Specific Taxes: the *canon eólico* and the Environmental Compensation Fund

The *canon eólico* was created in 2009 by the Galician regional government for the purpose of taxing negative externalities associated with the installation of wind farms. More specifically, this levy taxes the number of wind turbines and establishes a tax-exempt tax bracket and three other tax brackets with rates that increase with the number of wind turbines (see Table 4). Due to the quick development of wind energy in Galicia, the unit capacity of wind turbines is generally limited, which implies the installation of a high number of wind turbines per wind farm. Thus, most plants (65%) fall within the highest tax bracket, which accumulates 94.6% of the income collected through this tax. The *canon eólico* feeds the ECF, which distributes a significant part of its income to the municipalities affected by wind farms. These income flows are channelled as non-competitive subsidies, and therefore are not freely disposable. They are intended to implement actions related to production, employment generation, biodiversity conservation, the recreational use of natural resources, and the sustainable use of renewable energies [48].

Table 4. Characterisation and estimation of incomes derived from the *canon eólico* in Galicia (2017).

Canon eólico Tax Bracket (No. of Wind Turbines)	€ per Wind Turbine-Year	No. of Wind Farms	No. of Wind Turbines	Income (1000 euros)	Income (%)
Between 1–3	0	19	33	0	0
Between 4–7	2300	11	58	133	0.6
Between 8–15	4100	23	275	1128	4.9
More than 15	5900	101	3721	21,954	94.6
Total		154	4087	23,215	100

Source: Own elaboration from the SIDWEG.

After the *canon eólico* came into force, the total revenue collected through this new wind tax was estimated at around 22 million euros per year. This amount feeds the ECF, and the Galician regional government, which is responsible for managing this tax, determines that around one-third of those revenues (between 6–8 million euros per year) will go directly to the municipalities that have wind farms installed in their territories. In other words, the municipalities receive an income in the form of non-competitive subsidies according to the number of wind turbines and the metres of power evacuation lines in their territories. In the last years, ECF revenues have decreased (Figure 3), which has been mainly due to changes in the amounts allocated by the Galician government, and to a lesser extent, the reduction in the number of wind turbines caused by the repowering of wind farms [55]. In relation to the drop in EFC revenues allocated to municipalities with wind farms in their territories, there is no information on the reasons behind the regional government's reduction of the amounts in the last few years. In spite of this, it is worth highlighting that during the nine years in which the tax has been in force, the municipalities with wind farms and evacuation lines have received more than 63 million euros.

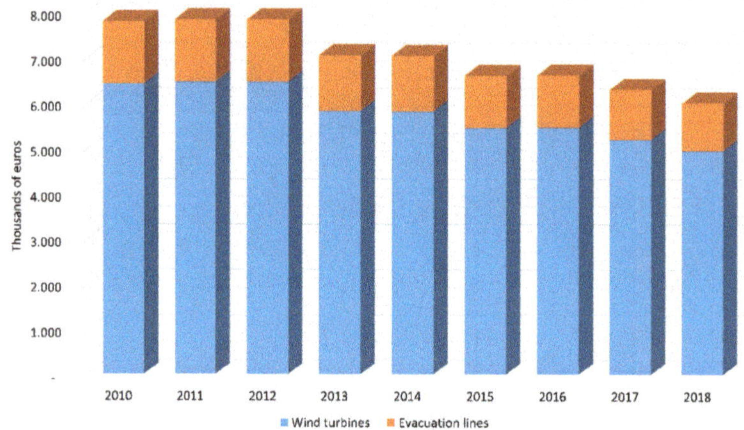

Figure 3. ECF non-competitive subsidies for municipalities with wind turbines and evacuation lines in their territories (thousands of euros). Source: Own elaboration from the SIDWEG.

With regard to the revenues from the *canon eólico* feeding the ECF, the Galician government has annually allocated, through competitive calls for proposals, between 1.3–2.8 million euros to environmental action programs to be developed in the municipalities. The use made of the remaining approximately 14 million euros is unknown, since there is no public information provided by the regional government regarding the management of these resources.

In 2017, the direct ECF revenues earned by municipalities with wind farms were estimated at 6.3 million euros, representing 35.4% of the total municipal incomes derived from wind power activity. From a geographical perspective, the Galician municipalities that received greater amounts from this tax were the smaller ones, which are dispersed across the region's territory (Figure 4). The economic impact of this tax on the municipalities located in the area of Serra do Xistral is worth highlighting, and among them, the case of Muras in particular, where 381 wind turbines are currently in operation. This municipality, which had a budget of about 1.7 million euros in 2016, has received, in the period during which the ECF has been active (2010–2018), more than 5.3 million euros in non-competitive subsidies for the number of wind turbines and evacuation lines located in its territory.

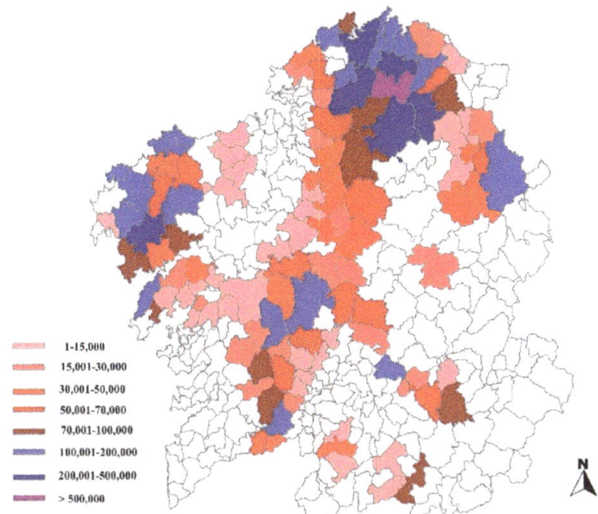

Figure 4. Geographical characterisation of ECF revenues for Galician municipalities with wind farms in their territories (2015) (euros). Source: Own elaboration from the SIDWEG.

4.2. Municipal Ownership of Wind Farms: Singular Wind Farms

In 2001, the specific category of singular wind farms was created to allow municipalities to promote this kind of installation. The capacity of each singular wind farm was limited to a maximum of 3 MW, which is a restriction that conditioned the generation of local revenues. Initially, the installation of this type of wind farm attracted the interest of many Galician municipalities. Up to 182 of the 315 municipalities in the region requested permits to build their own wind farms. However, this specific type of wind farm did not achieve the expected degree of development. Their limited expansion, despite the great interest they initially attracted among local governments, is mainly related to the legal uncertainty of the various calls for tenders. This uncertainty caused the cancellation of the largest call for tenders for this type of wind farm, and led to several subsequent court sentences that have paralysed the setting up of these singular wind plants [56].

The monitoring data of the administrative proceedings to install this type of wind farm indicate that in 2018, barely 13 of them were in operation (Figure 5). These wind farms have an aggregated installed capacity of 34.80 MW, which stands for only 1% of the total operating capacity.

Once the administrative authorisation is obtained, these farms can be managed through various mechanisms: (a) with the municipality' own resources; (b) by creating a joint venture controlled by the municipality; or (c) by issuing a call for tenders ending in an administrative concession to a company that will manage the wind farm. In Galicia, all the singular wind farms in operation were tendered by the local administration. From the information gathered and synthesised in the SIDWEG, it is possible to estimate the income per MW in operation earned by the municipalities. Thus, in 2017, the revenues per singular wind farm were estimated at between 19,000–47,000 euros per MW [56], with an average annual provision of 36,000 euros per MW, which is a small amount when compared to other sources of income. Thus, singular wind farms in Galicia contributed 1.2 million euros to local revenues (7% of the total). The collection and analysis of municipal wind farm tenders shows that the income flows usually combine an annual payment as a percentage of the wind farm's turnover (sometimes reaching 20%), and a fixed annual payment per installed capacity. Finally, it is worth mentioning that several singular wind farms awaiting administrative processing and not yet in operation are managed by joint ventures that will be interesting to observe in the near future, although at the moment it is not possible to compare the benefits of this mechanism with those of public tenders.

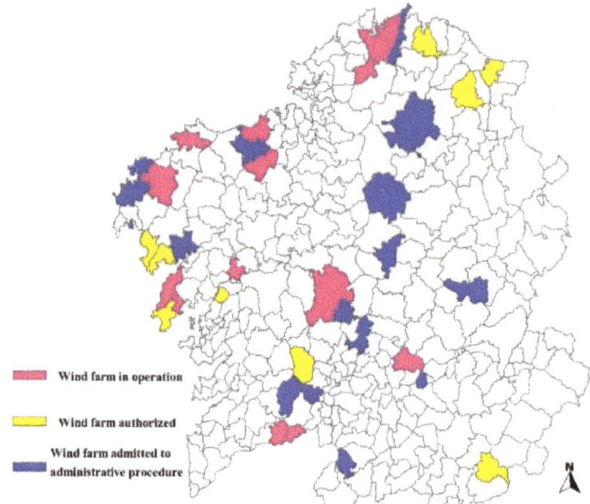

Figure 5. Geographical characterisation of municipally owned singular wind farms in Galicia (September 2018). Source: Own elaboration from the SIDWEG.

4.3. Other Incomes

This research work has identified two other wind-related income flows for local administrations, which nevertheless have little relevance on the total revenue (0.2%). The first source of income is linked to the signing of collaboration agreements between promoters and municipalities, which include the annual payment of some monetary amounts. For instance, the municipality of Cuntis signed an agreement with the company that developed the Monte Arca wind farm for the creation and maintenance of a new hiking trail. The agreement disaggregated the works to be implemented and the corresponding monetary amounts. In other countries, these income flows associated with specific agreements were framed within the concept of community benefits that are voluntarily provided by the promoters [57]. For instance, they are common in England, where they can be managed by local governments or social entities [31]. A second mechanism to obtain revenues for the local administration is to rent land belonging to the municipality for the installation of wind farms. However, in the case of Galicia, only 2% of the forest areas where the wind farms are usually located are municipally owned [58]. Consequently, only a few municipalities receive income via land leasing, according to the information available in the SIDWEG.

5. Discussion

5.1. Wind Revenues and Rural Development in Galicia

In the case of Galicia, wind power activity has created an important flow of direct income for the municipalities. If the estimated income earned by municipalities with wind farms in their territories in 2017 (17.8 million euros) is compared with the estimated turnover of the wind farms that same year, it is possible to observe how those revenues represent 3.17% of the estimated turnover of the wind farms. This percentage is relatively low, especially when considering the case of Greece, where local administrations collect the same percentage through only one tax (see Dimitropoulos and Kontoleon [24]). However, as underlined by Slattery et al., [10], these new sources of income gain further significance since the wind farms are installed in rural areas with little economic activity and important social and environmental problems. Thus, the revenues derived from wind power activity can be used to improve public services and/or the capacity of investment of the local economic network, as well as to implement activities linked to rural development and sustainability, as in Mexico [59] or Oklahoma [32].

As analysed before, taxes are the main direct wind revenue obtained by Galician municipalities; therefore, they present an opportunity to encourage local development through public expenditure [31]. However, in the case of incomes collected via conventional taxes (57.3% of the total in 2017), it is extremely difficult to identify which has been their final use due to at least two reasons: (a) tax conditions have changed during the period analysed, and (b) there is no disaggregated public information on the matter. In this sense, it is possible to affirm that wind power activity contributes to the economic growth of the affected areas, but it is not possible to determine with certainty the effects that it has on their rural development. In other regions, it has been proven that wind farms facilitate the maintenance of the provision of basic services for the population [32,33], and contribute to production diversification and the viability of agricultural activities [60]. In the case of Galicia, a singular example found in the SIDWEG is that of the municipality of Muras, which, ever since 2016, has allocated part of its wind revenues to helping local families pay their electricity bill.

On the other hand, ECF revenues (35.4% of the total in 2017) can be more easily monitored due to the limitations established in their regulatory framework. Thus, the financing of fire prevention, sanitation, improvement of infrastructures, and other environmental works and actions with ECF revenues is quite common. These expenditure policies foster the generation of economic activities and the creation of employment in rural municipalities with wind farms installed in their territories. For instance, in 2017, staff was hired with ECF revenues in the municipalities of Cedeira (seven workers), Ponteceso (nine workers), Aranga (eight workers), Cariño (two workers), and Muras (10 workers). These experiences prove that although municipalities are only granted a limited share of the amounts collected through the specific wind power tax (*canon eólico*), local revenues may be used to revive local employment and other economic activities in rural areas. If the taxes were directly controlled by the municipalities or if the amounts allocated were larger, the impact and the potential for revitalisation would notably increase, which would be of great relevance in the problematic context of these rural municipalities.

5.2. Regulatory Limitations to the Access, Control, and Generation of Wind Revenues

The opportunities, conditions, and barriers to local income generation greatly depend on the public policies that define and limit the spheres and possibilities [61]. In this sense, the analysis of the Galician regulatory framework shows how the current legislation has negatively conditioned the results obtained at a local level. First of all, it is necessary to bring into focus the limitations and barriers to entry imposed on the decision-making capacity of local administrations with regard to the planning and authorisation of wind farms. In contrast, the development of wind farms by a few private companies with few or no ties to the territory where the activity is being developed has been favoured, thus conditioning their local economic impacts [21]. Galician municipalities have witnessed the construction of wind farms in their own territories without actually having a say in the process. This situation is different in regions where there are governance processes associated with the installation of wind farms [45]. Thus, for instance, in the United States, local authorities have a crucial role in the planning and development of regulations, and they sometimes even decide the location of the wind farms [43]. In Sweden, they even have veto power over wind farm projects [44]. In line with what Fast and Mabee pointed out [60], the weakening of the decision-making capacity of Galician municipalities has taken away all the structural incentives for companies to negotiate with them.

Moreover, the regulatory framework has restricted the possibilities of building and managing municipally owned wind farms. This circumstance also makes the case of Galicia different from those of other regions where local ownership seems to be a relevant option [27]. In Denmark, for instance, cooperatives run by the local population control a great number of wind turbines (15%), while in Germany, community ownership is estimated at 20% [28]. From the information gathered in the SIDWEG, it is possible to observe how singular wind farms, which is the only category that allows local ownership, represented only 1% of the total operating capacity in 2017. Despite the initial interest awakened by this specific type of wind farm, the restrictions imposed on its capacity and the low degree of its development have resulted in a very limited income collection and a very low impact on the total revenue [56]. Thus, 3 MW per singular wind farm is a very low capacity considering the size of most farms in Galicia, the capacity of which is, in many cases, close to 50 MW; consequently, the farms' turnover and their positive impact at a local level are also conditioned by this limitation.

Third, the specific wind tax (*canon eólico*) associated with the ECF is levied on the number of wind turbines, which is in contrast with other compensation funds levied on the installed capacity, such as the one created in the region of Valencia (Spain), or on the plant's turnover, as in Greece [24] or Portugal [25]. In Galicia, this connection to the number of wind turbines puts the very logic of the tax and the ECF at stake. In theory, the ECF was designed to compensate for the negative environmental externalities generated by wind farms [62]. However, the definition of the tax, which is articulated into several tax brackets, hinders its own objectives. For instance, a wind farm with only three turbines built in an area where there were no previous installations would be exempt from paying the tax, despite the negative visual environmental impact that it would have on that specific geographical area [63]. In addition, the current repowering of Galician wind farms [64] is reducing the number of turbines, and consequently, the revenues derived from the ECF. One of the objectives of the *canon eólico* is to incentivise the repowering of wind farms by reducing the amounts to be paid if a large number of wind turbines with a low unit capacity are substituted by a smaller number of turbines with a higher unit capacity, or even by establishing an exempted bracket.

Both the design of the *canon eólico*, under non-local control, and the allocation of ECF revenues established by the Galician government reduce the economic impact of this tax on rural municipalities. The limited local impact is underlined by only between 26–36% of the 22 million euros that are annually collected, on average, through this specific wind tax benefiting the municipalities affected by wind farms. The analysis herein performed shows how, despite the creation of specific taxes, mechanisms, or categories such as the ECF or singular wind farms, the main revenue of municipalities with wind farms still comes from conventional taxes. These results confirm the shortcomings of specifically designed wind taxes and the need to advance in the development of public policies aiming at increasing the possibilities for economic and rural development in the municipalities where wind farms are built.

5.3. Limitations and Future Research Prospects

The absence of a public taxpayer register for wind-farming activities limits the availability of information about the wind revenues earned by the municipalities. Although we have tried to overcome this limitation through indirect calculations, future research works should aim at compiling in greater detail and at a higher level of disaggregation the amounts of local revenue derived from the development of wind farms. Hopefully, the regional government will also improve the level of transparency in the management of the wind power industry, which would allow arriving at more solid conclusions regarding the above-mentioned and currently lacking information about the destination of two-thirds of the annual revenues collected from the *canon eólico*. On the other hand, it will be interesting to address other issues related to the present research work, such as verifying whether the improvement of this industrial sector and of other services linked to wind power production generates any spillovers that may foster local development, as indicated in other works [65,66]. Investigating the role that local economic impact plays in the local processes of approval of wind farms in rural areas of Galicia will be another worthwhile line of research.

6. Conclusions and Policy Recommendations

The research work developed around the SIDWEG has allowed the identification and quantification of the direct local economic impacts of wind power activity in the rural municipalities of Galicia. Thus, several sources of local income have been identified: the taxation of wind farms through conventional taxes and through the ECF, the municipal ownership of wind farms, the leasing of municipal land to wind farm promoters, and the collaboration agreements drawn up between companies and municipalities. The income flows of municipalities with wind farms in their territories has been quantified as 17.8 million euros for 2017.

The main sources of local income are the IBI (37.7% in 2017) and the specific wind tax (*canon eólico*) (35.4%), followed by the IAE (19.6%) and the revenues from municipally owned wind farms (7.0%). In this sense, the present work shows how the public policies implemented have, through the regulatory framework, limited and conditioned the possibilities for increasing local revenues derived from wind power activity. The possibility of promoting municipally owned wind plants is only possible through the installation of singular wind farms. This option has been poorly developed, and proves to be very limiting in terms of installed capacity. The specific wind tax (*canon eólico*) and the ECF have a great potential to encourage local development through employment creation and the promotion of environmentally sustainable economic activities. However, the lack of control that local administrations have over these mechanisms and the limited amount of revenue (only around 30%) that is transferred to the municipalities reduces their potential. In brief, it is possible to affirm that wind power activity allows increasing local revenues, though limitedly, and that consequently, there is a need to continue researching the extent to which these new sources of income contribute to the sustainable development of rural municipalities beyond their economic growth.

Finally, considering our results and the currently intense relaunching of the wind power sector in Galicia, we include the following policy recommendations. First of all, our results make clear that, as pointed out in the literature [21], the restricted local ownership of wind farms and the limited negotiating capacity on the part of local authorities reduces the possible benefits and positive impacts of this activity. Therefore, it would be reasonable to implement a new policy aimed at increasing local ownership of renewable energy plants. This could be done through the relaunching of singular wind farms, eliminating the 3 MW per farm limitation, or through the design of a new mechanism that would favour the setting up of new locally owned wind plants. Secondly, understanding that taxes on wind power production may generate opportunities to encourage local development through public expenditure [31] and given the existence of a specific tax in Galicia (*canon eólico*), it would also be positive to hand over the total control over this tax to the local authorities in order to better support local projects of transformation, thus improving as well the level of energy justice [67] derived from progress in renewable energies.

Thirdly, it might be relevant to grant a binding decision-making capacity to the local governments in order to incentivise negotiation with power-promoting companies [60]. This would foster the drawing up of collaboration agreements between promoters and municipalities, which are common in other regions [57], and would lead to an increase of local revenues. These measures would not hinder the development of the wind power industry (many of them are still in force in countries such as Denmark or Germany [28], where wind power production is most extended), and would instead allow an increment of income flows for Galician rural municipalities, which in turn would facilitate the implementation of other measures favouring local development in a context of serious structural problems, including depopulation, a lack of employment opportunities, and environmental challenges such as forest fires.

Author Contributions: D.C. and X.S. designed the research plan and carried out the field work. D.C. systematised and estimated the collected data. D.C., X.S. and D.P.-N. analysed and discussed the results. D.C., D.P.-N. and X.S. jointly wrote the article. X.S. directed the research work.

Funding: This research received no external funding.

Conflicts of Interest: The authors declare no conflict of interest.

References and Notes

1. Intergovernmental Panel on Climate Change (IPCC). *Summary for Policymakers*; Edenhofer, O., Pichs-Madruga, R., Sokona, Y., Eds.; IPCC Special Report on Renewable Energy Sources and Climate Change Mitigation; Cambridge University Press: Cambridge, UK, 2011.

2. European Parliament. Directive 2009/28/EC of the European Parliament and of the Council, of 23 April 2009, on the Promotion of the Use of Energy from Renewable Sources and Amending and Subsequently Repealing Directives 2001/77/EC and 2003/30/EC. 2009. Available online: http://eur-lex.europa.eu/legal-content/EN/ALL/?uri=CELEX:32009L0028 (accessed on 20 November 2018).

3. Ertay, T.; Kahraman, C.; Kaya, İ. Evaluation of renewable energy alternatives using MACBETH and fuzzy AHP multicriteria methods: The case of Turkey. *Technol. Econ. Dev. Econ.* **2013**, *19*, 38–62. [CrossRef]

4. Kaplan, Y.A. Overview of wind energy in the world and assessment of current wind energy policies in Turkey. *Renew. Sustain. Energy Rev.* **2015**, *43*, 562–568. [CrossRef]

5. Michalak, P.; Zimny, J. Wind energy development in the world, Europe and Poland from 1995 to 2009. Current status and future perspectives. *Renew. Sustain. Energy Rev.* **2011**, *15*, 2330–2341. [CrossRef]

6. Kemmoku, Y.; Keiko, N.; Kawamoto, T.; Sakakibara, T. Life cycle CO_2 emissions of a photovoltaic/wind/diesel generating system. *Electr. Eng. Jpn.* **2002**, *138*, 14–23. [CrossRef]

7. Kayser, B. The future of the countryside. In *Beyond Modernization: The Impact of Endogenous Development*; van der Ploeg, J.D., Van Dijk, G., Eds.; Van Gorcum: Assen, Holland, 1995; pp. 179–190.

8. Ryser, L.; Halseth, G. Rural Economic Development: A Review of the Literature from Industrialized Economies. *Geogr. Compass* **2010**, *4*, 510–531. [CrossRef]

9. Varela-Vázquez, P.; Sánchez-Carreira, M.C. Socioeconomic impact of wind energy on peripheral regions. *Renew. Sustain. Energy Rev.* **2015**, *50*, 982–990. [CrossRef]

10. Slattery, M.C.; Lantz, E.; Johnson, B.L. State and local economic impacts from wind energy projects: Texas case study. *Energy Policy* **2011**, *39*, 7930–7940. [CrossRef]

11. Aceleanu, M.I.; Şerban, A.C.; Țîrcă, D.M.; Badea, L. The rural sustainable development through renewable energy. The case of Romania. *Technol. Econ. Dev. Econ.* **2018**, *24*, 1408–1434. [CrossRef]

12. Fouché, E.; Brent, A. Journey towards Renewable Energy for Sustainable Development at the Local Government Level: The Case of Hessequa Municipality in South Africa. *Sustainability* **2019**, *11*, 755. [CrossRef]

13. Benedek, J.; Sebestyén, T.; Bartók, B. Evaluation of renewable energy sources in peripheral areas and renewable energy-based rural development. *Renew. Sustain. Energy Rev.* **2018**, *90*, 516–535. [CrossRef]

14. Instituto Nacional de Estadística-INE. Indicadores de Estructura de la Población. Madrid. 2018. Available online: http://www.ine.es (accessed on 20 November 2018).

15. Instituto Galego de Estatística-IGE. Nomenclator. Santiago de Compostela. 2018. Available online: https://www.ige.eu (accessed on 20 November 2018).

16. Corbelle, E.; Crecente, R.O. Abandono de terras: Concepto teórico e consecuencias. *Revista Galega de Economía* **2008**, *17*, 47–62.

17. Fuentes-Santos, I.; Marey-Pérez, M.F.; González-Manteiga, W. Forest fire spatial pattern analysis in Galicia (NW Spain). *J. Environ. Manag.* **2013**, *128*, 30–42. [CrossRef] [PubMed]

18. Copena, D. Enerxía Eólica e Medio Rural: Unha Análise Aplicada dos Impactos Socioeconómicos dos Parques Eólicos no Mundo Rural Galego. Ph.D. Thesis, Universidade de Vigo, Vigo, Spain, 17 December 2015. Available online: http://www.investigo.biblioteca.uvigo.es/xmlui/handle/11093/658 (accessed on 16 April 2017).

19. Instituto Galego de Estatística-IGE. Poboación, Entidades e Densidade. Santiago de Compostela. 2017. Available online: https://www.ige.eu (accessed on 12 April 2018).

20. Copena, D.; Simón, X. Wind farms and payments to landowners: Opportunities for rural development for the case of Galicia. *Renew. Sustain. Energy Rev.* **2018**, *95*, 38–47. [CrossRef]

21. Delicado, A.; Figueiredo, E.; Silva, L. Community perceptions of renewable energies in Portugal: Impacts on environment, landscape and local development. *Energy Res. Soc. Sci.* **2016**, *13*, 84–93. [CrossRef]

22. Rand, J.; Hoen, B. Thirty years of North American wind energy acceptance research: What have we learned? *Energy Res. Soc. Sci.* **2017**, *29*, 135–148. [CrossRef]

23. Baxter, J.; Morzaria, R.; Hirsch, R. A case-control study of support/opposition to wind turbines: Perceptions of health risk, economic benefits, and community conflict. *Energy Policy* **2013**, *61*, 931–943. [CrossRef]
24. Dimitropoulos, A.; Kontoleon, A. Assessing the determinants of local acceptability of wind-farm investment: A choice experiment in the Greek Aegean Islands. *Energy Policy* **2009**, *37*, 1842–1854. [CrossRef]
25. Nadaï, A.; Krauss, W.; Afonso, A.I.; Drackié, D.; Hinkelbein, O.; Labussière, O.; Mendes, C. El paisaje y la transición energética: Comparando el surgimiento de paisajes de energía eólica en Francia, Alemania y Portugal. *Nimbus* **2010**, *25–26*, 155–173.
26. Nolden, C. Governing community energy—Feed-in tariffs and the development of community wind energy schemes in the United Kingdom and Germany. *Energy Policy* **2013**, *63*, 543–552. [CrossRef]
27. Yin, Y. An analysis of empirical cases of community wind in Oregon. Renew. Sustain. *Energy Rev.* **2013**, *17*, 54–73. [CrossRef]
28. Bauwens, T.; Gotchev, B.; Holstenkamp, L. What drives the development of community energy in Europe? The case of wind power cooperatives. *Energy Res. Soc. Sci.* **2016**, *13*, 136–147. [CrossRef]
29. Del Río, P.; Burguillo, M. An empirical analysis of the impact of renewable energy deployment on local sustainability. *Renew. Sustain. Energy Rev.* **2009**, *13*, 1314–1325. [CrossRef]
30. Paolo De Pascali, P.; Bagaini, A. Energy Transition and Urban Planning for Local Development. A Critical Review of the Evolution of Integrated Spatial and Energy Planning. *Energies* **2019**, *12*, 35. [CrossRef]
31. Munday, M.; Bristow, G.; Cowell, R. Wind farms in rural areas: How far do community benefits from wind farms represent a local economic development opportunity? *J. Rural Stud.* **2011**, *27*, 1–12. [CrossRef]
32. Castleberry, B.; Greene, J.S. Impacts of wind power development on Oklahoma's public schools. *Energy Sustain. Soc.* **2017**, *7*. [CrossRef]
33. Okkonen, L.; Lehtonen, O. Socio-economic impacts of community wind power projects in Northern Scotland. Renew. *Energy* **2016**, *85*, 826–833. [CrossRef]
34. Liebe, U.; Bartczak, A.; Meyerhoff, J. A turbine is not only a turbine: The role of social context and fairness characteristics for the local acceptance of wind power. *Energy Policy* **2017**, *107*, 300–308. [CrossRef]
35. Slattery, M.C.; Johnson, B.L.; Swofford, J.A.; Pasqualetti, M.J. The predominance of economic development in the support for large-scale wind farms in the U.S. Great Plains. *Renew. Sustain. Energy Rev.* **2012**, *16*, 3690–3701. [CrossRef]
36. Jami, A.A.N.; Walsh, P.R. The role of public participation in identifying stakeholder synergies in wind power project development: The case study of Ontario, Canada. *Renew. Energy* **2014**, *68*, 194–202. [CrossRef]
37. Mundaca, L.; Busch, H.; Schwer, S. 'Successful' low-carbon energy transitions at the community level? An energy justice perspective. *Appl. Energy* **2018**, *218*, 292–303. [CrossRef]
38. Upham, P.; García, J. A cognitive mapping approach to understanding public objection to energy infrastructure: The case of wind power in Galicia, Spain. *Renew. Energy* **2015**, *83*, 587–596. [CrossRef]
39. Breukers, S.; Wolsink, M. Wind power implementation in changing institutional landscapes: An international comparison. *Energy Policy* **2007**, *35*, 2737–2750. [CrossRef]
40. Johansen, K.; Emborg, J. Wind farm acceptance for sale? Evidence from the Danish wind farm co-ownership scheme. *Energy Policy* **2018**, *117*, 413–422. [CrossRef]
41. Krog, L.; Sperling, K.; Lund, H. Barriers and Recommendations to Innovative Ownership Models for Wind Power. *Energies* **2018**, *11*, 2602. [CrossRef]
42. Nadaï, A. "Planning", "siting" and the local acceptance of wind power: Some lessons from the French case. *Energy Policy* **2007**, *35*, 2715–2726. [CrossRef]
43. Jacquet, J.B. Landowner attitudes toward natural gas and wind farm development in northern Pennsylvania. *Energy Policy* **2012**, *50*, 677–688. [CrossRef]
44. Söderholm, P.; Ek, K.; Pettersson, M. Wind power development in Sweden: Global policies and local obstacles. *Renew. Sustain. Energy Rev.* **2007**, *11*, 365–400. [CrossRef]
45. Smith, A. Emerging in between: The multi-level governance of renewable energy in the English regions. *Energy Policy* **2007**, *35*, 6266–6280. [CrossRef]
46. Cassell, C.; Symon, G. *Essential Guide to Qualitative Methods in Organizational Research*; Sage Publications: London, UK, 2004; pp. 154–164.
47. McKim, C.A. The value of mixed methods research: A mixed methods study. *J. Mix Methods Res.* **2015**, *11*, 202–222. [CrossRef]

48. Presidencia. *Lei 8/2009, do 22 de Decembro, pola que se Regula o Aproveitamento Eólico en Galicia e se Crean o canon Eólico e o Fondo de Compensación Ambiental*; Diario Oficial de Galicia de 29 de decembro de 2009; Xunta de Galicia: Santiago de Compostela, España, 2009.
49. Saladié, S. *Impacte econòmic de les centrals eòliques en els pressupostos municipals a Catalunya*; Associació de Municipis Eòlics de Catalunya: La Granadella, Spain, 2014.
50. Iglesias, G.; del Río, P.; Dopico, J.A. Policy analysis of authorization procedures for wind energy deployment in Spain. *Energy Policy* **2011**, *39*, 4067–4076. [CrossRef]
51. The compensation for the generation of MWh by wind farms has changed during the expansion of the wind power industry. These changes have been implemented through various regulations. At the beginning of the wind power boom, the selling price of wind energy was fixed in relation to the electricity tariffs and the capacity installed, in addition to other complements. Later on, in 1998, a premium was introduced, the amount of which depended on various factors. After 2004, an incentive was created, in addition to the premium, for participation in the market; both were defined as a percentage of the baseline average electricity tariff. Three years later, two options were established for the sale of wind-generated electricity: (a) to transfer the electricity in exchange for a regulated tariff, which is expressed in euro cents per kWh, for all programming periods; (b) to sell the electricity in the electric energy production market. In the last case, the selling price of the electricity would be the one fixed at the regulated market or freely negotiated, which is later complemented by a premium in euro cents per kWh.
52. Flores, J.; Esteve, J. Una visión general de la fiscalidad de la actividad eléctrica en España. In *Los Tributos del Sector Eléctrico*; Becker, F., Cazorla, L.M., Martínez-Simancas, J., Eds.; Aranzadi Thomson Reuters: Cizur Menor, Spain, 2013; pp. 145–170.
53. Pérez de Ayala, M.A. La doctrina del Tribunal Supremo sobre la base del ICIO en la construcción de parques eólicos. *Estrateg. Financ.* **2010**, *276*, 70–74.
54. Blanco, M.I. The economics of wind energy. *Renew. Sustain. Energy Rev.* **2009**, *13*, 1372–1382. [CrossRef]
55. Santos-Alamillos, F.J.; Thomaidis, N.S.; Usaola-García, J.; Ruiz-Arias, J.A.; Pozo-Vazquez, D. Exploring the mean-variance portfolio optimization approach for planning wind repowering actions in Spain. *Renew. Energy* **2017**, *106*, 335–342. [CrossRef]
56. Copena, D.; Simón, X. Enerxía eólica e desenvolvemento local en Galicia: Os parques eólicos singulares municipais. *Revista Galega de Economía* **2018**, *27*, 31–48.
57. Ellis, G.; Cowell, R.; Warren, C.; Strachan, P.; Szarka, J.; Hadwin, R.; Miner, P.; Wolsink, M.; Nadaï, A. Wind Power: Is There A "Planning Problem"? Expanding Wind Power: A Problem of Planning, or of Perception? *Plan. Theory Pract.* **2009**, *10*, 521–547. [CrossRef]
58. Ministerio de Medio Ambiente y Medio Rural y Marino. Tercer Inventario Forestal Nacional. Madrid. 2008. Available online: https://www.miteco.gob.es/es/biodiversidad/servicios/banco-datos-naturaleza/informacion-disponible/ifn3.aspx (accessed on 25 June 2018).
59. Rodríguez-Pose, A.; Palavicini-Corona, E.I. Does local economic development really work? Assessing LED across Mexican municipalities. *Geoforum* **2013**, *44*, 303–315. [CrossRef]
60. Fast, S.; Mabee, W. Place-making and trust-building: The influence of policy on host community responses to wind farms. *Energy Policy* **2015**, *81*, 27–37. [CrossRef]
61. Del Río, P.; Burguillo, M. Assessing the impact of renewable energy deployment on local sustainability: Towards a theoretical framework. *Renew. Sustain. Energy Rev.* **2008**, *12*, 1325–1344. [CrossRef]
62. Wang, S.; Wang, S. Impacts of wind energy on environment: A review. *Renew. Sustain. Energy Rev.* **2015**, *49*, 437–443. [CrossRef]
63. Wolsink, M. Planning of renewables schemes: Deliberative and fair decision-making on landscape issues instead of reproachful accusations of non-cooperation. *Energy Policy* **2007**, *35*, 2692–2704. [CrossRef]
64. Villena-Ruiz, R.; Ramirez, F.J.; Honrubia-Escribano, A.; Gómez-Lázaro, E. A techno-economic analysis of a real wind farm repowering experience: The Malpica case. *Energy Convers. Manag.* **2018**, *172*, 182–199. [CrossRef]
65. Elola, A.; Parrilli, M.D.; Rabellotti, R. The Resilience of Clusters in the Context of Increasing Globalization. *Eur. Plan. Stud.* **2013**, *21*, 989–1006. [CrossRef]

66. Varela-Vázquez, P.; Sánchez-Carreira, M.C. Upgrading Peripheral wind sectors. *Technol. Anal. Strateg. Manag.* **2016**, *28*, 1152–1166. [CrossRef]
67. Sovacool, B.K.; Dworkin, M.H. Energy justice: Conceptual insights and practical applications. *Appl. Energy* **2015**, *142*, 435–444. [CrossRef]

© 2019 by the authors. Licensee MDPI, Basel, Switzerland. This article is an open access article distributed under the terms and conditions of the Creative Commons Attribution (CC BY) license (http://creativecommons.org/licenses/by/4.0/).

 *sustainability*

Article

Collective Energy Practices: A Practice-Based Approach to Civic Energy Communities and the Energy System

Nick Verkade * and Johanna Höffken

School of Industrial Engineering and Innovation Sciences, Eindhoven University of Technology, Room ATL 8.406, P O Box 513, 1600 MB Eindhoven, The Netherlands; j.i.hoffken@tue.nl
* Correspondence: n.verkade@tue.nl

Received: 10 May 2019; Accepted: 5 June 2019; Published: 11 June 2019

Abstract: Civic energy communities (CECs) have emerged throughout Europe in recent years, developing a range of activities to promote, generate, and manage renewable energy within the community. Building on theories of Social Practice, we develop the notion of Collective Energy Practice to account for the activity of CECs. This expands the practice-based understanding of energy, which thus far has mostly focused on energy practices of the home. Additionally, we build on earlier practice-based thinking to come to our understanding of a 'system of energy practices'. This view places the collective energy practices of CECs in a broader mesh of sites of practice, including policymaking, commercial activity, and grid management. Taking account of the enabling and/or restricting the influence of this broad system of energy practices is crucial in understanding the development of CECs' practices. We accomplish this through the qualitative analysis of our long-term empirical research of five Dutch CEC sites, but also draw on our earlier fieldwork on smart grid projects in the Netherlands.

Keywords: civic energy communities; community energy; local energy initiatives; grassroots innovation; energy transition; social practice theory; energy practices

---

1. Introduction

Civic Energy Communities (CECs) have been on the rise in many countries worldwide; specifically, in the Netherlands, CECs are showing steady growth since 2010. The fourth and most recent report by the sector's network organization counts 353 local energy cooperatives in the Netherlands [1], a country with 380 municipalities. CECs are local citizen organizations that aim to make their ways of using and generating energy environmentally, politically, and economically more sustainable. They achieve this through a range of activities, including: defining the visions of local sustainability, offering energy efficiency measures to the community, organizing collective buying of solar panels, and developing collectively owned renewable energy generation. CECs have the potential to make a significant contribution to the future energy system in Europe through these activities [2]. It is of importance to study the emergence of current and future activities of CECs in order to understand this transitioning system [3].

In this article, we focus on the activities of civic energy communities, specifically those activities through which they aim to generate and manage energy collectively. We develop two central arguments in this paper. First, we argue that the activities of community energy can be better understood through the conceptual lens of collective energy practices. In this argument, we bring together research regarding the development of community energy and work that conceptualizes energy from a social practice perspective. We find a useful distinction between three categories of collective energy practices:

promoting individual energy practices; developing collective energy generation; and, developing collective energy management.

Second, we argue that, to understand the (continuing) emergence of these collective practices, it is crucial to identify the linkages of these practices with the broader system of energy practices. Here, we work with the notion of a system of energy practices, building on theory and applications that were developed by Schatzki [4,5], Nicolini [6], and Watson [7]. This systemic view based within practice theory sees the energy system consisting of all the practices through which it is made and sustained. We will identify the crucial interlinkages with this system of energy practices that explain the emergence of collective energy practices for CECs and how these have been hindered and/or enabled thus far.

By developing these two arguments, we address the following research question: how can we understand the emerging practices of civic energy communities within a changing energy system?

Community Energy has been a topic of research for quite some time, being largely focused on renewable energy generation activities. Much of this research uses notions that are related to transition studies and the multi-level perspective. Energy communities [8] are often seen as 'niches' [9] within an unsustainable and obdurate regime [10] where 'grassroots innovations' [11,12] can develop. Although these innovations can be technical [13], CECs often work with existing products that are applied in new contexts, where non-technological aspects are the subject of innovation [3,14]. We build on this body of literature, but from a different perspective: we approach the activities of CECs from the perspective of social practice theory, in which the actual activities through which energy is generated and managed are the central object of study. This set of activities is still expanding as new material and non-material elements become available to CECs, and the place of community energy in a transitioning energy system is still in flux. The concept of collective energy practices will be mobilized to characterize the diverse activities of CECs.

It is argued that CECs are important in the development of local energy practices, because they pioneer initiatives to generate and manage energy [15,16]. In this innovative role, CECs actively challenge and question conventional practices through which the energy system is organized [17]. The challenges that are posed by local renewable energy generation to the grid infrastructure are becoming very palpable in several regions of the Netherlands [18]. CECs also challenge the conventional energy system by developing activities around managing the way that their community-generated energy is used. It is thus necessary to also take account of the practices of established energy system actors to understand the emergence and potential for CECs' energy practices. We do this by extending our concept of collective energy practice to be enmeshed in a broader system of energy practices, based on earlier system of practices work [7,19]. In this paper, we elaborate a practice-based perspective that captures both the specific practices of CECs as well as the practices of other sites in the energy system.

The following section reports on the methods that we used in our research into various sites of practice over a longer period of time. Section 3 is about collective energy practices, and it introduces our findings on three categories of collective energy practices of CECs. This is followed by a section where we take a more systemic perspective and show the various sites of practice that form the system of energy practices. We close with a discussion regarding the collective energy practices of CECs within the energy system, by highlighting their empirical and theoretical relevance.

2. Methodology

This paper is based on fieldwork at several sites of the Dutch energy system, as part of a longer ongoing research project on emerging energy practices in the Netherlands. The focal sites of energy practices for this paper are civic energy communities, and, in particular, innovative CECs that are, or were, early developers of collective energy practices. The CECs that were researched for this paper are exemplary for this innovative character because of projects that they are, or were, involved in. Five of these CECs were researched in-depth by following on-going projects and through semi-structured interviews with key individuals. These key individuals had been involved with initiating CECs, developing its practices, and interacting with other system actors. Interviews were performed at the home or workplace of the interviewees, and they lasted 60 to 90 min. The interviews were transcribed and analyzed by hand by the authors. The empirical material presented throughout this text in the form of telling quotes comes from these interviews with key individuals from the CECs. All five CECs are represented at least once through telling quotes, which was chosen to also reflect the expressions that were made in the interviews with other CECs. Our understanding of CEC's collective practices was expanded and triangulated by visiting events and utilizing reports from the Dutch community energy network organization HIER, and personal experience working with CECs. Table 1 provides some more information regarding the CEC names, locations, and the numbers that we use to reference them throughout the article. Thus, the analysis is built on a mixed method, formulating and deriving insights on the several 'sites of practice' from the interview data and site visits, and further substantiating these with literature on the energy sector. This table also forms an overview of the methods used to research the several practice sites.

Table 1. Overview of researched practice sites and mixed methodologies.

Site of Practice	Organization/Project (Location)	Method
CEC #1	Duurzame Energie Haaren (Haaren).	
CEC #2	Member of Brabant Provincial commission social innovation and initiator of several CECs.	
CEC #3	Morgen Groene Energie (Nuenen).	Semi-structured interviews.Sector events & reports.
CEC #4	Escozon (Heeten).	
CEC #5	Endona (Heeten).	
Grid management #1	Project SSmE ("Together energy smart"—Haaren).	Interviews with grid manager Enexis, other project partners and participants. Observing project meetings.
Grid management #2	Project JEM2 ("Your energy moment 2.0"—Breda).	
Local Policy		Based on literature & interviews at CEC sites.
Commercial		Based on literature, the project partners in two projects mentioned above & interviews at CEC sites.
National Policy		Based on literature.

Descriptive data on the public and commercial practices through which grid management is performed and changed were obtained in the authors' earlier fieldwork in two Dutch smart grid experiments (SSmE and JEM2: see also [20,21]). Our data about policy practices, which we collected through interviews with practitioners, is enriched by including relevant other research literature describing these practices.

3. Results

This section is divided into two larger parts. The first part deals with our notion and findings of the collective energy practices of CECs. The second part takes on a broader view, introducing our notion of a system of energy practices, and reporting on the several sites of practice that make up this system.

3.1. Collective Energy Practices

In this section, we focus on the activities of civic energy communities that affect the way that energy is used and produced in the Netherlands. We will do this from a perspective that is based on theories of social practice and introduce a notion that—as we will argue—better captures the activities of CECs. Practice theory [4,20] understands the social in terms of the actual practices of people, whose daily lives consist of engaging in socially prefigured, but still indeterminate and emergent, activities. This line of thinking has questioned the rationalist approach to how people relate to energy [21], and instead reframes energy consumption as an invisible, but often necessary, by-product of the meaningful practices that were performed in our daily life [22]. However, practice theory has also been used to understand recently emerging energy-related activity in the household. A range of research that is rooted in practice theory has been carried out to study the emerging energy practices of the home [23–27]. The concept of *energy practices* has been proposed as a particular set of practices through which " ... energy is highlighted, made visible, problematized, managed, stored or discussed, which in turn produces insights that can be used to shape [domestic] energy conditions [25]".

The focus in this line of research has mostly been placed on the activity of individual householders, within the arena of the home. The notion of these home energy management practices [28] is useful for this level of analysis, but it does not fully capture the phenomenon of energy collectives. Energy practices, as a concept, applies to much of the activity of CECs, as they are explicitly bringing energy to the foreground, highlighting and problematizing it, making it a matter of concern for the people that are involved and (as an aim) the wider community. However, the activities of a CEC have a more collective dimension than home energy practices: they only emerge after a group of people comes together and goes on to develop projects that they individually could or would not have. This distinctness from home energy practices is why we adopt the notion of *collective energy practices.*

Collective energy practices consist of the smaller practices that are performed by the community members, interrelated and united under the collective practice of e.g., operating a community-owned solar park. As a collective, by developing these collective practices, the CEC assembles the resources that are available within the community into actual practices, which would not emerge otherwise. An approach that sees the activity of a CEC as operating at a 'larger scale' than the individual is similar to how we might describe the activity of any other organization from the practice-theoretical perspective. This means that the recognizable activity that we observe within a collective, whether it is a CEC or a coal company, is a set of interrelated practices that are performed by multiple people at different times and places [29].

As will be seen in the description of the different collective energy practices, these are not being developed and performed exclusively by CECs. However, based on their founding principles, CECs aim to carry out their collective energy practices differently from the collective energy practices of other actors [30]. This means that this collective energy practice for a CEC should often include notions of democratic process, equality, local ownership, and the adequate scaling of technology. In the context of CECs, the term *collective* practices thus takes on an additional meaning, beyond the indication that they emerge from a multitude of individuals. The collective notion also holds that the activities that are developed by CECs often explicitly aim to benefit their community, work toward community-owned energy resources, and make sure that the community's wishes are in some way represented in the local development of the energy transition. Furthermore, it indicates that it is through these collective practices that the community they identify with is further established and shaped: the practices give meaning to the CEC, a reason to be organized as a collective.

We expand the toolkit of social practices to better understand the phenomenon emerging through the activities of CECs with the concept of collective energy practices. In the following section, we will provide a more detailed look into the collective energy practices of civic energy communities. We distinguish three collective energy practices of energy communities: promoting individual energy practices; developing collective energy generation; and, developing collective energy management. This distinction is initially based on the development path that we have seen for CECs, where these categories can be often seen as stages of development.

3.1.1. Promoting Home Energy Practices

The initial practices that most civic energy communities developed, apart from organizational practices, were aimed at promoting home energy practices among their participants. This distinct type of collective energy practices promotes and supports the growth of home energy practice that is organized at the community level. The CEC directly impacts the energy system by 'circulating' [20] technical and non-technical aspects of these home energy practices among the community members.

The widespread organization of community schemes for buying solar panels is the most visible of these collective energy practices: "The first few years, we have been very busy getting solar panels onto houses" (interview CEC #1 & #3). During the significant decrease in solar PV costs in recent years, many initiatives for collectively purchasing solar PV were started in the Netherlands. Up to the year 2018, at least 248 smaller and larger projects were counted [31]. This collective energy practice benefits the CEC by reaching out and recruiting members, whom themselves benefit by being "unburdened" in obtaining (cheaper) solar PV. Similar community schemes exist that promote home insulation, heat pumps, or energy monitoring devices. The community can also promote new knowledge and meanings regarding energy besides these ways of altering the material arrangement through which energy is used and produced within the home. Examples that were observed in our cases that help to achieve this include education on energy saving at schools, displaying individual 'sustainable' successes within the community, and promoting their vision and knowledge within political and civil society. In doing so, individual energy practices of energy monitoring and energy saving are promoted within the community.

However, these collective energy practices are not exclusive to CECs. In fact, of all the collective solar purchasing initiatives up to 2015 the majority (in number and capacity) of these initiatives was started by commercial actors, consumer organizations, or governmental programs [32]. Similarly, energy companies now offer easy access to solar panels and heat pumps to their customers, as well as a host of apps and devices to monitor and reduce energy usage. The way that CECs perform these collective practices is not very different from how this practice is performed by other societal or commercial actors (interview CEC #2). However, as a relatively easy to organize practice, which also offers clear benefits to the community members, it proved to be a successful first step in the development of many CECs.

3.1.2. Developing Collective Energy Generation

Following the promotion of individual energy practices, many CECs have moved on to another category of collective energy practice: developing collective energy generation. This practice, especially collective solar PV generation, has been widely picked up by CECs and grew fast in recent years: The amount of collectively owned solar projects grew from 277 in the year 2017, to 450 in 2018 [1].

The collective practice that was recognizable as collective energy generation consists of a host of smaller and diverse practices, related by their contribution to the collective practice. Smith, Hargreaves, Hielscher, Martiskainen, and Seyfang [11] lists many tasks within the set of practices that CECs needs to perform in setting up community energy: "Groups have to study technical information […], constitute themselves as legal entity, apply for grants, seek loans, raise money, think about insurance, permissions, marketing strategies" […]. This illustrates that developing collective generation is a more complex collective practice than the previous category of promoting individual energy practices.

In itself, the practice of generating energy through solar parks, wind turbines, or, for instance, biomass facilities is of course not exclusive to CECs. CECs often set out to apply a different (broader) range of elements, which are rooted in the principles on which the community operates and is built [30]. This becomes visible in the way that CECs shape their collective energy generation practices. As we have observed in our fieldwork, a democratic process and equality are basic principles for most Dutch CECs, which have legally established themselves as cooperative associations. Equality takes shape not only in the equal voting rights of cooperative members, but also in designing the investment structure of collective generation to be equally accessible for everyone (CEC #4). Rather than finding the largest investor for a local project, our case studies show how ownership is split into small shares that practically anyone can buy limited amounts of. Local ownership is not intrinsic to all collective solar generation by CECs, some projects allow for investment by anyone. However, local ownership has been a common goal for CECs, and over time it has become established as a standard form through changes in tax legislation. We will return to this topic in the next section. Lastly, the adequate scaling of collective generation means that whatever the type and size of the project that is developed by the CEC, it corresponds with what the community deems necessary (CEC #4). Thus, the leading principle might not simply be to develop the largest or cheapest project to have as much renewable energy as possible; the project should fit with the values and goals that a majority of the community finds important.

Such founding principles of CECs, which they aim to apply in the collective practices that they develop, are not always recognized in the way other actors might develop the same practice of collective energy generation. This is especially relevant in regard to solar parks in the Netherlands, where market based actors are also offering and developing solar projects within the municipality. In fact, private actors form the majority in developing collective generation, while CECs own a mere 2% of total solar power in the Netherlands [1]. The 'booming' growth of commercially developed solar parks has been increasingly met with discussion and opposition at the local level, because the values and principles of the community are often not included in these developments.

3.1.3. Developing Collective Energy Management

We distinguish a third category of collective energy practices, which are formed by the collective energy management practices that address how, when, and which energy is used within the community. Just like collective *generation*, the technologies through which collective energy *management* is achieved have a collective nature; people do not individually develop them, they are operated by and for a collective. Energy monitoring platforms for the community [25,33]; community energy storage [34]; the development of community virtual power plants [35]; and, operating micro-grids [36] are examples of this.

Energy management practices are not new, as the maintenance of the electricity system has always required careful control and balancing throughout the grid. However, a consequence of a renewables-based energy system is that there is a growing need to manage the dynamics of energy at the scale of the household and the community. A host of smart grid pilots and experiments has been conducted over the years to test how technologies (storage, smart appliances, and IT platforms), tariff schemes, regulations, and information feedback can be applied to create a demand response and manage energy usage at the local level. The palette of energy management practices through which the energy system is balanced and maintained grows as all of these elements are developed. Although grid management is, strictly speaking, not their responsibility in the current energy system, we observe that CECs are starting to explore how their communities might engage in these energy management practices.

One main reason for this is that it improves the amount of community-generated electricity that is actually used within the community. Besides any moral principles that the CEC might have on becoming self-sustainable, improving on this will likely also be financially rewarding. CECs anticipate changes in the degree to which their collective generation capacity has access to the grid infrastructure. The local grid is meeting constraints in its ability to accommodate all electricity generated by collective solar electricity generation (not just by CECs) in ever more parts of the Netherlands. Grid operators and the government are already exploring ways to limit the pressure that collective generation puts on the grid [37]. Measures, such as flexible taxation and/or pricing, lead to higher costs of using the grid, and the curtailment of solar parks is a flat loss of the generated energy [38]. These interventions are not unlikely (some DSOs see "curtailment as unavoidable" [18]) and they can be a push for the CEC as owners and users of the generated energy to decrease their demand of grid capacity, as this is likely to become limited and more expensive at times. Thus, collective energy management practices that achieve this will become a logical add-on to collective generation for CECs.

Another reason we find is that, even before these more critical measures have to be taken, energy management practices can support the growth and position of the CEC. "To maintain [the CEC] we must find a new business model for cooperatives, in which they perform tasks to unlock and retain as much value as possible (CEC #5)". Practices that unlock and extract value from flexible energy usage and storage are growing, often as a core practice of market-based aggregators. These practices generate value by reducing the energy usage, making use of fluctuating energy prices, and offering flexibility services for grid management. The CEC can strengthen its business model by performing these practices itself, by setting up and utilizing flexible capacity (a battery, aggregation) within the community.

3.2. Understanding Collective Energy Practices within a System of Energy Practices

The collective energy practices of CECs are developed and performed in deep relation with the energy system at large, which we understand as a socio-technical system. As Watson [7] argues, practices "are partly constituted by the socio-technical systems of which they are a part; and those socio-technical systems are constituted and sustained by the continued performance of the practices which comprise them". Collective practices of energy generation and management make use of, but also affect, the material infrastructures of the energy system. Furthermore, the devices that are applied in these practices are produced, supplied, and installed by other organizations. However, beyond the technical dimension, collective energy practices also relate to, make use of, and have an impact on the financial, legal, political, and cultural dimensions of the socio-technical energy system.

Just as the activity of CECs can be understood as collective energy practices, so can the activity of other organizations that contribute to the energy system. The "methods of planning and policy-making" that make and shape an infrastructure can be "considered practices in their own right" [39]. In this sense, the energy system is itself a larger constellation of practices that build it, feed it, regulate, use, and manage it. This system persists "through the routinized actions of actors throughout the system, as they perform the practices which reproduce the institutions and relations comprising the system". [7]. The 'system of energy practices' is thus defined "as a relatively stable configuration of linked practices and relations that together sustain a particular socio-technical mode of doing" [19], in this case generating energy and managing the electricity grid.

The system of energy practices is dispersed over many different sites of practice: the energy practices of an energy producer, a grid operator, or a policymaker all play a part in the broader system of energy practices. The particular socio-technical mode of undertaking energy generation and management is routinely re-constituted by these various sites of practice, and the activity of CECs is thus dependent and prefigured by the system of energy practices. However, at the same time, this does not presume that influence is exclusively in the hands of "professional and political practices" [39]; any one site of practice can, by doing things differently, challenge and change the system of energy practices. This change is affected by alterations in other sites of practice, such as the political and legislative arena, grid management and planning, or the market for technologies.

From our perspective on a system of energy practices, even though we have a particular interest in the energy practices of CECs, "appreciating the relations between practices—not just interdependent but also competitive relations—is in fact essential to understanding the dynamics within practices" [7]. This means that, to understand development and change in collective energy practices, it is imperative to follow their linkages into the places and practices of national policymaking, grid management, local government, and commercial activity in the energy system. This means that not only the collective practices of CECs have to be studied, but also the sites of practice that are important for their emergence, growth, or failure.

We can identify the crucial enabling or hindering linkages to other sites of practice throughout the system of energy practices by studying the practices of CECs. This process of studying one focal practice and tracing the important relations to other sites of practice is similar to the 'zooming' approach that was presented by Nicolini [40]. Some of these other sites of practice have been an object of our own empirical work, while we refer to in-depth studies of others where available for other practices. We will report on the crucial enabling and hindering relations to other sites of practice that we have identified in studying the collective energy practices of CECs in the remainder of the section.

3.2.1. National Policy Practices

Several national policies figure strongly in the emergence of collective energy practices. First of all, the collective practices that promote individuals to acquire solar panels were enabled by a national governmental policy that made solar panels more economically attractive. Net-metering policy for individual households' self generated solar electricity has had major consequences in this regard, and the VAT rebate on solar systems has been another supporting policy. The net-metering policy was made as a general governmental policy to promote individual solar energy generation, without the reference or influence of CECs. It merely acts as a 'prerequisite' to the emergence of this particular collective energy practice, which we have shown has been a very common first step for CECs. The net-metering policy was intended as a temporary support mechanism, but it has been extended several times and it will last until the year 2023, when it will be replaced by a different policy [41,42].

A second relation to the sites of national policy practice becomes apparent by zooming in on the development of collective energy generation. The investment and subsidy structure of collective energy generation can take different forms, but most of the community-based energy generation is shaped by a particular national policy instrument: the 'postal code rose' (PCR) arrangement. CECs that develop energy generation capacity in shared ownership, for instance, by wind turbines or larger solar installations, have argued that this self-supplied energy should not be taxed in the same way as grid-supplied energy. Before a policy instrument ultimately came about, the tax regime in place heavily limited the economic feasibility of community energy. Repeated attempts by policymakers and local energy representatives to fight for a different tax regime for community energy throughout recent years in the Netherlands have been documented by [43]. This process shows how energy transition at the local level intersects with the practices of the national government, where the concerns of CECs were constantly balanced against reduced tax incomes.

Ultimately, this process resulted in the postal code rose (PCR) arrangement, which exempts the energy supplied to owners of collective generation within the same and adjacent postal code areas from taxation. This PCR arrangement, in turn, became a steering element for the further development of CECs' collective energy generation practices, because it only applied to the specific legal form 'cooperative association'. Already before the PCR arrangement was established, many CECs organized themselves as cooperatives [1], but this national tax policy has further entrenched the cooperative form.

The net-metering and PCR arrangement are only temporary measures, and it is not completely certain for individual users and collectives what future policy might look like. The temporary aspect of national legislation is of influence on the collective energy practices of CECs, because it introduces further risks: "what are the parameters, who is going to ensure our continuity?" (CEC #3). CECs can look for support from the European level: recent rulings by the European Parliament on the 'Clean Energy for All Europeans' package strives for national policies within the European Union (EU) to further recognize and accommodate the generation and management practices of CECs [44,45]. The tax related legislative changes are one example of this accommodation. Another example is the "Experimenting arrangement" that was first established by the national government in 2014 [46]. Under this arrangement, cooperatives can deviate from the "Elektriciteitswet" (Electricity Bill), meaning that they can take on roles and responsibilities that they are normally excluded from.

3.2.2. Local Policy Practices

Sites of local governmental practices are another site with an important relation to the development of collective practices of CECs [47]. The practices of local government have proved a valuable source of support and "strong facilitation" for the development of many CECs (CEC #2) although the municipality only acts at the fringes of the system of energy practices. This support can consist of funds to organize events, physical space on (municipal) roofs or lands for energy generation, or signaling the importance of the CEC to other organizations and inhabitants. The link between CEC and municipal practices works both ways, because the municipality is also often dependent on the collective practices that were developed by CECs. The sustainability of the energy system is rising on most municipal agendas, since "Paris and Groningen have become serious issues (Refers to the Paris climate agreement, and to earthquakes resulting from gas production in the Dutch province Groningen)" (CEC #3). However, the municipal capacity to pursue this sustainable agenda is still limited [48], especially in smaller municipalities: "there simply is no capacity or quality to bring it into practice" (CEC #3). Changing this takes a long time due to the long and fluctuating cycles of local government practices, such as elections, setting a policy agenda, developing programs, and eventually appointing civil servants. CECs fill this gap by developing collective energy practices that promote sustainable action by citizens of the municipality [1].

CECs' collective energy generation practices can develop at a scale that exceeds the boundaries of individual municipalities. In some areas these municipalities have remained relatively small, so that, in deploying and connecting energy generation, CECs quickly run into these boundaries. In these cases, officials and planners from other municipalities also get involved, with their own visions and practices, complicating the process for the CEC. "But energy does not follow these boundaries . . . I think, regarding energy management and generation, the municipal boundaries in this particular area are completely outdated" (CEC #3). Thus, the physical limits of local governmental practice are not always where the practices of CECs want to end.

3.2.3. Commercial Practices

The practices of commercial organizations in the energy system also prominently figure in the development of collective energy practices of CECs. The sites of practice are dispersed over a range of different businesses, such as commercial energy generators, suppliers of energy generation and management technology, and organizations that offer support services to CECs. Commercial sites of practice in the energy system relate to the collective practices of CECs in various ways, from supporting to competing with CECs.

For instance, the CEC often collects offers from a few suppliers who will sell and install the technology in question in organizing the promotion of individual energy practices. The relation between the CEC and these suppliers are described as "ambivalent" (CEC #3): on the one hand, by promoting solar panels, heat pumps, or home insulation, the CEC basically organizes a market for the supplier. In this way, the CEC and the supplier can support each other's practices. On the other hand, the commercially driven actors also perceive the CEC as a competitor and as costly, because the CEC usually wants a small discount for its members. From the supplier's point of view, the collective practice of promoting individual energy practices can also be performed by the supplier itself. This ambivalent and possibly competitive relation between CECs and commercial actors is a theme that we find in the development of all CECs' collective energy practices.

The relation between CEC's and commercial practices can have a supportive character: the annual report of CEC network organization HIER [1] signals a "growing willingness of commercial market parties to cooperative with local communities". This cooperative stance is, for instance, visible among some of the commercial energy generating companies. No CEC by itself has the capacity (and very few the license) to supply its members-clients with energy at any time. If a CEC wants to directly sell energy its members, then it needs to enter into a partnership with an established and licensed energy supplier. A few commercial and sustainable energy suppliers, such as Greenchoice, have positioned themselves as supporting community energy [1]. These companies often also offer additional administrative services to the still inexperienced CEC besides being the buyer and co-supplier of energy community generated energy. By taking care of these administrative practices, the commercial actor prevents a lot of overhead administrative costs for the CEC. By offering services to automate and professionalize the administrative processes that come with collective energy generation and postal code rose projects, commercial actors enable the further growth of community energy.

The routine practices of commercial actors can also be of hindrance to the CEC's development of collective energy practices, despite a willingness to cooperate with CECs. We find that commercial organizations that the CEC has to work with or rely on in developing their collective practices are not attuned to this new player on the field. The variety of technological, legislative, and financial elements that need to be integrated into these collective practices are not readily on offer; instead, they come from a range of organizations that each offer partial solutions based on the limits of their expertise and responsibilities. Collective generation, but especially collective management practices, involve "all these incredible technological parts, that ultimately need to be integrated into one solution. There is no party that I know that offers these integrated solutions" (CEC #3). Thus, these elements need to be integrated by the CEC, a complex task that does not make further growth of collective energy practices easier.

Furthermore, our research on CECs developing collective energy generation practices also shows the competitive side of the relation between CECs and the sites of commercial energy practices. Developing collective energy generation has also become an attractive business opportunity with the ongoing decrease in costs and substantial governmental subsidies. This has led to strong growth in solar park developments that are not in the hands of communities, and that are not postal code rose projects. These commercial practices of energy generation at the local level compete with CEC energy generation practices. Especially with the growing criticism of land-based solar parks, the amount of space for energy generation that they compete over is limited at the community level. "The community takes this step first, before the big companies start moving, too. Subsequently, it is the question if you want to keep doing that, as a small organization" (CEC #3). Commercial actors can now offer complete packages for solar parks to a land-owning farmer or municipality, with which it is difficult for a slower-moving, volunteer-based CEC project to compete on economic terms.

3.2.4. Grid Management Practices

Yet another site of energy system practices are the sites where the electricity grid is produced and managed. Especially, the regional distribution grid operators (DSO) in the Netherlands have been quite visible by setting up or joining experimental projects to develop and test mostly technical, measures that will help in managing the electricity grid at the local level. However, the practices of grid management have something of a dual nature, which also reflects on how they relate to the collective practices of CECs.

DSOs have developed innovative elements that are in close collaboration with CECs that are developing collective energy management practices. Our own research [20] has covered a DSO-led project, in which (some of) the tools for individual household energy management practices are introduced to a CEC, with the goal "to explore the potential of social cohesion for energy management". In recent years, DSOs are increasingly faced with communities who want to take up responsibilities in managing their energy. A cVPP project by a CEC in Loenen [49], and the GridFlex project with the CEC in Heeten [50] who want to experiment (under aforementioned Experimenting arrangement) with a local energy market and offer flexibility services are examples of this. These collective energy management practices are very relevant for DSOs, because these can support the DSOs task of local grid management, which might need 'smarter' solutions than what the conventional grid management practices entailed thus far.

On the other hand, by far the largest task that grid operators work on is maintaining a stable and reliable grid at all costs. It is this core practice on which they are judged, and this leads to a very defensive and risk-averse stance versus the uncertainties of working with collective energy management practices by CECs (CEC #3). Bids that are offered on the market for flexibility services (made mostly by commercial businesses such as aggregators) need to satisfy strict demands on reliability, availability, and response time. In this sense, the collective energy management practices of CECs will likely be held to the same standards. This development is still young and whether CECs are able to meet these standards in their grid management practices remains to be seen.

Besides the physical space that is needed for land-based solar parks, commercial and CEC energy generation practices also increasingly compete for capacity on the grid. In particular, in less populated areas that have been laid out with relatively low grid capacity, the limitations of the grid to accommodate collective energy generation are becoming visible [1]. In these areas, but perhaps also more generally in the near future, this limited grid capacity can be a barrier to the growth of collective energy generation. Furthermore, the nature of the practices of grid construction and planning is such that lifting this barrier by grid expansion will take a long time. Especially, the planning procedures can take several years, because this part of the process is also entrenched in governmental practices [18].

4. Discussion and Conclusions

The three categories of collective energy practices that CECs engage in are in quite different stages of development. The promotion of individual energy practices is ubiquitous among CECs, collective energy generation is growing fast, while collective energy management is currently undertaken by a fraction of CECs. We want to stress that not all CECs develop all of these collective energy practices, and there is not a fundamental ordering to these three sorts of energy practice. However, we observe that these collective practices are logically consecutive and increasingly complex steps in the development of CECs. Furthermore, we find that CECs become more enmeshed with the broader energy system of practices with each step of developing these practices. The collective energy practices of CECs become more impactful on the existing energy system and the practices through which it is maintained.

In each of our proposed categories of collective energy practices, CECs (still) take up a small share when compared to public and market-based actors. However, we do not wish to discount the role of the collective energy practices of CECs within the energy transition that is unfolding throughout the system of energy practices. As Watson [7] says, a practice based perspective on the energy system holds that changes to the system can result from changes in any site of practice: "if small interventions initiate or give momentum to positive feedback effects in desirable processes of recruitment and defection, their cumulative effects on the overall system can be substantial [7]". CECs developing collective energy practices ahead of others, precisely because they do not purely operate from business as usual, commercial position, and can thus contribute to the growth of these collective practices overall.

As their development progresses, CECs are increasingly performing tasks that touch upon, both positively and negatively, the practices of other actors in the energy system, in particular, those of grid management and market-based actors, such as energy generators and aggregators. As we noted, CECs often are started from a fundamentally different position regarding sustainable energy as compared to the current socio-technical system [30]. The organization of ownership and decision-making, the distribution of benefits, and the scale of technology are aspects of collective energy practices in which CECs aim to be distinctive. However, as Hicks and Ison [51] find, in reality, that these aspects are actually continuums of choices between the community 'ideal type' and the business as usual way of designing collective energy practices. This means that 'community energy' is an ambiguous term and that CEC energy practices come in many different forms. This highlights how it is likely that the collective energy practices of CECs, as they develop, will be required to conform to the practices of other sites of the energy system. This is already visible in the fast development of collective energy generation, where CECs compete for resources and space on economic terms with market-based actors. The question is whether CECs in this dynamic can keep bringing their basic principles into the way that they design their collective energy practices.

We have shown that the relation between sites of practice is varied: they can be enabling or hindering, and this relation can change with time. An enabling relation is seen, for example, in commercial energy suppliers that facilitate CECs to supply energy to their members as a reseller, or offer administrative services in support of collective energy generation. Conversely, the other sites of practices also hinder collective energy practices. This is seen in the mundane practices of grid management (coupled with booming commercial energy developments), which have culminated in the current situation of grid-constrained areas with no more grid-space for collective energy generation. Over time, the practices in these different sites of the energy system can also change their relation to the practices of CECs. The postal code rose arrangement is a striking example of this, which changed the site of taxation practices from a hindrance to collective energy generation to one that supports this practice for CECs. This observation is line with Macrorie's [19] earlier definition of a system of practices as "a relatively stable configuration of linked practices and relations that together sustain a particular socio-technical mode of doing", in that the particular mode of 'doing' linked energy practices is relatively stable, but is always open to change. We might say that, with time, the relatively stable configuration of practices at each site of the energy system needs to change to accommodate challenging practices, such as those of CECs.

In this article, we approached the topic of community energy and its role in a transitioning energy system from the perspective of social practice theory. We introduced the notion of collective energy practices: sets of energy practices beyond those of the individual, which emerge from some form of collective organization. These collective energy practices work towards collectively defined goals and, by being performed, also shape and maintain the collective. Collective energy practice is a useful concept, because it adds to the thus far individual-focused conceptual toolkit of practice-based understandings of sustainable energy and the energy transition. Conversely, understandings of community energy can be approached from a practice-based understanding with this concept.

A main characteristic of a practice-based perspective is that, ontologically, it places the practice itself center stage, rather than whoever is performing that practice. The collective that performs the practice can be a civic energy community, but also another organization, such as a market-based company, a public or government institution, or an organization from civil society. The point of focusing on the practice and 'leaving open' who the performing actor is, is not only inherent to practice theory, but it allows a perspective in which different organizations develop their own, sometimes competing, versions of collective energy practices.

We also extended the view on practices outwards, building on earlier work on systems of practice by Watson [7] and Macrorie [19] to come to the notion of an energy system of practices. Our descriptions of several collective energy practices are powerful examples that their emergence must be understood as embedded and shaped by other sites of practice within this system of practices.

The research question that was formulated for this article was: how can we understand the emerging practices of civic energy communities within a changing energy system? In this article, we empirically and theoretically answer this question. Our current energy system is changing and we have shown that the emerging activities of CECs play a role in these dynamics. We theorize this empirical phenomenon, in which a range of energy practices is collectively organized, by using the notion of collective energy practices and we show how such practices unfold within the broader energy system. Thereby, we highlight the empirical and theoretical importance of collective energy practices, as it gives us guideposts for both understanding and shaping the transition of our energy system.

Author Contributions: Conceptualization, N.V. and J.H.; Methodology, N.V.; Investigation, N.V. Writing—original draft preparation, N.V.; Writing—review and editing, N.V. and J.H.; Supervision, J.H.

Funding: This research is fully funded by The Netherlands Organization for Scientific Research NWO.

Conflicts of Interest: The authors declare no conflict of interest.

Abbreviations

CEC	Civic Energy Community.
cVPP	Community Virtual Power Plant
DSO	Distribution System Operator.
PCR (arrangement)	Postal-code Rose (arrangement).

References

1. HIER. Lokale Energie Monitor 2018. In *Lokale Energie Monitor*; Schwencke, A.M., Ed.; HIER opgewekt: Utrecht, The Netherlands, 2018.
2. Kampman, B.; Blommerde, J.; Afman, M. *The Potential of Energy Citizens in the European Union*; CE Delft: Delft, The Netherlands, 2016.

3. De Vries, G.W.; Boon, W.P.C.; Peine, A. User-Led Innovation in Civic Energy Communities. *Environ. Innov. Soc. Transit.* **2016**, *19*, 51–65. [CrossRef]

4. Schatzki, T.R. *Site of the Social: A Philosophical Account of the Constitution of Social Life and Change*; Penn State Press: State College, PA, USA, 2002.

5. Schatzki, T. Where the Action Is (on Large Social Phenomena Such as Sociotechnical Regimes). Available online: https://www.semanticscholar.org/paper/Where-the-Action-Is-(On-Large-Social-Phenomena-Such-Schatzki/f3ec3188fdff5fb14c1d346d34c6b3f8a55bb265 (accessed on 4 June 2019).

6. Nicolini, D. Is Small the Only Beautiful? Making Sense of 'Large Phenomena'from a Practice-Based Perspective. In *The Nexus of Practices*; Routledge: Abingdon-on-Thames, UK, 2016; pp. 110–125.

7. Watson, M. How Theories of Practice Can Inform Transition to a Decarbonised Transport System. *J. Transp. Geogr.* **2012**, *24*, 488–496. [CrossRef]

8. Dóci, G.; Vasileiadou, E.; Petersen, A.C. Exploring the Transition Potential of Renewable Energy Communities. *Futures* **2015**, *66*, 85–95. [CrossRef]

9. Seyfang, G.; Hielscher, S.; Hargreaves, T.; Martiskainen, M.; Smith, A. A Grassroots Sustainable Energy Niche? Reflections on Community Energy in the UK. *Environ. Innov. Soc. Transit.* **2014**, *13*, 21–44. [CrossRef]

10. Geels, F.W. Regime Resistance against Low-Carbon Transitions: Introducing Politics and Power into the Multi-Level Perspective. *Theory Cult. Soc.* **2014**, *31*, 21–40. [CrossRef]

11. Smith, A.; Hargreaves, T.; Hielscher, S.; Martiskainen, M.; Seyfang, G. Making the Most of Community Energies: Three Perspectives on Grassroots Innovation. *Environ. Plan. A Econ. Space* **2015**, *48*, 407–432. [CrossRef]

12. van der Waal, E.; van der Windt, H.; van Oost, E. How Local Energy Initiatives Develop Technological Innovations: Growing an Actor Network. *Sustainability* **2018**, *10*, 4577. [CrossRef]

13. Ornetzeder, M.; Rohracher, H. Of Solar Collectors, Wind Power, and Car Sharing: Comparing and Understanding Successful Cases of Grassroots Innovations. *Glob. Environ. Chang.* **2013**, *23*, 856–867. [CrossRef]

14. Haxeltine, A.; Avelino, F.; Wittmayer, J.; Kemp, R.; Weaver, P.; Backhaus, J.; O'Riordan, T. Transformative Social Innovation: A Sustainability Transitions Perspective on Social Innovation. In Proceedings of the Social Frontiers: The Next Edge of Social Innovation Research, London, UK, 14–15 November 2013.

15. Oteman, M.; Kooij, H.-J.; Wiering, M. Pioneering Renewable Energy in an Economic Energy Policy System: The History and Development of Dutch Grassroots Initiatives. *Sustainability* **2017**, *9*, 550. [CrossRef]

16. Hewitt, R.J.; Bradley, N.; Baggio Compagnucci, A.; Barlagne, C.; Ceglarz, A.; Cremades, R.; McKeen, M.; Otto, I.M.; Slee, B. Social Innovation in Community Energy in Europe: A Review of the Evidence. *Front. Energy Res.* **2019**, 7. [CrossRef]

17. Proka, A.; Hisschemöller, M.; Loorbach, D. Transition without Conflict? Renewable Energy Initiatives in the Dutch Energy Transition. *Sustainability* **2018**, *10*, 1721. [CrossRef]

18. Solar Magazine. Enexis En Tennet Maken Plannen Voor Netaansluitingsproblematiek Zonneparken. *Solar Magazine*, 1 March 2019.

19. Macrorie, R. Reconstructing Low-Energy Housing Using 'Systems of Practice'. Ph.D. Thesis, University of East Anglia, Norwich, UK, 2016.

20. Verkade, N.; Höffken, J. Is the Resource Man Coming Home? Engaging with an Energy Monitoring Platform to Foster Flexible Energy Consumption in The Netherlands. *Energy Res. Soc. Sci.* **2017**, *27*, 36–44. [CrossRef]

21. Verkade, N.; Höffken, J. The Design and Development of Domestic Smart Grid Interventions: Insights from the Netherlands. *J. Clean. Prod.* **2018**, *202*, 799–805. [CrossRef]

22. Shove, E.; Pantzar, M.; Watson, M. *The Dynamics of Social Practice. Everyday Life and How It Changes*; Sage: London, UK, 2012.

23. Shove, E. Beyond the Abc: Climate Change Policy and Theories of Social Change. *Environ. Plan. A* **2010**, *42*, 1273–1285. [CrossRef]

24. Naus, J. The Social Dynamics of Smart Grids. Ph.D. Thesis, Wageningen University, Wageningen, The Netherlands, 2017.

25. Hargreaves, T.; Nye, M.; Burgess, J. Making Energy Visible: A Qualitative Field Study of How Householders Interact with Feedback from Smart Energy Monitors. *Energy Policy* **2010**, *38*, 6111–6119. [CrossRef]

26. Smale, R.; van Vliet, B.; Spaargaren, G. When Social Practices Meet Smart Grids: Flexibility, Grid Management, and Domestic Consumption in the Netherlands. *Energy Res. Soc. Sci.* **2017**, *34*, 132–140. [CrossRef]

27. Røpke, I. The Roles of Households in the Smart Grid. 2013. Available online: http://vbn.aau.dk/files/77312030/KU_Smart_grid_4.pdf (accessed on 25 May 2019).
28. Strengers, Y. *Smart Energy Technologies in Everyday Life: Smart Utopia?* Palgrave Macmillan: London, UK, 2013.
29. Smale, R.; Spaargaren, G.; van Vliet, B. Householders Co-Managing Energy Systems: Space for Collaboration? *Build. Res. Inf.* **2019**, *47*, 585–597. [CrossRef]
30. Schatzki, T.R. *The Timespace of Human Activity: On Performance, Society, and History as Indeterminate Teleological Events*; Lexington Books: Lanham, MD, USA, 2010.
31. Van der Schoor, T.; Lente, H.v.; Scholtens, B.; Peine, A. Challenging Obduracy: How Local Communities Transform the Energy System. *Energy Res. Soc. Sci.* **2016**, *13*, 94–105. [CrossRef]
32. PolderPV. Inkoop Acties Zonnepanelen in Nederland. Available online: http://www.polderpv.nl/inkoopacties_Nederland.htm (accessed on 9 May 2019).
33. The Solar Future Vii—3. Inkoopacties. Available online: http://www.polderpv.nl/nieuws_PV126.htm#6jun2015_TSF3_inkoopacties (accessed on 9 May 2019).
34. Kloppenburg, S.; Boekelo, M. Digital Platforms and the Future of Energy Provisioning: Promises and Perils for the Next Phase of the Energy Transition. *Energy Res. Soc. Sci.* **2019**, *49*, 68–73. [CrossRef]
35. Koirala, B.P.; van Oost, E.; van der Windt, H. Community Energy Storage: A Responsible Innovation towards a Sustainable Energy System? *Appl. Energy* **2018**, *231*, 570–585. [CrossRef]
36. Van Summeren, L.; Wieczorek, A. Defining Community-Based Virtual Power Plant. Available online: http://www.nweurope.eu/projects/project-search/cvpp-community-based-virtual-power-plant/ (accessed on 3 May 2019).
37. Hirsch, A.; Parag, Y.; Guerrero, J. Microgrids: A Review of Technologies, Key Drivers, and Outstanding Issues. *Renew. Sustain. Energy Rev.* **2018**, *90*, 402–411. [CrossRef]
38. Netbeheer Nederland. *Position Paper—Aansluiten Duurzaam Op Land*; Netbeheer Nederland: The Hague, The Netherlands, 2019.
39. Bird, L.; Lew, D.; Milligan, M.; Carlini, E.M.; Estanqueiro, A.; Flynn, D.; Gomez-Lazaro, E.; Holttinen, H.; Menemenlis, N.; Orths, A.; et al. Wind and Solar Energy Curtailment: A Review of International Experience. *Renew. Sustain. Energy Rev.* **2016**, *65*, 577–586. [CrossRef]
40. Shove, E.; Watson, M.; Spurling, N. Conceptualizing Connections: Energy Demand, Infrastructures and Social Practices. *Eur. J. Soc. Theory* **2015**, *18*, 274–287. [CrossRef]
41. Nicolini, D. *Practice Theory, Work, and Organization: An Introduction*; Oxford University Press: Oxford, UK, 2012.
42. Energeia. Wiebes: Consumenten Kunnen Tot 2021 Nog Salderen. Available online: https://energeia.nl/energeia-artikel/40072302/wiebes-consumenten-kunnen-tot-2021-nog-salderen (accessed on 20 March 2019).
43. Rijksoverheid. *Salderingsregeling Verlengd Tot 2023*; Rijksoverheid: The Hague, The Netherlands, 2019.
44. Kooij, H.-J.; Lagendijk, A.; Oteman, M. Who Beats the Dutch Tax Department? Tracing 20 Years of Niche–Regime Interactions on Collective Solar Pv Production in The Netherlands. *Sustainability* **2018**, *10*, 2807. [CrossRef]
45. REScoop.eu. European Parliament Puts Its Final Stamp on New Rules to Empower Citizens Energy Communities. ReScoop.eu. Available online: https://www.rescoop.eu/blog/european-parliament-puts-its-final-stamp-on-new-rules-to-empower-citizens (accessed on 28 March 2019).
46. Ministry of Economic Affairs. *Besluit Experimenten Decentrale Duurzame Elektriciteitsopwekking*; Ministry of Economic Affairs: Den Haag, The Netherlands, 2015.
47. Hoppe, T.; Graf, A.; Warbroek, B.; Lammers, I.; Lepping, I. Local Governments Supporting Local Energy Initiatives: Lessons from the Best Practices of Saerbeck (Germany) and Lochem (The Netherlands). *Sustainability* **2015**, *7*, 1900–1931. [CrossRef]
48. Elzenga, H.; Schwencke, A.M. Lokale Energiecoöperaties: Nieuwe Spelers in De Energie. Available online: http://asisearch.nl/wp-content/uploads/2016/11/2015-Artikel-PBL-Tijdschrift-Bestuurskunder-2015-artikel-Lokale-energiecoo%CC%88peraties-Elzenga-Schwencke.pdf (accessed on 4 June 2019).
49. DPL. Community-based Virtual Power Plant Loenen. Available online: https://duurzaamloenen.nl/community-based-virtual-power-plant-loenen (accessed on 11 June 2019).

50.   GridFlex. Gridflex Heeten. Available online: https://gridflex.nl/ (accessed on 11 June 2019).
51.   Hicks, J.; Ison, N. An Exploration of the Boundaries of 'Community' in Community Renewable Energy Projects: Navigating between Motivations and Context. *Energy Policy* **2018**, *113*, 523–534. [CrossRef]

© 2019 by the authors. Licensee MDPI, Basel, Switzerland. This article is an open access article distributed under the terms and conditions of the Creative Commons Attribution (CC BY) license (http://creativecommons.org/licenses/by/4.0/).

Article

Remediation of Potential Toxic Elements from Wastes and Soils: Analysis and Energy Prospects

Alberto González-Martínez [1], Miguel de Simón-Martín [1], Roberto López [2], Raquel Táboas-Fernández [1] and Antonio Bernardo-Sánchez [3,*]

[1]   Department Area of Electrical Engineering, School of Mining Engineering, University of León (Spain), Campus de Vegazana s/n, 24071 León, Spain; alberto.gonzalez@unileon.es (A.G.-M.); miguel.simon@unileon.es (M.d.S.-M.); raqueltaboas@gmail.com (R.T.-F.)
[2]   Department Area of Physical Chemistry, Faculty of Biological and Environmental Sciences, University of León (Spain), Campus de Vegazana, s/n, 24071 León, Spain; rlopg@unileon.es
[3]   Department Area of Mines Engineering, School of Mining Engineering, University of León (Spain), Campus de Vegazana s/n, 24071 León, Spain
*   Correspondence: antonio.bernardo@unileon.es; Tel.: +34-987-29-35-54

Received: 28 April 2019; Accepted: 13 June 2019; Published: 15 June 2019

Abstract: The aim of this study is to evaluate the application of the main hazardous waste management techniques in mining operations and in dumping sites being conscious of the inter-linkages and inter-compartment of the contaminated soils and sediments. For this purpose, a systematic review of the literature on the reduction or elimination of different potential toxic elements was carried out, focusing on As, Cd and Hg as main current contaminant agents. Selected techniques are feasible according to several European countries' directives, especially in Spain. In the case of arsenic, we verified that there exists a main line that is based on the use of iron minerals and its derivatives. It is important to determine its speciation since As (III) is more toxic and mobile than As (V). For cadmium (II), we observed a certain predominance of the use of biotic techniques, compared to a variety of others. Finally, in mercury case, treatments include a phytoremediation technique using *Limnocharis flava* and the use of a new natural adsorbent: a modified nanobiocomposite hydrogel. The use of biological treatments is increasingly being studied because they are environmentally friendly, efficient and highly viable in both process and energy terms. The study of techniques for the removal of potential toxic elements should be performed with a focus on the simultaneous removal of several metals, since in nature they do not appear in isolation. Moreover, we found that energy analysis constitutes a limiting factor in relation to the feasibility of these techniques.

Keywords: hazardous waste; contaminated soil; potential toxic elements; removal; mine waste

---

1. Introduction

Waste is the unusable result of a material after it has been used to develop a job or operation [1]. In the EU, total waste production amounted to 2.5 billion tons every year [2]. The increase in waste generation produced over last 50 years is the main reason for environmental legislation enacting. Actually, the main guideline in EU environmental policy is the 7th Environment Action Program, which promotes the protection, conservation, and enhancement of the natural capital Member States as one of the main objectives for 2020.

The transposition of EU directives on waste matter to Spanish legislation promoted the enactment of Law 22/2011, 28th July, on waste and contaminated soils. Based on this set of new laws, waste can be defined as any substance or object that its holder discards or intends to dispose of [3].

Law 22/2011, 28th July, aims to "regulate the management of waste by promoting measures to prevent its generation and mitigate adverse impacts on human health and the environment associated with its generation and management, improving efficiency in the use of resources and regulating the legal regime of contaminated soils". The importance of waste management lies in the fact that, if it is not done, the environment could deteriorate irreversibly.

A waste generation comparison of European countries between 2004 and 2014 is shown in Figure 1. The circle line represents the waste amount generated in 2004, while the solid circle represents the waste generation in 2014. In Spain, as in other European countries, waste generation has been linked to economic growth, so the trend of recent years has been to reduce the volume of generated waste (e.g., reduction factor of 0.69 in Spain and 0.49 in Portugal). However, some countries, as Latvia, Sweden, Netherland, and Norway, increased their waste generation to a factor of 2.09, 1.82, 1.43, and 1.42 in a decade [4]. EU annual report on waste statistics shows the main reasons of this increment to be the high shares of major mineral wastes (because their relatively sizeable mining and quarrying activities) and construction and demolition activities of that countries [5].

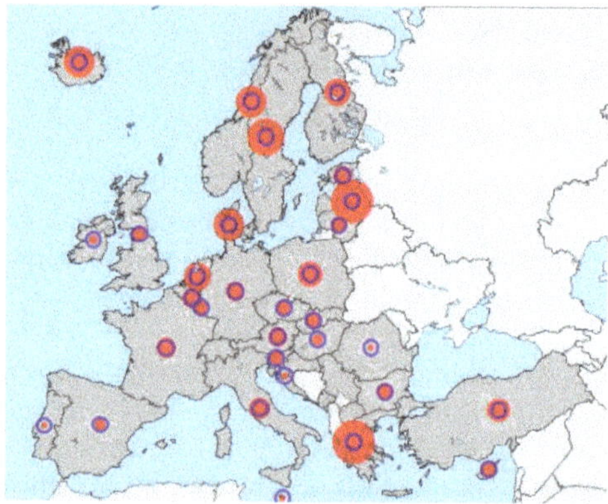

Figure 1. Waste generation comparison in EU Member States for the 2004 (circle line) to 2014 (solid circle) period. Source: [4].

The contribution of the different activities to the generation of waste in Spain in 2014 are presented in Figure 2. More than half of the waste generated in 2014 came from the services, construction and mining sectors. The waste generated in industry and in the services sector are different in nature, while in mining and construction waste is mostly of a mineral kind [6].

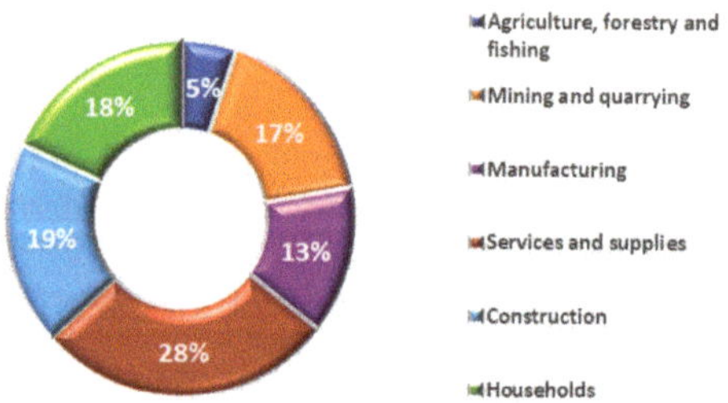

Figure 2. Generation of waste by economic activity in Spain in 2014. Adapted from: [6].

The extractive industry obtains minerals through mining techniques, which include drilling and blasting operations, among others. In addition, for commercial use, these extracted minerals must benefit from mineral processing techniques. In both stages of the process, waste is generated. In this sense, mining, or extractive industries wastes are the solid, aqueous, or paste residues that remain after the investigation and use of a geological resource, provided that such material constitutes waste as it is defined in Law 22/2011, on waste and contaminated soils [3].

The Environmental European Agency defines percolation as a process which concerns the movement and filtering of fluids through porous materials [7]. In soil contamination, this term is associated with the movement of polluted water solutions across different land layers. These solutions, also called *leachates*, are characterized to displace large quantities of heavy metals, resulting in clean soils contamination. As a consequence, leaching is found to be one of the worst environment problems caused by mining wastes. These leachates cause serious environmental damage, triggering the contamination of large amounts of lands with dangerous metals at a long distance away from the contamination source. In Carnoulès (France), mining installations were abandoned 55 years ago. An acid stream with pH between 2.5–4.7 and 50–350 mg/L of arsenic flowed into the Amous river, with a very high environmental impact [8]. On the other hand, in Giant mine (Canada), 237,000 tons of arsenic trioxide dust is currently safe, but when the rock at Giant Mine was crushed and mined out, the arsenic was exposed to the environment. Large sections of the underground mine were backfilled with waste rock and tailings. The arsenic concentration in these sources was hundreds of times lower than in the arsenic trioxide waste in the storage chambers. However, the large volumes meant they also could contaminate the surrounding groundwater [9].

One of the most important problems in relation to potential toxic elements is that they are not biodegradable. [10]. As a consequence, heavy metals become toxic at very low concentrations. This toxicity depends on the concentration, the chemical form, and the persistence in which they occur, since in trace concentrations they may be indispensable for living beings (Na, K, Mg, Ca, V, Mn, Fe, Co, Ni, Cu, Zn, and Mo) [11].

Metals, as those cited before, coexist with other cations and anions in acid water solutions because their high insolubility in alkaline pH [12]. It is also true that acid solutions are frequently found in wastewater produced in industries such as mining, metal processing, electroplating, textiles, tanning, and oil refining, as well as in leaching from some dumpling sites. As a result, mining waste is characterized to produce severe environmental damage when an efficient treatment is not applied to these emissions.

Khalid et al. found mercury (Hg), lead (Pb), cadmium (Cd), and arsenic (As) to be the main metal(-oids) produced by metalliferous mining and smelting activities [13]. In addition, the World Health Organization identify the same metals/metalloids to be the major public health concerns [14]. As a consequence, these are the four metals/metalloids most widely studied in the literature. Between them, mercury, cadmium, and arsenic have been found to have high significant presence in mining waste products and dumping sites leaching [15]. This is the reason why this work is focused on the available disposal techniques for their remediation.

The choice of a suitable toxic elements' remediation technique considers the properties of both the elements and the contaminated site. Any treatment is associated with the modification of toxicity, mobility and volume of the waste, whether by physical, chemical, or biological processes. These treatments can be classified according to several factors. Classification according to the type of treatment may be described as follows [16]:

Biological treatments or bioremediation: Based on the use of the metabolic activity of organisms like plants, fungi, or bacteria for the transformation or elimination of the pollutants.

Physic-chemical treatments: Based on the use of the physical and/or chemical properties of the contaminants or the medium to remove, separate, or contain them.

Thermal treatments: Based on the use of heat to increase volatilization or melt contaminants.

The main advantages and disadvantages of these technologies are summarized in Table 1.

Table 1. Advantages and disadvantages of biological, physical-chemical, and thermal treatments for arsenic, cadmium, and mercury remediation. Adapted from [16] and [17].

Treatment	Advantages	Disadvantages
Biological	Economical Sustainable Removes contaminants Minimal or no further treatment needed	Longer treatment time Need for toxicity verification Only suitable if soil allows microbial growth Very sensitive to process conditions changes
Physic-chemical	Economical (intermediate) Short application periods Easier to control	Need for waste generated treatment Recovery systems for extraction fluids
Thermal		Energy and engineering needs Possible contaminant gaseous emissions Expensive

The remediation of environments contaminated with potential toxic elements by chemical methods can be excessively expensive due to the specificity of potential toxic elements reactivity. This specificity can make in-situ treatments to be more difficult to carry on when different metals must be removed from the area. On the other hand, biological methods can be applied both in-situ and ex-situ with high elimination specificity due to (1) the specific conditions in which metabolic organisms' reactions take place, and (2) the wide range of organisms with different metabolic routes suitable to soil bioremediates [18].

A controversy about scientific evidence for environmental damages caused by the use of bioremediation continues today. On one hand, Swannwell et al. reported some information about field evaluations of marine oil spill bioremediation [19]. In the *Exxon Valdez* incident description, they showed evidence that bioremediation causes less environmental damage than conventional techniques. However, several works indicated that microbes used in bioremediation may produce metabolites and other residues (i.e., dead biomass) which could make soils be even more toxic after bioremediation than they were before [20,21]. As a consequence, (1) high volume of solid waste should be managed at the end of the remediation process [22], and (2) soil characteristics must be studied wide enough before deciding to apply this remediation technique as many factors (degree of exposure, metal concentration, temperature, salinity) can completely change whether bioremediation is suitable or not to be used as a specific soil treatment. Furthermore, it has been a wide agreement on the interest in biological techniques, mainly characterized by its safety and low cost. In addition, they can be applied at the contaminated site. However, bioremediation processes produce residues [22]. Carrying out on-site treatment involves other advantages such as reducing exposure and avoiding problems associated with accidental discharges during transport. The increase in the use of biological techniques in relation to traditional physic-chemical techniques is due to the fact that the latter involve the alteration or elimination of the ecosystem and its physic-chemical properties, despite the fact that these technologies are faster and more effective [23,24].

This paper itself is innovative in terms of conducting a systematic review of the most common contaminant agents considering not only the remediation effects, but also analyzing the involved energy needs and prospects, which will make the different treatments more or less sustainable. It will represent the leading work for a more efficient and sustainable design of remediation techniques than can be applicable not only to hazardous wastes from mining or dumping sites, but also from environmental disasters or intensive potential toxic elements activities. The systematically conducted review will constitute a significant contribution and, from the author's point of view, a nexus between environmental researchers and energy researchers, which is new in the literature in this field.

Finally, the better the treatment techniques are known and optimized, the better locally available energy sources can be used to feed them, especially considering renewable energy sources. The authors observe that a high energy consumption of certain treatments can constitute a real barrier for their effective application as it can make them unaffordable due to the associated costs for the energy supply and the impossibility of supplying the required energy just by using renewable energy sources, such as photovoltaic systems, with an average specific energy capacity of about 170 W/m^2. Thus, the local remediation techniques that can take advantage of locally generated renewable energy sources have become of high interest, reducing the overall lifecycle impact on the environment. Thus, this paper fits as an outstanding application of locally available energy sources and sustainability. The final target is to remediate the human impact on the environment without producing, or at least minimizing the production, of greenhouse gas emissions, reducing the dependence on imported fuels, and creating economic development and improving local employment, i.e., in a sustainable way.

1.1. Arsenic

Arsenic is a chemical element that belongs to the group of metalloids or semi-metals, due to its intermediate behavior between metals and non-metals. It can be found in four states of oxidation: +5 (arsenate), +3 (arsenide), 0 (arsenic) and −3 (arsine), often as metal sulphides, arsenides, or arsenates. In surface environments, the predominant forms are trivalent [As (III)] and pentavalent [As (V)] arsenic, which occur as oxyanions. There are also methylated derivatives of arsenic such as monomethylarsonic acid (MMAA), dimethylarsinic acid (DMAA), and trimethylarsine oxide (TMAO) [25]. In water, it is mostly presented as arsenate (+5), but under anaerobic conditions, it is presented as arsenide (+3).

The toxicity of arsenic depends on the state of oxidation in which it is found and/or the chemical compound of which it is part. The inorganic species are always more toxic than the organic ones. The most toxic form is arsine (AsH$_3$), followed by arsenic trioxide (As$_2$O$_3$), arsenide [As (III)], and arsenate [As (V)].

The main route of arsenic exposure is ingestion through water and food. Water pollution is a global problem, as the World Health Organization's recommended maximum concentration limit for drinking water (10 μg As/L) has been exceeded in many places [26].

The presence of arsenic in the environment can cause serious risks to health, such as neuropathy, conjunctivitis, diabetes, renal damage, an enlarged liver, bone marrow depression, high blood pressure, cardiovascular disease or skin, lung and liver cancer [15].

1.2. Cadmium

Cadmium is a metal found in the Earth's crust with an abundance in soil of 0.16 ppm. Its most common oxidation state is +2. Most of this element is produced as a by-product of zinc smelting, although it can also be produced as greenockite (CdS) [27]. It is a potential toxic element that causes environmental problems and can be accumulated in the human body through the food chain [28]. This accumulation may cause skeletal damage, evidenced by low bone mineralization. In addition, cadmium is also carcinogenic if inhaled [15].

Environmental contamination by cadmium occurs frequently in areas surrounding zinc-, lead- or copper-smelting factories [29]. Scenarios such as the incineration of waste and other materials containing cadmium as a pigment or stabilizer, the burning of fossil fuels, and the use of phosphate fertilizers can also be sources of soil or air pollution [30].

The World Health Organization recommends a maximum concentration limit of cadmium in water for human consumption of 3 μg Cd/L [26]. The US Agency for Toxic Substances and Disease Registry (ATSDR) lists cadmium as the sixth-most dangerous metal according to its prevalence and severity due to the intoxication it causes.

1.3. Mercury

In these substances, mercury is in an oxidation state of +2, +1 and smaller. It is usually found as cinnabar (HgS), which may contain drops of metallic mercury, metallic mercury, mercury vapor, inorganic mercury (I) and (II), alkylmercury, and phenylmercury [25].

Mercury is one of the most polluting elements for humans and many animals due to its high toxicity. The different species of mercury present different characteristics in terms of environmental behavior, bioavailability (as rate of which a toxin is absorbed) [31], metabolism, and effects on organisms. It must be in ionic forms to show harmful effects. The World Health Organization defined mercury to be responsible for abnormal brain development, resulting in a lower IQ, and consequently a lower earning potential. In addition, mercury affects other aspects in brain and produces cardiovascular diseases [15]. Mercury salts exhibit high toxicity when dissolved in water. Alkylmercury has irreversible effects on the nervous system, and methylmercury has serious teratogenic, carcinogenic, and mutagenic effects [25]. In addition, mercury is easily absorbed by biota (especially in its organic forms), which leads to its accumulation in the food chain.

The maximum allowable limit of Hg (II) ions in drinking water is 1 µg Hg/L [32], and the maximum limit recommended for inorganic mercury, by the World Health Organization, is 6 µg Hg/L [26].

2. Systematic Review

A systematic review is defined as "an observational and retrospective research design, which synthesizes the results of multiple primary investigations" [33]. In this case, such a review has been carried out in order to assess the state of the art in the field of mining-waste management and the remediation of contaminated soils and leaches. The following conditions have been considered in the review:

Sources: Journal articles and technical review reports have been included, provided that they have been indexed and involved a peer-review process.
Language: English.
Period of publication: Research articles published between 2004 and 2017 have been included.
Requirements: All studies analyzed are freely accessible.
Keywords: "Removal mining", "arsenic", "cadmium", "mercury".
Type: Original research contributions and review works have been considered.

The systematic review was performed after the search conditions were established. 210 documents on arsenic removal, 190 on cadmium, and 137 on mercury were consulted; 537 documents were thus analyzed in total. A distribution of number or documents studied by year is shown in Figure 3.

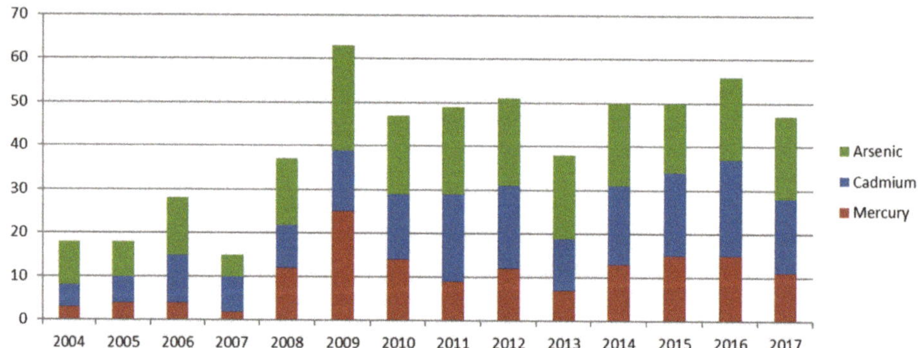

Figure 3. Number of references, organized by publication year, resulting from the systematic review of studies on potential toxic elements removal techniques (As, Cd, and Hg) used in the preparation of this article.

An initial search was conducted in bibliographic databases, applying the conditions established above. Through this process, Table 2 results were obtained. These results refer to general techniques for eliminating each of the potential toxic elements considered in this work: Arsenic, cadmium, and mercury (keywords).

Table 2. Number of references from the systematic review of potential toxic elements removal techniques.

Metal/Metalloid	Articles	Reviews
Arsenic	210	15
Cadmium	190	7
Mercury	137	9
Total	**537**	**31**

The previous search was filtered according to the specific elimination techniques analyzed, through which the results shown in Table 3 were obtained.

Table 3. Number of references from the systematic review of specific techniques for the removal of potential toxic elements (As, Cd, and Hg).

Metal/Metalloid	Articles	Reviews
Arsenic		
Adsorption	48	4
Electrocoagulation	2	-
Precipitation	47	2
Bioleaching	3	-
Others	110	9
Cadmium		
Adsorption	38	2
Sequential soil washing	1	-
Biosynthesis of metal chalcogenides	1	-
Others	150	5
Mercury		
Phytoremediation	6	1
Nanobiocomposite hydrogel	3	1
Others	141	3
TOTAL	**537**	**31**

From the evaluated references, a 5.24%, 3.16%, and 2.19% of them face the problem of arsenic, cadmium, and mercury removal from the leachates, respectively. This sort of references has increased significantly in the recent years.

2.1. Arsenic

In relation to techniques for reducing arsenic, there is a predominant line based on the use of iron. Among these techniques, some of them should be highlighted: Co-precipitation and natural adsorption by iron minerals [34,35], electrocoagulation using iron electrodes [36], use of iron-rich sludge from coal treatment [37], sorption using coconut shell and iron oxide-coated sand [38], and biogenic ferric precipitates [39]. There are also biotic arsenic removal techniques, such as bioleaching using *Acidothiobacillus thiooxidans*.

2.1.1. Co-Precipitation and Natural Adsorption by Iron Minerals

Arsenic co-precipitation can be performed by natural attenuation, produced in leaching mine process, or by using a specific installation to develop it.

Natural attenuation is based on ferrihydrite and schwertmannite formation during acid mine drainage. Arsenic is incorporated into their chemical structure during their formation, causing a solid precipitate in aqueous solution. The natural attenuation of arsenic presents in acid mine effluents occurs principally through the co-precipitation of As (V) with schwertmannite. The main advantages of this method are its low cost and high efficiency, making arsenic concentrations in water negligible [34]. However, this process can only be produced in environments where ferrihydrite and schwertmannite are formed.

Ferrihydrite and schwertmannite formation can also produce goethite. Goethite has a high specific active surface [35], which adsorbs arsenic. This phenomenon is called natural adsorption because it is produced without the need of a designed installation.

At last, arsenic can be precipitated by using ferric salts in a designed reactor. Figure 4 shows a schematic model of a co-precipitation arsenic system.

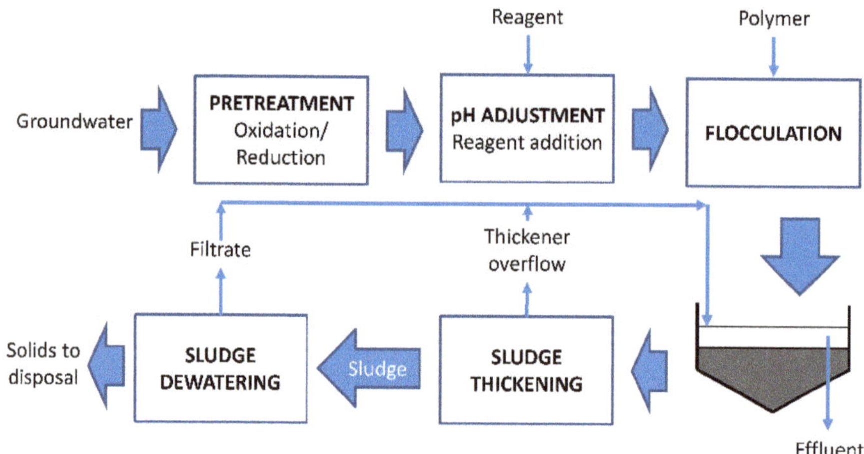

Figure 4. Schematic diagram a co-precipitation arsenic system. Adapted from: [40].

In a reagent-based system, ferric salts are added to residual water, making arsenic form a solid matrix in the aqueous solution. Afterwards, ferric-arsenic salts are separated from water and conducted to a sludge treatment. Solubility of As (V) is lower than As (III). As a consequence, sludge stability over time is also responsible for the oxidation of As (III) before neutralization. These are the main reasons why a pretreatment of As (III) oxidation is sometimes included.

2.1.2. Hydrometallurgical Process Based on Arsenic Precipitation with Lime

One of the most important technology for arsenic removal is neutralization with lime and the recirculation of precipitation sludge to enhance their properties (stability, water content). Arsenic is precipitated as calcium arsenite and/or calcium arsenate at pH 11–12. The ratio of Fe/As and the proportion of As (III)/As (V) are important parameters that influence the performances of As removal and sludge stability [36].

As a global method, lime neutralization is a relatively economic process for arsenic immobilization. However, the precipitates show poor long-term stability and therefore must be deposited in hazardous landfills. [36].

2.1.3. Electrocoagulation Using Iron Electrodes

Electrocoagulation with iron electrodes is a promising technique in comparison to those which need reagents addition because the limitation of oxidizers and absorbents reaction and absorption capabilities.

Both As (III) and As (V) are present in arsenic contaminated groundwater. When electric current passes through iron electrodes, Fe (II) dissolves. In the presence of dissolved oxygen, Fe (II) oxidizes to Fe (III) and forms a solid matrix, $FeAsO_4$ and $Fe(OH)_3$ onto which As can be sorbed. Thus, arsenic is removed from the aqueous phase by separating the precipitates. The optimum pH for As (III) removal is 7, using low intensity, in absence of other substances. In addition, it has also been determined that increasing concentrations of phosphate, silicate, and natural organic matter reduces the efficacy of electrocoagulation [41]. On the other hand, bicarbonate, nitrate, sulphate, and chloride do not affect the elimination efficiency.

Electrocoagulation can be applied in situ and also ex situ. In ex situ technology, contaminated water passes between electrodes. The current causes arsenic to migrate toward the electrodes, and also modifies the pH and the redox potential of water causing arsenic to precipitate. Afterwards, solids are removed from the water by a filtration stage [40].

Arsenic electrocoagulation removal has been studied under various conditions of electricity intensity, pH, electrode connectivity, dissolved oxygen, temperature, and sedimentation. Several designs have been developed, but further research is necessary to better understand this process and make possible a significative increasing in arsenic removing [41], specially, the dependence of pH in function of the presence of other substances in the aqueous solution. The system pH affects the oxidation of Fe (II) and As (III). The presence of phosphate and silicate also has an effect by reducing the removal of arsenic, but it is not known how this process works.

2.1.4. Removal of As (III) and As (V) Using Iron-Rich Sludge

Amorphous or crystalline iron oxides have a high arsenic adsorption capacity. Removal of As (III) with a typical adsorbent is more difficult than the removal of As (V) due to the positive charge of the adsorbent surface and the neutral charge of As (III) in neutral conditions of pH. This is the reason why its extraction is pH dependent: the adsorption of arsenic decreases at high pH or in the presence of phosphates [42].

Most coal-mining drainage contains high proportions of iron and its treatment systems produce large quantities of iron-rich sludge whose disposal and reuse cause major environmental problems, so its use as an adsorbent has environmental and economic advantages.

The iron-rich sludge employed consists of goethite and calcite from the drainage treatment of coal mines; it provides an arsenic adsorption of almost 70%, after one hour of contact. After 8 h of contact, the adsorption increases up to 95%. However, after 48 h, the treatment improves just an extra 2–5%. The maximum adsorption capacity of As (III) and As (V) is in the range 67–72 and 21.5–24.6 mg/g adsorbent, respectively, at 293–323 K [37].

2.1.5. Fixed-Bed Sorption System Using Coconut Shells and An Iron Matrix

The removal of different potential toxic elements from gold-mine effluent can be carried out in fixed downflow bed columns using coconut shells and different iron solid matrix, such as iron-oxide-coated sand, ferric hydroxide-coated newspaper pulp, granular ferric hydroxide, or iron filings mixed with sand [38,40].

There are few studies in which agricultural and natural materials are used in continuous systems for the removal of potential toxic elements from effluents, so it is necessary to carry out pilot-scale research to demonstrate the applicability of biosorption processes in wastewater treatment on an industrial scale. Acheampong and Lens [38] showed the sorption system based on coconut shells' capacity for Cu (II) sorption. This method can be applied to remove As (V) in continuous and non-continuous bed conditions, by reusing in multiple stages of sorption and desorption. Figure 5 shows a schematic diagram of an arsenic adsorption pilot plan.

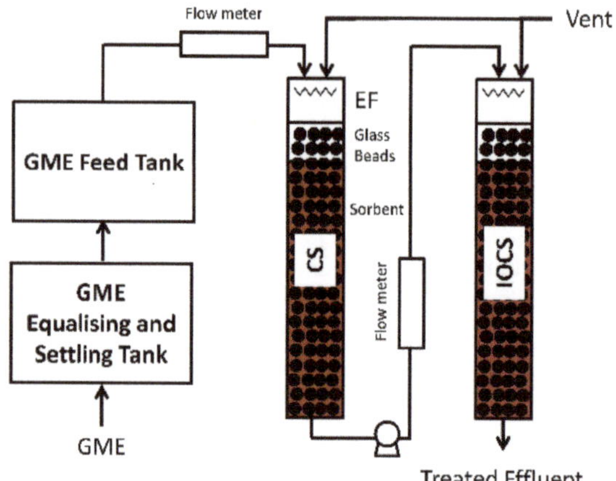

Figure 5. Schematic diagram of the two-stage pilot plant set-up (CS: coconut shell; IOCS: iron oxide coated sand; EF: effluent distributor; GME: gold mining effluent; FM: flow meter). Adapted from [38].

Coconut shell with a particle size between 0.5–1.4 mm is introduced in the first column, filling a bed height of 150 cm. First column manages to remove up to 100% of Cu and Fe, while the As remains in the effluent. In the second column, iron-oxide-coated sand with a particle size between 1.0–3.0 mm is introduced, filling a bed height of 150 cm. In both cases, the sorbent is placed between a layer of beads of glass on the top and a sieve placed on the bottom of the columns. The influent is pumped into the columns and passes through them. After the treatment, the concentrations of the different potential toxic elements obtained are measured. After 3 cycles in both columns, only As is still detected in the effluent. The optimum number of cycles is found to be 8. Other variables optimized are length of the bed, contact time, number of columns, sorption sites, and volume of gold mining effluent, and As is satisfactorily removed from contaminated water.

This installation demonstrates the ability of the coconut shell and iron-oxide-coated sand to remove arsenic and copper, as well as other heavy metals (Pb, Fe, and Zn) in order to comply with actual normative in the Netherlands. In addition, it demonstrates the capacity of the pilot plant for the treatment of large volumes of contaminated effluents.

2.1.6. Arsenic Removal with Biogenic Ferric Precipitates

Removal of arsenic can be carried out in situ by precipitating the iron present in the contaminated water. Iron oxides, such as jarosite, schwertmannite, or goethite can be applied in the removal of arsenic from mining effluents. A study of a biological treatment for acidic solutions containing iron [Fe (II)] and arsenic [As (III)] has been performed by Ahoranta et al. [39].

A diagram of the process is shown in Figure 6. It consists on a fluidized bed with biological iron oxidation reactor, followed by an iron precipitator for the subsequent separation of precipitates by gravity in a settler. Furthermore, experimental conditions have been optimized for arsenic removal. In this sense, the highest iron oxidation and precipitation rates were 1070 and 28 mg As/L for 5 and 7 h of retention time respectively, achieving a solid precipitate with 96–98% of Fe (II) present in affluent at pH 3.0. These conditions, after the treatment process, yielded to reach a 99.5% of arsenic removal efficiency.

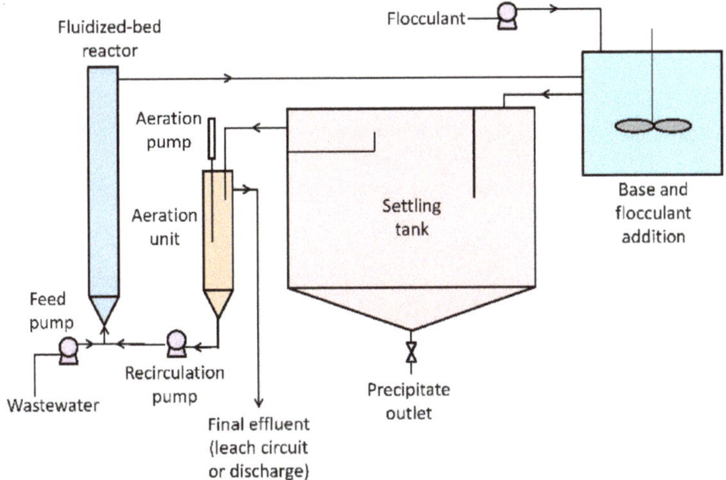

Figure 6. Schematic diagram of the experimental system. Adapted from [39].

2.1.7. Bioleaching of Arsenic Using *Acidithiobacillus thiooxidans*

In-situ or ex-situ bioleaching technology applied to contaminated sediments uses acidophilic and chemotropic bacteria to oxidize sulphur and Fe (II) under acidic conditions. This oxidation makes solubilization of potential toxic elements contained in the sulfides of the sediments possible.

Bioleaching using *Acidithiobacillus thiooxidans* is an effective method for the reduction of arsenic concentration from mine tailings with high concentrations of arsenic (approximately 34,000 mg/kg) [39].

The studies reported by Ahoranta et al. [39] were performed to determine the optimum conditions to bioleach the potential toxic elements present in batteries. The initial experimental conditions were: pH 1.8–2.2, temperature between 25 and 40 °C and solid concentration in the interval of 0.5–4.0%. Optimal conditions were set to pH value of 1.8, 30 °C, and 0.5% solids concentration. These conditions yielded to a maximum As removal efficiency of 47%.

The results showed that the leaching efficiency was similar regardless of pH. A higher leaching velocity was observed with low initial pH (1.8), which could be due to a greater initial fixation of the cells to the tailings. Furthermore, it was observed that changes in temperature and solids concentration affect leaching. Leaching efficiency decreased with an increase in temperature due to the lower bacterial growth. The highest initial leaching ratio was obtained at 25 °C. On the other hand, efficiency increased with a decrease in solids concentration. Researchers also observed that jarosite presented in tailings after leaching was higher when solids concentration was also higher due to the lower efficiency of arsenic bioleaching.

Experimental installation is usually based on a packed-bed column reactor where residual water with arsenic is pumped. When column saturates, arsenic is stripped and reactor is biologically regenerated. Some studies performed at The Center for Bioremediation at Weber State University have studied this process on a bench scale reactor. In this system, bacteria produce sulfuric, nitric, and organic acids that are important in process optimization. In addition, surfactants are produced because of the biological activity, enhancing arsenic leaching in the system [40].

2.2. Cadmium

Between all techniques studied, there is now a certain predominance towards the use of biotic techniques for its extraction, as opposed to other types of techniques such as soil washing, the use of activated and non-activated adsorbents, or absorption with zeolites.

2.2.1. Microorganism Mediated Biosynthesis of Cadmium Chalcogenides

This removal method consists in nanocrystals production of metal chalcogenides (CdS, CdSe, CdTe) based on the metabolic activity of living organisms [39]. Nanocrystal produced, such as Q-dots, are characterized to be semiconductors and, hence, they are suitable for developing cutting edge technologies including optical devices (optical storage, light-emitting diodes), solar energy conversion, or signaling of in vivo process as fluorescent labels [43]. In this process, bioremediation is based on the minimization of cadmium bioavailability because microorganisms do not decompose potential toxic elements.

Microorganisms cells surface contains chelates that allow cadmium to cross cell walls. Once inside the cell, cadmium is combined with S^{2-}, Se^{2-} or Te^{2-} to produce Q-dots chalcogenides. These particles can be separated from cellular solution by magnetic methods [43]. Biosynthesis is achieved in a modular reactor based on microorganisms trapped in mineral matrices that combine cadmium depletion and chalcogenides nanoparticles production.

The main disadvantage of this environmental-remediation technique is the difficulty with (i) Q-dots metal chalcogenides recovering, and (ii) biological manipulating. In this way, further research is needed in order to design bioreactors to minimize a possible environment contamination because of the low recovery of toxic Q-dots and to make biological manipulation safer.

2.2.2. Removal Using Dead Biomass

Bio-treatment for potential toxic elements removing has demonstrated to be an interesting environmental technology due to its low operating costs and high detoxification efficiency [44]. The use of dead biomass is proposed in order to overcome disadvantages detected in living biomass studies, such as the addition of nutrients and the maintenance of microorganisms, since they are sensitive to water quality (pH, oxidation-reduction state), but there are other problems of application in real scale, such as the difficulty of separating biomass from water after treatment and low mechanical resistance with long-term use. If the dead biomass is immobilized in a polymer matrix, its presence confers high resistance on chemical environments and provides other advantages such as efficient regeneration and ease of solution separation [45].

The effectiveness of immobilized dead *Bacillus drentensis* sp. use in a polymer matrix for the bio-sorption of metals in acid waters from underground mines has been studied [46]. Figure 7 shows the configuration of the pilot plant feasibility test. One of the most important issue to be considered is the immobilization matrix material, as it determines the mechanical strength, stiffness, and porosity characteristics [47]. Alginate, polyacrylamide, and polyvinyl alcohol has been used as bio-carrier packed layer but, recently, the use of polysulphone has provided good results for Pb (II) and Cu (II) removal, reporting maximum adsorption capacities of 0.3332 and 0.5598 mg/g, respectively. Hence, poysulphone material reduces the process costs [48].

Column tests were performed by adding different amounts of biomass to an artificial solution with different initial concentrations of several potential toxic elements (Cd, Cu, As, Zn, Pb, Fe) at pH 3.0. Polysulphone has a porous structure that adsorbs cadmium from the solid form at the boundary between the biomass and the matrix and/or the interior of the biomass. Elimination efficiency in lead and copper column tests is over 87%. The amount of lead absorbed per gram of bio-carrier is 1553 g, which shows the high sorption capacity of the elements used for the removal of potential toxic elements. In the pilot prototype, 80 tons of groundwater were contaminated with potential toxic elements for 40 days. The disposal capacity was, at least, 1098 tons per kilogram of bioculture.

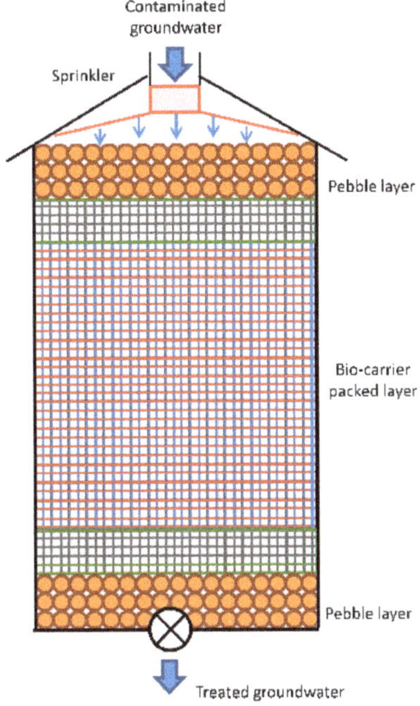

Figure 7. Configuration of the pilot plant feasibility test. Adapted from [39,46].

2.2.3. Removal by Nostoc Muscorum Cyanobacterium

There are many works on the removal of a single metal in aqueous solutions using cyanobacteria, but recent works such the one presented in [12] proposes the biosorption of potential toxic elements of a multicomponent system consisting of Cu (II), Zn (II), Pb (II) and Cd (II) present in an aqueous solution using *Nostoc muscorum*.

The choice of cyanobacteria is based on the fact that they are organisms that require a very low nutritional contribution and can survive under strict environmental conditions. The study was carried out using the statistically valid Plackett–Burman factorial design to better understand the biosorption of present metals. As main factors, the concentration levels of the four potential toxic elements (Cu (II), Zn (II), Pb (II) and Cd (II)) were chosen in the study.

The results show that Pb (II) is eliminated with high efficiency (96%), followed by Cu (II) (96%), Cd (II) (80%) and Zn (II) (71%). Lead shows an inhibitory effect on the removal of other metals, which is attributed to its small ionic radius in comparison with other metals. In addition, the values obtained from sorption show that the cyanobacteria *Nostoc muscorum* must be further studied because its potential in the disposal of contaminated waste water. Maximum elimination rates after 60 incubation hours for Pb (II), Cu (II), Cd (II) and Zn (II) are 96.3%, 96.42%, 80.04%, and 71.3% respectively. These results could be improved with a better process knowledge.

2.2.4. Adsorption with Activated and Non-Activated Carbons Prepared by Pyrolysis of Oily Sludges

The adsorption of potential toxic elements present in aqueous solutions with active carbon has been demonstrated to be an efficient and low-cost adsorption process [24]. The adsorbents, porous carbons obtained from oily solid pyrolysis wastes [49], were physically and chemically activated to improve their effectiveness in the removal of potential toxic elements from wastewater [50]. Carbon activation contributes to reduce the sludge volume and reduces the environmental costs and problems associated with the disposal of contaminated waste.

The adsorption of cadmium ions with active and non-active carbons has been achieved, reaching average adsorption rates between 78% and 98%. The most significant factors in cadmium removal were the initial metal concentration and the adsorbent dose used, 500 ml with 10 mg Cd/L and pH 6. Is has also been reported that oily sludge is a potential resource for the generation of carbonaceous adsorbents for the treatment of wastewater, but certain economic, technical and environmental aspects must be improved if it is to be used on a large scale.

2.2.5. Adsorption by Zeolites Synthesized from Rice Husks

Zeolites are effective for purifying water and gases [51–53]. There are numerous works on their production using coal ash [54–58]. However, there is hardly any research on the production of zeolites from the silica content of ash from rice husks. Investigations have been conducted on the use of synthesized zeolites from rice husk ash, the white part of which has a high content of silica that can be used as a precursor of zeolites for cadmium removal in aqueous solution.

The elimination capacity of Cd (II), was studied in [28] with a concentration between 50 and 500 mg/L. The maximum removal capacity of the cadmium ion of the Na-A zeolite is 736.38 mg/g and that of the Na-X zeolite is 684.46 mg/g. Maximum rates for zeolites Na-A and Na-X elimination are 98% and 97% respectively, for 9.2 g/L dosage and pH of 9.0. Na-A zeolite exhibits higher cadmium removal than Na-X zeolite because of the higher cation exchange capacity. These results were compared with those obtained with a commercial synthetic zeolite and an elimination capacity close to that of commercial zeolites was reported.

2.2.6. Sequential Soil Washing Technique

Soil washing is an ex-situ technique that is suitable for the removal of potential toxic elements from the ecosystem due to its high efficiency and profitability. A considerable amount of literature has been published on soil washing technique applied to potential toxic elements removal with a single reagent and/or in a single washing step [59–61]. In these studies, the fact of several metals coexisting in contaminated areas have pointed to the insufficiency of a single reagent and/or a single washing step as a successful contamination elimination.

A triplicate sequential soil wash test using a typical chelating agent (Na_2EDTA), an organic acid (oxalic acid) and a weak inorganic acid (phosphoric acid) has been carried out in [62] to evaluate how to clean soil contaminated with potential toxic elements (As and Cd) in the surroundings of a mining zone.

The results analysis shows that the primary components of arsenic and cadmium in the soil are arsenic residues (O-As) and exchangeable fraction, which represents 70% of cadmium and 60% of total arsenic. The different status as which potential toxic elements are present in soil makes a sequence of soil washing necessary. In Wei et al. [62], the optimum sequence of soil washing is phosphoric acid, oxalic acid, and Na_2EDTA with the following experimental conditions: Agitation speed, 150 rpm; liquid/solid ratio, 15/1; washing time, 30 min; room temperature; agent concentrations of 0.075 M for Na_2EDTA, 0.075 M for oxalic acid and 0.05 M for phosphoric acid, and the original pH of the washing agents. The results demonstrate that the tested six influencing factors except temperature had a marked effect on the efficiency of heavy metal removal.

The good elimination efficiency (41.9% for arsenic and 89.6% for cadmium) and the minimization of harmful effects of mobility and bioavailability of the present potential toxic elements were pointed to be under consideration for further study.

2.3. Mercury

In the case of mercury, three conventional and two innovative technologies are reported here. As conventional methods, thermal desorption, chemical extraction, and solidification/vitrification are explained. As innovative methods, both a biological technique based on the use of an aquatic plant,

Limnocharis flava—present in wetlands, and a physical/chemical low-cost technique based on the use of a natural adsorbent—a nanobiocomposite hydrogel modified with glutaraldehyde, are reported.

2.3.1. Thermal Treatment Based on Mercury High Temperature Thermal Desorption

Thermal desorption is a no destructive technology based on the volatilization of water in land also with the mercury it contains. It is a non-destructive technology because neither water molecules nor metals or organics pollutants are destroyed during operation. In addition, mercury is not usually oxidized in this system. It is characterized to be an in-situ or ex-situ treatment. In ex-situ case, land must be excavated and, afterwards, it must be transported to the equipment installation location [63].

As it is explained by the United States Environmental Protection Agency, land is introduced into a heated desorber, where both a carrier gas or a vacuum system can be used to transport the volatilized mercury produced by heating to the gas treatment located downstream. This treatment consists in a simple particle removal system, such as a baghouse or a wet scrubber, followed by a mercury condensation step and a sulfur-impregnated activated carbon adsorption to adsorb the residual mercury. Finally, residual gas is emitted to the air. After being treated by thermal desorption, the soil is cooled by spray water and it is returned back to the site [64].

The mains design parameters to be controlled in this technology are the bed (desorber) temperature profile and residence time of the carrier gas.

The most common equipment used to heat the soil in desorber are screw units and rotary driers. Screw units transport the land along the process unit while a hot oil or steam stream circulates through small pipes located between the screw hollows. On the other hand, rotary driers are slightly inclined cylinders that can be directly or indirectly fired. In the directly fired option, fire is applied directly upon soil surface while the indirectly fired option is characterized by heating the heater surface by the use of flames located in a reactor placed in contact with the rotary drier. In both screw units and rotary driers, land is heated to 320–500 °C. The technology has proven it can produce a final contaminant concentration below 2 mg/kg, in some cases even lower than 0.05 mg/kg [63,64]. After soil sieving, land particles are introduced into an extractor. HCl (acid) is used as extractant agent when mercury is mainly present as Hg(0), while organic compounds, such as acetone, hexane, methanol, or amines are used when mercury is found to be bonded to other organic structures.

In in-situ operation mode, thermal desorption method is applied simultaneously by heating the soil with resistors (usually 400–600 °C) and using a vacuum for a gas emission. The soil may be heated with a steam high-flow at high temperatures through injection wells.

2.3.2. Chemical Extraction Mercury Removement

Chemical extraction consists in the use of an extractant agent (acid or organic compound) to separate mercury from soil because the high solubility of the metal in the extractant agent. Thus, the volume of the hazardous waste that must be treated is reduced. It is an ex situ and a non-destructive method.

Before chemical extraction, a physical soil separation into different particle sizes is often used: Thus, fine and coarse fractions are usually treated by separately. This step may result in the enhancing of extraction kinetics [65].

Residence time of soil in the extraction unit is generally on the range between 10 and 40 minutes, but it depends on the soil type and mercury concentration. The soil-extractant mixture is continuously pumped out of the extractor tank, and the soil and extractant are separated using hydro cyclones. Afterwards, soil is rinsed with water to remove entrained extractant agent and mercury. Finally, extraction solution and rinse water are regenerated by using NaOH and flocculants to remove mercury. Thus, the extractant agent is reformed and mercury is concentrated in a precipitate potentially suitable to be recovered.

Chemical extraction efficiency is usually over 90% and it could reach 99% if metals are not strongly bonded to other structures.

2.3.3. In situ Solidification/Vitrification Method

The objective of this method is to reduce the mercury mobility in the environment in both chemical and physical meanings by trapping it within its "host" medium. For the purpose of mercury remediating, an electric current is applied in order to melt soil at very high temperatures (1600–2000 °C). Consequently, once the soil is cooled down, mercury is immobilized in a vitrified glass and crystalline mass. After applying heat, water vapour, some mercury amount previously volatilized, and other compounds from organics pyrolysis are captured in the off-gas collection hood. Afterwards, off-gas is treated by removing particulates, the volatilized mercury, and other pollutants from the gas. Finally, vitrified soil is found to be chemically stable, leach-resistant, glass, and crystalline material similar to obsidian or basalt rock.

Soil composition, especially humidity, determines whether this technology is suitable to be applied or not. If silt and clay contents are high enough, water release between soil particles can be very difficult. In addition, a high soil porosity can make necessary a previous land compaction.. Additionally, humidity reduces the technology efficiency [65].

The main advantage of the solidification/vitrification method is the high mercury "removal", over 99%, but it has several disadvantages, such as (1) the land depth in which the treatment is suitable to be applied is small in comparison with other technologies (2) high cost, and (3) the land reutilization is limited as soil characteristics are completely modified [66].

2.3.4. Removal of Mercury from Gold Mine Effluents Using Limnocharis Flava In Constructed Wetlands

The mechanisms of potential toxic elements removal in horizontal artificial wetlands are complex and include binding to sediments and soils, precipitation and co-precipitation of metals as insoluble salts, plant absorption, and, to a lesser extent, microbial metabolism [67]. A schematic view of experimental pilot plant is shown in Figure 8.

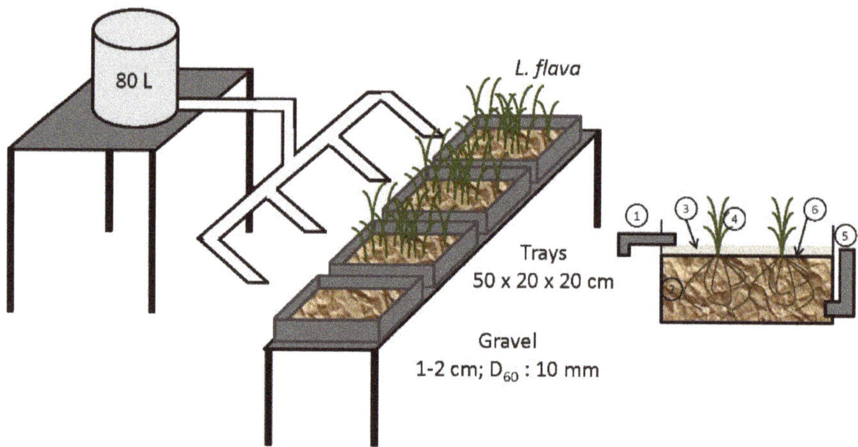

Figure 8. Schematic view of experimental setup in the pilot-scale constructed wetland system. Lateral section of planted systems *Limnocharis flava*: 1. Supply pipe; 2. Distribution area (coarse gravel); 3. Gravel 1-2 cm; D60: 10 mm; 4. *Limnocharis flava*; 5. Adjustable output pipe to control water level; 6. Water level. Adapted from [68].

The potential of the *Limnocharis flava* aquatic plant for the phytoremediation of water contaminated with mercury in a constructed wetland with horizontal underground flow was evaluated in [62] through a pilot experiment using an effluent from a gold mine in El Alacrán (Colombia) with a mercury content of 0.11 ± 0.03 mg/mL enriched with $HgNO_3$ (1.50 ± 0.09 mg/mL) [68].

The results obtained show a mercury concentration reduction only in the wetlands in which *Limnocharis flava* was used. The most important variable for mercury removing is the exposure time. Mercury concentration was reduced from 1.50 ± 0.13 to 0.15 ± 0.04 mg/mL after 30 days. With exposure time optimization, removal yield increases up to nine times relative to the control experiment.

The *Limnocharis flava* macrophyte has a high potential for the phytoremediation of water contaminated with mercury, reducing its concentration in contaminated water by 90%. The construction of this type of wetland is a viable option for the treatment of effluents from gold mines rich in mercury. Construction of this system is planned in El Alacrán (Colombia) in order to optimize it and evaluate its performance for several years before its complete construction.

2.3.5. Removal of Hg (II) Ions from An Aqueous Environment Using Glutaraldehyde Crosslinked Nanobiocomposite Hydrogel Modified by TETA and β-cyclodextrin

A study of Hg (II) ions removal from aqueous solutions using a new nanobiocomposite hydrogel modified by triethylene tetramine (TETA) and β-cyclodextrin has been reported to be an alternative and cost-effective technique for the remediation of mining wastewater [69].

The nanobiocomposite hydrogel is composed of chitosan, vegetable gum, montmorillonite, and zinc oxide nanoparticles crosslinked by glutaraldehyde and modified by TETA and β-cyclodextrin for an effective removal of Hg (II) ions from wastewater. This reduction is based on the availability of amine and hydroxyl residues for enhancing their interaction with glutaraldehyde.

The maximum elimination values of Hg (II) achieved by TETA and β-cyclodextrin are 97.9% and 70.1%, respectively. In absorption case, results reported by TETA and β-cyclodextrin are 407.9 mg/g and 292.1 mg/g, respectively. The main experimental used parameters are: pH, 6.0; contact time, 6 h; initial mercury concentration, 250 mg/L; and biomass dosage, 0.6 g/L.

Results analysis reported a higher adsorption of Hg (II) yield using TETA-cyclodextrin because of its greater surface roughness, lower compaction, and greater availability of functional groups. In addition, maximum mercury removal from mine wastewater was set at 80%. This result was obtained with a maximum column height of 12 cm, a minimum flow rate of 1 ml/min, and maximum dilution of 50%.

3. Discussion on the Systematic Review

All the techniques studied present adequate results for their application at a pilot scale. Some of them have pilot-scale projects that support the results obtained in the laboratory. The studies at pilot scale are necessary in order to identify the adjustments for their real-scale implementation. Table 4 shows a summary of the main studies considered in this work.

Table 4. Summary table of the reviewed studies.

Metal/Metalloid	Treatment	Articles	References
Arsenic	Adsorption	2	[34,35]
	Electrocoagulation	1	[41]
	Precipitation	1	[36]
	Bioleaching	5	[38–42]
Cadmium	Adsorption	11	[24,28,50–58]
	Sequential soil washing	4	[59–62]
	Biosynthesis of metal chalcogenides	7	[12,39,43–47]
Mercury	Phytoremediation	2	[62,67]
	Nanobiocomposite hydrogel	1	[69]
	Others	4	[63–66]

Between the techniques here explained, very efficient methodologies for arsenic, cadmium, and mercury removal can be found. However, in each case, not only a specific determination as to the applicability of the principles and provisions of each technology must be made, but also economic issue is an important factor to determine final decision. To make a closer discussion about this issue, Table 5 reports a costs summary of different remediation technologies for metals removing based on the USEPA studies [63]. The methodology followed in USEPA studies allows to classify the technologies into two scenarios: large and small sites. Hence, the scale up effect can be observed.

Table 5. Costs summary ($€/m^3$) of different remediation technologies. Source: [63].

Technology	Large Site Cost	Small Site Cost	Average Cost	In-Situ/Ex-Situ
Chemical extraction	320	1468	894	"ex situ"
Phytoremediation	214	1312	763	"ex situ"
Adsorption	303	405	354	"ex situ"
Bioleaching	105	145	125	"ex situ"
Solidification	58	178	118	"in situ"
Soil washing	62	166	114	"ex situ"
Thermal desorption	63	148	105	"ex situ"
Electrocoagulation	83	104	93	"in situ"
Precipitation	76	91	84	"ex situ"
Bioremediation	27	89	58	"in situ"

From inspection of Table 5, as it was expected, ex situ treatments present, on average, higher associate costs than in situ ones. One of the main reasons of this finding is associate to the costs implied in land excavation and soil transportation. For the sake of comparison, technologies can be classified by treatments groups as thermal, physical/chemical, and biological treatments. A closer observation of Table 4 shows that bioremediation emerges as the best interesting methodology to metals remove. By contrast, phytoremediation is observed to be one of the most expensive methods. Most of physical/chemical treatments costs are situated in a compromise cost place (e.g., soil washing, adsorption, electrocoagulation, precipitation) but chemical extraction is catalogued to be the most expensive treatment because of the need of a specific treatment for the extracting agent to be used afterwards. Finally, thermal methodologies (e.g., solidification, thermal desorption) can be interesting from an economic point of view, when application is possible, because the in situ application.

3.1. Arsenic

In arsenic removal, it is important to determine the arsenic's speciation, since As (III) is more toxic, more mobile, and more difficult to eliminate by using many of the treatment technologies previously explained than As (V) is.

In the systematic review carried out in the previous section, two predominant lines have been evaluated; the most common techniques are those based on the use of iron compounds. However, those that eliminate arsenic through the use of bacteria have also been found interesting for mining contaminated waters.

There are few studies that compare co-precipitation and adsorption by iron minerals. Also, few studies evaluate the design of arsenic's mechanism of natural attenuation.

If adsorption-co-precipitation rates are observed, the rate of arsenic adsorption is found to be 0.90 mM higher than arsenic-ferrihydrite co-precipitation. No significant differences are found with schwertmannite use. The adsorption with goethite reports better results than co-precipitation. The reason is currently still unknown, and so more research is required. Arsenic adsorption by schwertmannite seems to be more effective than for the other iron minerals [34]. The As adsorption by schwertmannite included exchange reactions between sulfate and As (V) [70].

The elimination of As (III) by electrocoagulation with iron electrodes depends on Fe (II) oxidation and Fe (III) precipitation, so it also depends on the operating parameters affecting Fe (II) oxidation, such as current intensity, pH, initial concentration of arsenic, or the presence of other ions in the aqueous solution [41].

Another disadvantage of arsenic removal by electrocoagulation is the only partial oxidation of Fe (II) to Fe (III). In addition, the presence of phosphates and silicates adversely affects arsenic removal in Fe (II) and Fe (III) co-precipitation, similar to electrocoagulation systems with iron electrodes [41]. However, one advantage of electrocoagulation with iron electrodes is that As (III) conversion to As (V) can be achieved without the use of any chemical reagent.

Arsenic [III and V] removal by using iron rich sludge can be achieved by a sludge composition of goethite and calcite, according to X-ray diffraction (XRD) analysis. It also contains other iron oxides [37]. The possibility of reuse the adsorbent in several consecutive cycles has been also studied. The adsorbent adsorption-desorption capacity decreased by about 10% in each cycle. As a result, iron-rich sludge technology can be used effectively at an average of three times [37].

Adsorption systems based on agricultural and natural fixed-beds have been successfully used for zinc and nickel removal. In this case, a low effectivity was detected in water treatment from electroplating using sugarcane bagasse, probably due to electrostatic repulsion forces between the cationic surface of the bagasse and the metal ions [71]. However, when coconut shell was used, this situation was not detected because it was negatively surface charged at the work pH of 6.5. As a consequence, potential toxic elements with positive charge were removed successfully because of attractive forces between the sorbent and the sorbate [38].

Removal of arsenic present in acid solutions by biogenic ferric precipitate has also been carried out, but the stability or storage of arsenic-containing precipitates was not studied [38]. The main inconvenience in using this technique is the discrepancy observed between the results for the arsenic present in the liquid phase and for that present in the solid phase due to an overestimation of the total concentration of arsenic in the effluent [39]. However, this inconvenience could be avoided by the use of a graphite furnace or inductively coupled plasma emission spectrometry (ICP-ES) since the detection limits are lower [39].

Finally, bioleaching of arsenic using *Acidithiobacillus thiooxidans* make solubilization of potential toxic elements contained in the sulfides of the sediments possible. In addition, solids concentration was observed to be inversely proportional to the elimination efficiency and directly proportional to the time at which the leaching begins [72]. However, it was found some corrosion problems of mine tailings produced by bacterium [72].

3.2. Cadmium

Removal of cadmium from a contaminated matrix can be carried out through a variety of biotic or abiotic techniques. These techniques must be studied in each particular situation because their high sensitivity to environment conditions.

Conventional techniques for the treatment of cadmium contaminated water are not characterized to be economically profitable for small and medium sized industries (see, for example, chemical extraction for small sites in Table 4). This is the reason why innovative techniques with natural materials are being studied for decontamination.

A very little amount of literature has been published on the metal chalcogenides biosynthesis using encapsulated microorganisms for cadmium removal. However, findings reported by Vena et al. [43] suggest this technique to be interesting in this field, for its ability to reuse waste as functional nanomaterials, so further research must be carried on.

The results obtained from the extraction of potential toxic elements with *Bacillus drentensis* sp. indicate the convenience of using polysulphones as a bio-carrier packed layer adsorbent because its high porosity and low cost. As in the metal chalcogenides case, there is little amount of literature published on the item of cadmium removal with *Bacillus drentensis* sp. [46]. The findings of this study suggest that cadmium removal by *Bacillus drentensis* sp. could be a good solution for water treatments. In this sense, further research in the field is necessary.

The kinetics of elimination of potential toxic elements with the bacterium *Nostoc muscorum* conforms to the second-order kinctics described in the existing literature. [12] and given as follows:

$$\frac{1}{Q_t} = \frac{1}{k_2 Q_e^2} + \frac{1}{Q_e} t \tag{1}$$

where k_2 (g^{-1} mg^{-1} min^{-1}) is the rate constant of the pseudo second-order sorption and Q_e and Q_t denote the amounts of metal ions sorbed at equilibrium and at time t (mg g^{-1}), respectively. The metal removal system in a multicomponent system is complex and unique for each combination of metal concentration and microbial strain employed in the process [12]. For this reason, the study of one of these metals and microorganisms cannot predict its elimination in a multicomponent system [73].

Removal of cadmium present in aqueous solutions with carbonaceous adsorbents are more efficient when activated carbons (96%) are used than in the case of non-activated carbon adsorbents used (20%) [50]. The values predicted by the Taguchi strongly differ from those obtained in experiments performed for both adsorbents, with an error of 11% for non-activated carbon and 7% for activated carbon. Therefore, the confirmatory experiments were carried out under the optimum expected conditions and the percentage of cadmium adsorption obtained in the confirmatory experiments was adjusted to the expected amounts. The preparation of porous carbons from oily solid waste contributes to reduction of sludge volume, costs, and environmental problems associated with its elimination. Adsorption cadmium efficiencies up to 98.28% can be obtained with active carbons use.

Cadmium adsorption by zeolites synthesized from rice husk eliminates Cd (II) by varying its concentration between 50 and 500 mg/L, while temperature (30 °C), solution pH (7.0) and cadmium dose (2.0 g/L) are keeping constant. Results show a cadmium removal efficiency of 97–98% after the previous treatment. Both Na-A and Na-X zeolite present similar results of that obtained with synthetic zeolites, such as 3A [38]. However, adsorption capacity of the Na-A and Na-X zeolites is much higher than that of other low-cost adsorbents such as fly ash [74], Egyptian kaolin clay [75], or petroleum shale ash [76]. As a result, cadmium removal by zeolites synthesized from rice rusk are presented to be a promising technique for cadmium sustainable reducing.

Finally, the results obtained for removing cadmium by soil sequential washing suggest that cadmium bioavailability in soil must be optimized to enhance cadmium removal. Some studies point that metals increasing solubility due to washing can be attributed to the dissolution of the remaining fraction and the transformation of other metal fractions. However, this result differs from previous studies in which amorphous Fe/Al oxide dissolution increases the bioavailability of cadmium in soils [77]. Cadmium removal efficiency near 90% suggests soil washing to be a promising technique to be optimized with further investigation.

3.3. Mercury

Five technologies for mercury removing were commented on in previous sections: Three could be classified as *conventional technologies* (thermal desorption, chemical extraction, and solidification/ vitrification) as their used has been more broadly tested. In contrast, two *innovative techniques* were also commented on: The use of the *Limnocharis flava* macrophyte in constructed wetlands for the treatment of gold mine effluents, and the use of a new hydrogel nanobiocomposite.

Thermal desorption is considered to be a good method for soils remediation as the final metal concentration can reach a value even lower than 0.05 mg/kg. However, thermal techniques are not usually the most economic ones between those possible to be used. As it is shown in Table 4, average thermal desorption costs can be established at 105 €/m³ (landfilling of activated carbon particles included).

Chemical extraction is characterized by a high efficiency of Hg removal (over 90%) but this efficiency is very sensitive on the different Hg combinations with other compounds present in soil. In addition, from Table 4 it can be concluded that chemical extraction is a very expensive technology.

In this sense, using chemical extraction is only reserved for situations where other technology cannot possibly be applied.

Solidification/vitrification presents a very high efficiency for Hg removing, as this is usually over 99% of initial Hg concentration, but the main disadvantage it presents is the change of the soil characteristics that can hinder some further uses.

Constructed wetlands are considered a viable alternative because of their low maintenance and operation costs. In addition, mercury concentrations are reduced by 90% if *Limnocharis flava* is used. These results are similar to those obtained by other authors who used species such as *Typha domingensis* [78], *Myriophylhum aquaticum*, *Ludwigna palustris*, and *Mentha* aquatic, but other authors report smaller efficiencies: 49% [79], 78,2% [80], and between 25% and 50% [81].

Mercury absorption by aquatic plants is affected by various factors, such as pH, temperature, mass flow, solar radiation, presence of chlorides, sulphates and phosphates, dissolved oxygen, biological oxygen demand, total organic carbon, total dissolved solids, and total suspended solids [68]. In the analyzed article [68], only the concentration of mercury at which *Limnocharis flava* is exposed and the exposure time have been studied in detail. The transfer coefficient (TC) for background mercury concentration (0.04 ± 0.01 μg/g) was 0.02 mL/g, and a maximum value of 53.44 mL/g was reached after an exposure of 30 days, corresponding to a concentration of Hg of 8.02 ± 1.14 μg/g in the aerial part of the plant. On the other hand, using *Typha domingensis*, a TC of $7{,}751 \pm 570$ mL/g and an initial concentration of Hg of 9.0 ± 0.4 mg Hg/L were obtained after 27 days [78].

The effectiveness of the removal of Hg (II) ions from the aqueous medium using a nanobiocomposite hydrogel crosslinked with glutaraldehyde modified by triethylene tetramine (TETA) and β-cyclodextrin through the realization of six cycles of metal removal has been demonstrated. The effectiveness decreases as the cycles are repeated because the surface of the biosorbent becomes damaged. It is necessary to carry out experiments on the biosorbents reuse and regeneration, so that they can be used on an industrial scale in a sustainable way [69]. Finally, although bioremediation presents a lower efficiency than conventional techniques, the reduction in process costs (e.g., 118 €/m^3 for solidification vs. 58 €/m^3 for bioremediation) makes it more attractive for implementation and further investigation. The new nanobiocomposite hydrogel modified by TETA and β-cyclodextrin has proved to be an ideal and profitable agent for the remediation of contaminated water. When the hydrogel is modified with TETA-cyclodextrin, it has better elimination efficiencies than when modified with β-cyclodextrin. The optimization of these removal characteristics must be further investigated because of its good results.

4. Future Prospects

In order to guarantee the sustainability of human life on Earth, research on applications of waste treatments are mandatory. In particular, the EU is especially concerned about the importance of the transition towards a circular economy model and, thus, it has been developing a legislative framework for promoting it. Not only environmental consequences are involved, but a significant impact on the EU members' economies must also be considered. Such a transition presents an opportunity to transform the economy and generate new and sustainable competitive advantages for Europe [82].

In a number of end-uses, waste minimization involves using renewable material and energy [83]. Moreover, as [84] points out, by promoting the use of energy efficient equipment and renewable energy technologies, and also adopting measures for reduction of carbon footprint, the concern for hazardous wastes is also addressed in direction of long-term sustainability. Following the aforementioned, this paper intends to bridge the gap between the main techniques for arsenic, cadmium, and mercury disposal from hazardous wastes and locally available energy sources and sustainability.

In this sense, novel techniques involving biological treatments seem promising. Apart from the bioleaching or biosynthesis described in the previous section, microbial fuel cells (MFCs) and microbial electrolysis cells (MECs) can also be applied as effective treatments or as a complementary treatment with success.

A MEC is a device capable of converting the chemical energy contained in wastewater into hydrogen while reducing its organic load with an input of electricity [85]. In a MEC, electrochemically active bacteria oxidize organic matter and generate CO_2, electrons and protons in such a way the bacteria transfer the electrons to the anode and the protons are released to the solution. The electrons then travel through a wire to a cathode and combine with the free protons in solution thanks to an externally supplied voltage (higher than 0.2 V) and biologically assisted conditions [86].

On the other hand, MFCs can be defined as devices able to transform chemical to electrical energy via electrochemical reactions involving biochemical pathways [87]. To do that, these devices use electroactive microbes to degrade organics and produce electricity. These microbes are more sustainable and more durable compared to selective enzymes used in EFCs (enzymatic fuel cells), although they have worse electrochemical catalytic performance [87]. To produce electricity, a bacterial metabolic activity (the microorganisms donate electrons to an anode while oxidizing organic/inorganic waste) is produced in the anodic compartment while, separated by a membrane, electron acceptor conditions are shown in the cathode. In the oxidation reaction HCO_3^-, H^+, or CH_4 can be produced as co-products [88,89].

MECs and MFCs have shown a great potential for complementing costly wastewater treatment systems, being self-sustainable or even having a net positive energy output while pollutants are removed. Thus, interest is MECs and MFCs has been growing exponentially since the beginning of the 21st century although very few industrial applications exist nowadays due to the complexity of the maintenance of the proper conditions for the microbial population and the very low current rates. Although this technology is still not at an industrial-scale phase, some authors [90–93] point out its capacity to remove a mixed concentration of potential toxic elements from a contaminated influent, as the metal electroplate in the cathode chamber, with the advantage that they have very low energy needs. Moreover, as cells operate as MFCs, they can have a positive energy balance or, in the case of MECs, they can produce hydrogen or methane that can be used as bio-fuels [85,94]. Nevertheless, some large-scale applications are shown in [95,96].

These systems can be applied not only to remove organic waste from waste waters, but they have also shown good performance for electrochemical struvite precipitation [97], recovery of cobalt [90,91], or other heavy metals such as Cu or Ni [92].

In [96] a costs and revenues analysis both for MECs and MFCs is provided. The authors state that for a MEC, the revenue per kg of COD (chemical oxygen demand) removed can be considered constant if constantly applied voltage and constant cathodic Coulombic efficiency remained, and the costs decreased with decreasing internal resistance. For MFCs, however, the costs decrease with increasing current density. However, the revenues decrease with increasing current density for constant internal resistance. Thus, the highest revenues are obtained at the lowest internal resistance as the cell voltage is higher compared to higher internal resistances. In general terms, authors in [96] estimate the revenues of both systems in 0.35 €/kg (of COD removed) for the treatment of wastewater.

5. Conclusions

In this work, the application of the main hazardous waste management techniques in mining operations and in dumping sites has been reviewed. The systematic review has focused on the elimination of As, Cd, and Hg which are of particular interest of several European countries' directives, especially in Spain.

The development of biological methods is fundamentally due to the fact that conventional methods have higher costs, generate sludge, and may be inefficient. The use of biomass is increasingly being studied because it is an ecological, efficient, and technically feasible technique.

There is still hard work to do in the management of hazardous waste. What the treatments described have in common is that they are expensive, need long treatment times, and demand high rates of energy and/or chemical products. Moreover, they entail lots of human, machine, and energy resources in extraction and transport and in the restoration of the contaminated land fields.

In particular, future research lines must be focused not only on the improvement of the treatments' removal performance, but also on their energy requirements, as this can be one of the most limiting factors in carrying out treatment in a feasible way. Thus, apart from the improvement of current treatments, engineers and scientists must aim to remove potential toxic elements and other poisonous substances directly from leaches, as they constitute the main contamination vector. This objective is particularly relevant to the case of dumping sites, where leaches are extracted with efficacy thanks to drainage systems, but sophisticated post-treatments, energy-costly drying processes, and special storage management are needed.

The prospective analysis points out that biological treatments seem to be promising because of their effectiveness and their low energy consumption. In this item, the use of microorganisms for effluents decontamination is a promising and reversible technique that needs improvements—for example, in reactors design—for its further implantation. Moreover, the integration of MFCs or MECs as treatments to remove potential toxic elements may add other advantages such as the production of bio-fuels or positive energy balances.

Author Contributions: This paper is the result of the joint work of the authors. Conceptualization, A.G.-M. and A.B.-S.; Review, R.T.-F. and M.d.S.-M.; Analysis, R.L. and M.d.S.-M; Investigation, R.T.-F. and A.G.-M.; Methodology, A.B.-S., R.L. and M.d.S.-M.; Supervision, A.B.-S. and A.G.-M.; Validation, A.B.-S. and R.L.; Writing—original draft, A.G.-M. and M.d.S.-M.; Writing—review & editing, A.G.-M.

Funding: The APC was funded by Laboratorio de Inspección Técnica de la Escuela de Minas (LITEM), Universidad de León (Spain) and MDPI.

Acknowledgments: The authors want to thank all contributors to the project, and to the editors and reviewers for their valuable comments to increase the overall quality of the manuscript.

Conflicts of Interest: The authors declare no conflict of interest. The founding sponsors had no role in the design of the study; in the collection, analyses, or interpretation of data; in the writing of the manuscript, and in the decision to publish the results.

References

1. RED ESPAÑOLA DE COMPOSTAJE. *Residuos agrícolas*; Ediciones Paraninfo: Madrid, Spain, 2014; ISBN 978-84-8476-698-8.
2. European Commission 7th Environment Action Programme. Available online: http://ec.europa.eu/environment/action-programme/index.htm (accessed on 1 June 2018).
3. B.O.E. *Ley 22/2011, de 28 de Julio, de Residuos y Suelos Contaminados*; B.O.E.: Madrid, Spain, 2011.
4. EUROSTAT Total Amount of Waste Generated in EU. Available online: http://ec.europa.eu/eurostat/tgm/mapToolClosed.do?tab=map&init=1&plugin=1&language=en&pcode=ten00106&toolbox=types (accessed on 1 June 2018).
5. Waste Statistics—Statistics Explained. Available online: https://ec.europa.eu/eurostat/statistics-explained/index.php?title=Waste_statistics (accessed on 31 May 2019).
6. EUROSTAT Generation of Waste by Economic Activity. Available online: http://ec.europa.eu/eurostat/tgm/table.do?tab=table&plugin=1&language=en&pcode=ten00106 (accessed on 1 June 2018).
7. Water Glossary—European Environment Agency. Available online: https://www.eea.europa.eu/themes/water/glossary/ (accessed on 31 May 2019).
8. Casiot, C.; Bruneel, O.; Personné, J.-C.; Leblanc, M.; Elbaz-Poulichet, F. Arsenic oxidation and bioaccumulation by the acidophilic protozoan, Euglena mutabilis, in acid mine drainage (Carnoulès, France). *Sci. Total Environ.* **2004**, *320*, 259–267. [CrossRef] [PubMed]
9. Branch, G. Arsenic Trioxide and Underground Issues at Giant Mine. Available online: https://www.aadnc-aandc.gc.ca/eng/1100100027413/1100100027417 (accessed on 3 November 2018).
10. Prieto, M.J.; González, R.C.A.; Román, G.A.D.; Prieto, G.F. Contaminación y fitotoxicidad en plantas por metales pesados provenientes de suelos y agua. *Trop. Subtrop. Agroecosyst.* **2009**, *10*, 29–44.
11. Navarro-Aviñó, J.; Alonso, A.I.; López-Moya, J. Aspectos bioquímicos y genéticos de la tolerancia y acumulación de metales pesados en plantas. *Ecosistemas* **2007**, *16*, 10–25.

12. Roy, A.S.; Hazarika, J.; Manikandan, N.A.; Pakshirajan, K.; Syiem, M.B. Heavy Metal Removal from Multicomponent System by the Cyanobacterium: Kinetics and Interaction Study. *Appl. Biochem. Biotechnol.* **2015**, *175*, 3863–3874. [CrossRef] [PubMed]

13. Khalid, S.; Shahid, M.; Niazi, N.K.; Murtaza, B.; Bibi, I.; Dumat, C. A comparison of technologies for remediation of heavy metal contaminated soils. *J. Geochem. Explor.* **2017**, *182*, 247–268. [CrossRef]

14. WHO International Programme on Chemical Safety: Ten Chemicals of Major Public Health Concern. Available online: http://www.who.int/ipcs/assessment/public_health/chemicals_phc/en/ (accessed on 1 June 2018).

15. European Commission. *Science for Environment Policy. Soil Contamination: Impacts on Human Health*; European Commission: Brussels, Belgium, 2013; pp. 1–29.

16. Volke, T.; Velasco, J. *Tecnologías de remediación para suelos contaminados*; Instituto Nacional de Ecología: Mexico, 2002.

17. García-Carmona, M.; Romero-Freire, A.; Sierra Aragón, M.; Martínez Garzón, F.J.; Martín Peinado, F.J. Evaluation of remediation techniques in soils affected by residual contamination with heavy metals and arsenic. *J. Environ. Manag.* **2017**, *191*, 228–236. [CrossRef] [PubMed]

18. Vullo, D.L. Microorganismos y metales pesados: Una interacción en beneficio del medio ambiente. *Química Viva* **2003**, *2*, 93–104.

19. Swannell, R.P.J.; Lee, K.; Mcdonagh, M. Field Evaluations of Marine Oil Spill Bioremediation. *Microbiol. Rev.* **1996**, *60*, 342–365.

20. Igiri, B.E.; Okoduwa, S.I.R.; Idoko, G.O.; Akabuogu, E.P.; Adeyi, A.O.; Ejiogu, I.K. Toxicity and Bioremediation of Heavy Metals Contaminated Ecosystem from Tannery Wastewater: A Review. *J. Toxicol.* **2018**, *2018*, 2568038. [CrossRef]

21. Bioremediation of Polluted Waters Using Microorganisms | IntechOpen. Available online: https://www.intechopen.com/books/advances-in-bioremediation-of-wastewater-and-polluted-soil/bioremediation-of-polluted-waters-using-microorganisms (accessed on 31 May 2019).

22. Cortón, E.; Viale, A. Solucionando grandes problemas ambientales con la ayuda de pequeños amigos: Las técnicas de biorremediación. *Rev. Ecosistemas* **2006**, *15*. Available online: https://www.revistaecosistemas.net/index.php/ecosistemas/article/view/499 (accessed on 1 November 2018).

23. Moreno, B. Estrategias de recuperación de suelos contaminados por tricloroetileno basadas en el uso de vermicompost de alperujo y especies vegetales con potencial fitorremediador. Ph.D. Thesis, Universidad de Granada, Granada, Spain, 2009.

24. Song, B.; Zeng, G.; Gong, J.; Liang, J.; Xu, P.; Liu, Z.; Zhang, Y.; Zhang, C.; Cheng, M.; Liu, Y.; et al. Evaluation methods for assessing effectiveness of in situ remediation of soil and sediment contaminated with organic pollutants and heavy metals. *Environ. Int.* **2017**, *105*, 43–55. [CrossRef] [PubMed]

25. González, D.R. *Utilización de lodos rojos de bauxita en la contención e inactivación de residuos tóxicos y peligrosos*; Universidade de Santiago de Compostela: Santiago, Spain, 2008. Available online: https://minerva.usc.es/xmlui/handle/10347/2492 (accessed on 14 June 2019).

26. WHO. *Guidelines for Drinking-Water Quality*; WHO: Geneva, Switzerland, 2011.

27. Greenwood, N.N.; Earnshaw, A. *Chemistry of the Elements*; Elsevier: Amsterdam, The Netherlands, 2012; ISBN 978-0-08-050109-3.

28. Santasnachok, C.; Kurniawan, W.; Hinode, H. The use of synthesized zeolites from power plant rice husk ash obtained from Thailand as adsorbent for cadmium contamination removal from zinc mining. *J. Environ. Chem. Eng.* **2015**, *3*, 2115–2126. [CrossRef]

29. Baird, C.; Cann, M. *Química Ambiental*; Reverté: Barcelona, Spain, 2015; ISBN 978-84-291-7915-6.

30. Navarrete López, R.E. Riesgo por cadmio en un tiradero abandonado en el municipio de Sahuayo, Michoacán. Ph.D. Thesis, Tesis de Maestría en Ciencias de la Salud Ambiental, CUCBA, Universidad de Guadalajara, Zapopan, México, 2004.

31. Chow, S.-C. Bioavailability and Bioequivalence in Drug Development. *Wiley Interdiscip. Rev. Comput. Stat.* **2014**, *6*, 304–312. [CrossRef] [PubMed]

32. WHO. *Guidelines for Drinking-Water Quality*; WHO: Geneva, Switzerland, 2004.

33. Beltrán, G.; Óscar, A. Revisiones sistemáticas de la literatura. *Revista colombiana de gastroenterología* **2005**, *20*, 60–69.

34. Park, J.H.; Han, Y.-S.; Ahn, J.S. Comparison of arsenic co-precipitation and adsorption by iron minerals and the mechanism of arsenic natural attenuation in a mine stream. *Water Res.* **2016**, *106*, 295–303. [CrossRef] [PubMed]

35. Han, J.; Katz, L.E. Capturing the variable reactivity of goethites in surface complexation modeling by correlating model parameters with specific surface area. *Geochimica et Cosmochimica Acta* **2019**, *244*, 248–263. [CrossRef]

36. Nazari, A.M.; Radzinski, R.; Ghahreman, A. Review of arsenic metallurgy: Treatment of arsenical minerals and the immobilization of arsenic. *Hydrometallurgy* **2017**, *174*, 258–281. [CrossRef]

37. Yang, J.-S.; Kim, Y.-S.; Park, S.-M.; Baek, K. Removal of As(III) and As(V) using iron-rich sludge produced from coal mine drainage treatment plant. *Environ. Sci. Pollut. Res.* **2014**, *21*, 10878–10889. [CrossRef]

38. Acheampong, M.A.; Lens, P.N.L. Treatment of gold mining effluent in pilot fixed bed sorption system. *Hydrometallurgy* **2014**, *141*, 1–7. [CrossRef]

39. Ahoranta, S.H.; Kokko, M.E.; Papirio, S.; Özkaya, B.; Puhakka, J.A. Arsenic removal from acidic solutions with biogenic ferric precipitates. *J. Hazard. Mater.* **2016**, *306*, 124–132. [CrossRef]

40. United States Environmental Protection Agency. *Arsenic Treatment Technologies for Soil, Waste, and Water*; United States Environmental Protection Agency: Washington, DC, USA, 2008; pp. 1–132.

41. Banerji, T.; Chaudhari, S. Arsenic removal from drinking water by electrocoagulation using iron electrodes- an understanding of the process parameters. *J. Environ. Chem. Eng.* **2016**, *4*, 3990–4000. [CrossRef]

42. Nicomel, N.R.; Leus, K.; Folens, K.; Van Der Voort, P.; Du Laing, G. Technologies for Arsenic Removal from Water: Current Status and Future Perspectives. *Int. J. Environ. Res. Public Health* **2016**, *13*, 62. [CrossRef] [PubMed]

43. Vena, M.P.; Jobbágy, M.; Bilmes, S.A. Microorganism mediated biosynthesis of metal chalcogenides; a powerful tool to transform toxic effluents into functional nanomaterials. *Sci. Total Environ.* **2016**, *565*, 804–810. [CrossRef] [PubMed]

44. Song, H.; Yim, G.-J.; Ji, S.-W.; Neculita, C.M.; Hwang, T. Pilot-scale passive bioreactors for the treatment of acid mine drainage: Efficiency of mushroom compost vs. mixed substrates for metal removal. *J. Environ. Manag.* **2012**, *111*, 150–158. [CrossRef] [PubMed]

45. Zawierucha, I.; Kozlowski, C.; Malina, G. Immobilized materials for removal of toxic metal ions from surface/groundwaters and aqueous waste streams. *Environ. Sci. Process. Impacts* **2016**, *18*, 429–444. [CrossRef] [PubMed]

46. Kim, I.; Lee, M.; Wang, S. Heavy metal removal in groundwater originating from acid mine drainage using dead *Bacillus drentensis* sp. immobilized in polysulfone polymer. *J. Environ. Manag.* **2014**, *146*, 568–574. [CrossRef]

47. El-Naas, M.H.; Al-Muhtaseb, S.A.; Makhlouf, S. Biodegradation of phenol by Pseudomonas putida immobilized in polyvinyl alcohol (PVA) gel. *J. Hazard. Mater.* **2009**, *164*, 720–725. [CrossRef]

48. Seo, H.; Lee, M.; Wang, S. Equilibrium and Kinetic Studies of the Biosorption of Dissolved Metals on *Bacillus drentensis* Immobilized in Biocarrier Beads. *Environ. Eng. Res.* **2013**, *18*, 45–53. [CrossRef]

49. Hu, G.; Li, J.; Zeng, G. Recent development in the treatment of oily sludge from petroleum industry: A review. *J. Hazard. Mater.* **2013**, *261*, 470–490. [CrossRef]

50. Mohammadi, S.; Mirghaffari, N. Optimization and Comparison of Cd Removal from Aqueous Solutions Using Activated and Non-activated Carbonaceous Adsorbents Prepared by Pyrolysis of Oily Sludge. *Water Air Soil Pollut.* **2015**, *226*, 2237. [CrossRef]

51. Wingenfelder, U.; Hansen, C.; Furrer, G.; Schulin, R. Removal of Heavy Metals from Mine Waters by Natural Zeolites. *Environ. Sci. Technol.* **2005**, *39*, 4606–4613. [CrossRef] [PubMed]

52. Elabed, A.; El Khalfaouy, R.; Ibnsouda, S.; Basseguy, R.; Elabed, S.; Erable, B. Low-Cost Electrode Modification to Upgrade the Bioelectrocatalytic Oxidation of Tannery Wastewater Using Acclimated Activated Sludge. *Appl. Sci.* **2019**, *9*, 2259. [CrossRef]

53. Wang, P.; Sun, Q.; Zhang, Y.; Cao, J. Synthesis of Zeolite 4A from Kaolin and Its Adsorption Equilibrium of Carbon Dioxide. *Materials* **2019**, *12*, 1536. [CrossRef] [PubMed]

54. Du Plessis, P.; Ojumu, T.V.; Fatoba, O.O.; Akinyeye, R.O.; Petrik, L.F. Distributional Fate of Elements during the Synthesis of Zeolites from South African Coal Fly Ash. *Materials* **2014**, *7*, 3305–3318. [CrossRef] [PubMed]

55. Izidoro, J.d.C.; Fungaro, D.A.; dos Santos, F.S.; Wang, S. Characteristics of Brazilian coal fly ashes and their synthesized zeolites. *Fuel Process. Technol.* **2012**, *97*, 38–44. [CrossRef]

56. Shoumkova, A.; Stoyanova, V. Zeolites formation by hydrothermal alkali activation of coal fly ash from thermal power station "Maritsa 3", Bulgaria. *Fuel* **2013**, *103*, 533–541. [CrossRef]

57. Kunecki, P.; Panek, R.; Koteja, A.; Franus, W. Influence of the reaction time on the crystal structure of Na-P1 zeolite obtained from coal fly ash microspheres. *Microporous Mesoporous Mater.* **2018**, *266*, 102–108. [CrossRef]

58. Wu, D.; Sui, Y.; Chen, X.; He, S.; Wang, X.; Kong, H. Changes of mineralogical–chemical composition, cation exchange capacity, and phosphate immobilization capacity during the hydrothermal conversion process of coal fly ash into zeolite. *Fuel* **2008**, *87*, 2194–2200. [CrossRef]

59. Oustan, S.; Heidari, S.; Neyshabouri, M.; Reyhanitabar, A.; Bybordi, A. Removal of heavy metals from a contaminated calcareous soil using oxalic and acetic acids as chelating agents. *Int. Conf. Environ. Sci. Eng. IPCBEE* **2011**, *8*, 152–155.

60. Qiu, R.; Zou, Z.; Zhao, Z.; Zhang, W.; Zhang, T.; Dong, H.; Wei, X. Removal of trace and major metals by soil washing with Na$_2$EDTA and oxalate. *J. Soils Sediments* **2010**, *10*, 45–53. [CrossRef]

61. Yin, X.; Chen, J.-J.; Lü, C. Impact of compounded chelants on removal of heavy metals and characteristics of morphologic change in soil from heavy metals contaminated sites. *Huanjing Kexue/Environ. Sci.* **2014**, *35*, 733–739.

62. Wei, M.; Chen, J.; Wang, X. Removal of arsenic and cadmium with sequential soil washing techniques using Na2EDTA, oxalic and phosphoric acid: Optimization conditions, removal effectiveness and ecological risks. *Chemosphere* **2016**, *156*, 252–261. [CrossRef] [PubMed]

63. Federal Remediation Technologies Roundtable. Available online: https://frtr.gov/ (accessed on 1 June 2019).

64. Chang, T.C.; Yen, J.H. On-site mercury-contaminated soils remediation by using thermal desorption technology. *J. Hazard. Mater.* **2006**, *128*, 208–217. [CrossRef] [PubMed]

65. Khan, F.I.; Husain, T.; Hejazi, R. An Overview and Analysis of Site Remediation Technologies. *J. Environ. Manag.* **2004**, *71*, 95–122. [CrossRef] [PubMed]

66. Mulligan, C.N.; Yong, R.N.; Gibbs, B.F. Remediation technologies for metal-contaminated soils and groundwater: An evaluation. *Eng. Geol.* **2001**, *60*, 193–207. [CrossRef]

67. Galletti, A.; Verlicchi, P.; Ranieri, E. Removal and accumulation of Cu, Ni and Zn in horizontal subsurface flow constructed wetlands: Contribution of vegetation and filling medium. *Sci. Total Environ.* **2010**, *408*, 5097–5105. [CrossRef]

68. Marrugo-Negrete, J.; Enamorado-Montes, G.; Durango-Hernández, J.; Pinedo-Hernández, J.; Díez, S. Removal of mercury from gold mine effluents using Limnocharis flava in constructed wetlands. *Chemosphere* **2017**, *167*, 188–192. [CrossRef]

69. Varghese, L.R.; Das, N. Removal of Hg (II) ions from aqueous environment using glutaraldehyde crosslinked nanobiocomposite hydrogel modified by TETA and β-cyclodextrin: Optimization, equilibrium, kinetic and ex situ studies. *Ecol. Eng.* **2015**, *85*, 201–211. [CrossRef]

70. Fukushi, K.; Sasaki, M.; Sato, T.; Yanase, N.; Amano, H.; Ikeda, H. A natural attenuation of arsenic in drainage from an abandoned arsenic mine dump. *Appl. Geochem.* **2003**, *18*, 1267–1278. [CrossRef]

71. Sousa, F.W.; Sousa, M.J.; Oliveira, I.R.N.; Oliveira, A.G.; Cavalcante, R.M.; Fechine, P.B.A.; Neto, V.O.S.; de Keukeleire, D.; Nascimento, R.F. Evaluation of a low-cost adsorbent for removal of toxic metal ions from wastewater of an electroplating factory. *J. Environ. Manag.* **2009**, *90*, 3340–3344. [CrossRef]

72. Lee, E.; Han, Y.; Park, J.; Hong, J.; Silva, R.A.; Kim, S.; Kim, H. Bioleaching of arsenic from highly contaminated mine tailings using Acidithiobacillus thiooxidans. *J. Environ. Manag.* **2015**, *147*, 124–131. [CrossRef] [PubMed]

73. Gikas, P. Single and combined effects of nickel (Ni(II)) and cobalt (Co(II)) ions on activated sludge and on other aerobic microorganisms: A review. *J. Hazard. Mater.* **2008**, *159*, 187–203. [CrossRef] [PubMed]

74. de C. Izidoro, J.; Fungaro, D.A.; Abbott, J.E.; Wang, S. Synthesis of zeolites X and A from fly ashes for cadmium and zinc removal from aqueous solutions in single and binary ion systems. *Fuel* **2013**, *103*, 827–834. [CrossRef]

75. Ibrahim, H.S.; Jamil, T.S.; Hegazy, E.Z. Application of zeolite prepared from Egyptian kaolin for the removal of heavy metals: II. Isotherm models. *J. Hazard. Mater.* **2010**, *182*, 842–847. [CrossRef] [PubMed]

76. Shawabkeh, R.; Al-Harahsheh, A.; Hami, M.; Khlaifat, A. Conversion of oil shale ash into zeolite for cadmium and lead removal from wastewater. *Fuel* **2004**, *83*, 981–985. [CrossRef]

77. Drahota, P.; Grösslová, Z.; Kindlová, H. Selectivity assessment of an arsenic sequential extraction procedure for evaluating mobility in mine wastes. *Anal. Chim. Acta* **2014**, *839*, 34–43. [CrossRef] [PubMed]

78. Gomes, M.V.T.; de Souza, R.R.; Teles, V.S.; Araújo Mendes, É. Phytoremediation of water contaminated with mercury using Typha domingensis in constructed wetland. *Chemosphere* **2014**, *103*, 228–233. [CrossRef] [PubMed]

79. Zheng, S.; Gu, B.; Zhou, Q.; Li, Y. Variations of mercury in the inflow and outflow of a constructed treatment wetland in south Florida, USA. *Ecol. Eng.* **2013**, *61*, 419–425. [CrossRef]

80. Chavan, P.V.; Dennett, K.E.; Marchand, E.A.; Gustin, M.S. Evaluation of small-scale constructed wetland for water quality and Hg transformation. *J. Hazard. Mater.* **2007**, *149*, 543–547. [CrossRef]

81. Kröpfelová, L.; Vymazal, J.; Švehla, J.; Štíchová, J. Removal of trace elements in three horizontal sub-surface flow constructed wetlands in the Czech Republic. *Environ. Pollut.* **2009**, *157*, 1186–1194. [CrossRef] [PubMed]

82. European Commission. *Closing the Loop—An EU Action Plan for the Circular Economy*; European Commission: Brussels, Belgium, 2015; p. 21.

83. Pinto, V.N. E-waste hazard: The impending challenge. *Indian J. Occup. Environ. Med.* **2008**, *12*, 65–70. [CrossRef] [PubMed]

84. Kaur, A.; Kaur, S. Study Paper on E-waste: A Hazardous Waste. *Int. J. Innov. Res. Comput. Commun. Eng.* **2016**, *4*, 12–15.

85. Escapa, A.; Gil-Carrera, L.; García, V.; Morán, A. Performance of a continuous flow microbial electrolysis cell (MEC) fed with domestic wastewater. *Bioresour. Technol.* **2012**, *117*, 55–62. [CrossRef] [PubMed]

86. Kadier, A.; Simayi, Y.; Abdeshahian, P.; Azman, N.F.; Chandrasekhar, K.; Kalil, M.S. A comprehensive review of microbial electrolysis cells (MEC) reactor designs and configurations for sustainable hydrogen gas production. *Alex. Eng. J.* **2016**, *55*, 427–443. [CrossRef]

87. Santoro, C.; Arbizzani, C.; Erable, B.; Ieropoulos, I. Microbial fuel cells: From fundamentals to applications. A review. *J. Power Sources* **2017**, *356*, 225–244. [CrossRef] [PubMed]

88. Slate, A.J.; Whitehead, K.A.; Brownson, D.A.C.; Banks, C.E. Microbial fuel cells: An overview of current technology. *Renew. Sustain. Energy Rev.* **2019**, *101*, 60–81. [CrossRef]

89. Kondaveeti, S.; Mohanakrishna, G.; Lee, J.-K.; Kalia, V.C. Methane as a Substrate for Energy Generation Using Microbial Fuel Cells. *Indian J. Microbiol.* **2019**, *59*, 121–124. [CrossRef] [PubMed]

90. Jiang, L.; Huang, L.; Sun, Y. Recovery of flakey cobalt from aqueous Co(II) with simultaneous hydrogen production in microbial electrolysis cells. *Int. J. Hydrog. Energy* **2014**, *39*, 654–663. [CrossRef]

91. Huang, L.; Yao, B.; Wu, D.; Quan, X. Complete cobalt recovery from lithium cobalt oxide in self-driven microbial fuel cell – Microbial electrolysis cell systems. *J. Power Sources* **2014**, *259*, 54–64. [CrossRef]

92. Luo, H.; Liu, G.; Zhang, R.; Bai, Y.; Fu, S.; Hou, Y. Heavy metal recovery combined with H2 production from artificial acid mine drainage using the microbial electrolysis cell. *J. Hazard. Mater.* **2014**, *270*, 153–159. [CrossRef]

93. Venkata Mohan, S.; Velvizhi, G.; Vamshi Krishna, K.; Lenin Babu, M. Microbial catalyzed electrochemical systems: A bio-factory with multi-facet applications. *Bioresour. Technol.* **2014**, *165*, 355–364. [CrossRef] [PubMed]

94. Escapa, A.; Gómez, X.; Tartakovsky, B.; Morán, A. Estimating microbial electrolysis cell (MEC) investment costs in wastewater treatment plants: Case study. *Int. J. Hydrog. Energy* **2012**, *37*, 18641–18653. [CrossRef]

95. Rozendal, R.A.; Hamelers, H.V.M.; Rabaey, K.; Keller, J.; Buisman, C.J.N. Towards practical implementation of bioelectrochemical wastewater treatment. *Trends Biotechnol.* **2008**, *26*, 450–459. [CrossRef] [PubMed]

96. Sleutels, T.H.J.A.; Ter Heijne, A.; Buisman, C.J.N.; Hamelers, H.V.M. Bioelectrochemical Systems: An Outlook for Practical Applications. *ChemSusChem* **2012**, *5*, 1012–1019. [CrossRef] [PubMed]

97. Cusick, R.D.; Ullery, M.L.; Dempsey, B.A.; Logan, B.E. Electrochemical struvite precipitation from digestate with a fluidized bed cathode microbial electrolysis cell. *Water Res.* **2014**, *54*, 297–306. [CrossRef] [PubMed]

© 2019 by the authors. Licensee MDPI, Basel, Switzerland. This article is an open access article distributed under the terms and conditions of the Creative Commons Attribution (CC BY) license (http://creativecommons.org/licenses/by/4.0/).

 *sustainability*

Article

A System Dynamics Model to Assess the Effectiveness of Governmental Support Policies for Renewable Electricity

Huilu Yu [1,*, Youning Yan [1] and Suocheng Dong [2]**

[1] College of Resources and Environmental Engineering, Ludong University, Yantai 264025, China; yuluous@aliyun.com

[2] Institute of Geographic Sciences and Natural Resources Research, Chinese Academy of Sciences, 11A Datun Road, Anwai, Chaoyang District, Beijing 100101, China; dongsc@igsnrr.ac.cn

* Correspondence: huiluyu73@ldu.edu.cn

Received: 6 March 2019; Accepted: 30 May 2019; Published: 21 June 2019

Abstract: China's support policy for renewable electricity belongs to a feed-in tariffs scheme. With the rapid development of renewable electricity industries, this set of policies brought about a heavy fiscal burden for the government. The exploration of whether current support policy provided excessive subsidies for renewable electricity is of great practical significance. We hold an idea that the internalization of positive externality is the only criterion for the government to support the development of a renewable electricity industry. The problem of whether the current policy provides excessive subsidies for renewable electricity industry can be solved by assessing whether its positive externality is internalized, as renewable electricity industry has a characteristic of externality. Our study object is an assumed biomass power plant in Jingning County, Gansu Province. A system dynamics model was built. Applying the environmental cost accounting method and net present value analysis method, we connected the techno-economic analysis of the biomass power plant with the measurement of positive externality of biomass power generation together. In this system dynamics model, we developed an indicator to reveal whether the subsidies provided by governmental policies can compensate the positive externality generated by the assumed biomass power plant. This study mainly draws the following conclusions: Firstly, China's current support policy does provide excessive subsidies for the renewable power industry. The subsidies received by biomass power plants from the government are higher than the positive externality generated by them; secondly, the positive externality measurement of the biomass power industry is influenced by many regional factors; thirdly, without governmental policy support, biomass power plants cannot compete with traditional power companies; fourthly, as biomass power generation is concerned, the current price subsidy intensity is about US$0.0132 higher per kWh than a reasonable level. Furthermore, the parameters frequently applied in the calculation of the prices of pollutant emission reduction in Chinese research papers are relatively small, which is only half of their actual values. Jingning County, situated in inland west-northern China, lacks typicality. There is a limitation in judging whether the government's support policy for renewable electricity is reasonable through a feasibility analysis of investment in a biomass power generation project. This may be the main drawback of this study.

Keywords: biomass power generation; positive externalities; support policy; apple branches; Jingning

---

1. Introduction

1.1. Background and Literature Review

In order to expand the market share of renewable electricity, many countries in the world have implemented policies to encourage the development of renewable energy industries [1,2]. Price-based feed-in tariff (FIT) and market-based renewable portfolio standards (RPSs) schemes are the two most frequently-used support schemes [3,4]. Countries which have implemented FIT schemes include Germany, Denmark, and Spain; they set fixed prices or a price premium over the market price of electricity for a specified time period [5]. The RPS scheme is represented by the UK, the US, and Australia, which require electricity producers to acquire a certain percentage of renewable electricity. Tradable green certificates (TGCs) are issued for all renewable electricity produced.

At present, renewable electricity production in China is supported by the FIT scheme. Under this scheme, the installed renewable energy capacity has significantly increased. The portion of the FIT higher than the market price coming from the Renewable Energy Development Fund (introduced by the government) has created a great financial burden on the government [6]. According to the relevant research, the financial gap of the Renewable Energy Development Fund had reached US$18.13 billion by 2018 [7].

According to the Chinese government plan [8], China's renewable energy policy will ultimately transform from an FIT scheme to an RPS scheme, which means that the subsidy funds that renewable electricity producers had received from the government before will mainly come from the TGC market in the future.

It is known that the reason that the FIT scheme for renewable energy has brought a heavy fiscal burden on to China's government is the higher subsidy intensity. Thus, a question has arisen: What kind of subsidy intensity is rational?

The current theoretical research related to the FIT focuses on assessing its effectiveness. Scholars have usually discussed effectiveness from the aspects of social welfare, technological innovation, and installation growth of renewable energy [9]. Scholars have seldom paid attention to the rationality and accuracy of the FIT subsidy intensity.

Compared with traditional fossil energy, renewable energy is conducive to environmental protection and resource conservation—namely, it has positive externalities. Positive externality refers to the beneficial effects of economic activities of one economic entity on other economic entities; the recipients of positive externalities do not need to pay any cost [10]. The internalization of positive externality transforms the external benefits generated by the positive externality maker into private benefits of the positive externality maker in different ways [11]. It can solve the social optimal supply shortage caused by a lack of incentives, thus overcoming the efficiency loss caused by positive externalities and re-achieving the Pareto optimality [10]. Compared with traditional power, renewable electricity has positive externalities. It can increase social benefits and reduce pollutant emissions. According to the principle of modern economics, there is a phenomenon of insufficient resource allocation in industries with positive externalities. The internalization of positive externalities is the fundamental way to solve this type of problem. That the government formulates policies to support the development of renewable electricity is a form of the internalization of positive externality.

The positive externality of renewable electricity is the main reason why the government provides a subsidy to encourage its development. Biomass power generation is a classic type of renewable energy, for convenience, we take it as an example to demonstrate our research route.

Generally speaking, biomass power generation can be viewed as a type of agricultural circular economy [12]. The question of how to scientifically determine the subsidy intensity of biomass power generation is equivalent to the question of how to regulate the development of a circular economy accurately.

A circular economy has obvious positive externalities [13,14]. Despite the academic disagreement over the impact of the development of a recycling economy on the performance of a business, most scholars approve of the viewpoint that the development of a circular economy needs the support of the government [15,16]. Many countries have begun to pay more attention to the role of the circular economy in achieving sustainable development [17,18].

The internalization of externalities is an effective way to promote the sustainable development of industries with external characteristics. Therefore, whether the externalities have been compensated for can provide a reference standard for evaluating the effect of regional transformation from a traditional economy to a circular economy, and can provide a quantitative standard for evaluating the developmental performance of a regional circular economy, as long as the externality of the circular economy can be monetized. After decades of attempts, scholars have successfully established a method for an environmental economic loss assessment, based on the market value approach; although the assessment of environmental losses is a very difficult problem [11,19]. The application of environmental loss assessment techniques and methods has led to the development of relevant empirical studies [20,21]. Research on the measurement of economic losses caused by environmental pollution in China has also achieved rich results [22,23].

At present, scholars in the field of environmental management have studied the rationality and optimization of industrial policies from the perspective of the quantitative measurement of externalities [24,25]. For example, Ding et al. (2008) [26] proposed a reverse logistics investment valuation model to calculate the number of subsidies the government needs in order to provide theoretical guidance and a decision-making basis for the government to regulate and control an enterprise's investment in environmental protection projects or technologies through the implementation of incentive policies. Ding et al. (2014) [13] built a quantitative assessment model of the internalization of externalities with a life cycle assessment and net present value analysis to explore the reasonable subsidy space and promote the development of new energy vehicles.

Relatively speaking, there have been few achievements in the study of the internalization of externalities from the perspective of green products or enterprises. This type of study is in its infancy.

Present studies have paid attention to the negative environmental externalities of economic activities, while few scholars have focused on the positive externalities of a circular economy in the social and environmental fields.

China supports the development of a circular economy, but its support for a circular economy lacks a scientific standard of judgment [9]. In the field of biomass power generation, China has formulated detailed supportive policies, including tax breaks and price subsidies. However, whether the government's support policy for biomass power generation is reasonable is a problem to which few scholars have paid close attention.

1.2. Comprehensive Utilization of Agricultural By-Products

An important component of the development of an agricultural circular economy is to comprehensively utilize agricultural by-products. According to estimates, the dry weight of apple tree branches pruned per hectare ranges from 2.4 to 6.6 tons every year [27,28]. For primary apple production areas, an apple tree branch is an important type of agricultural by-product; the total production of apple tree branches is quite substantial, and has considerable potential utilization value. However, in practice, the resource utilization of apple tree branches has not caught people's attention. Currently, apple tree branches are used mainly as firewood for cooking or heating (as is the case for straw, also), which not only produces waste but also pollutes the environment. In some apple production areas, apple tree branches are smashed and bagged to produce edible fungus [29,30]. Few apple production regions use apple tree branches as feedstock for biomass power generation. These regions are apt to obtain a type of clean energy named biogas, by building methane-generating pits and constructing an industrial circular economy chain, named "grass-livestock-biogas-fruit," centered on biogas [31,32]. In rural areas, biogas is an important kind of clean energy, which generally uses livestock excrement as feedstock. This is the reason why the development of biogas utilization typically accompanies the development of livestock breeding industries. The sustainable utilization of biogas cannot be separated from livestock excrement (having a high heating value) as the fill for a methane-generating pit. Both apple orchard management and livestock breeding belong to a labor-intensive industry; however, it is difficult for an orchard operator to undertake the dual work of both managing orchards and caring for livestock, as rural labor is insufficient when viewed against the backdrop of rapid urbanization. For instance, the areas of apple tree cultivation in Jingning County of the Gansu Province account for 68.7% of the total area of arable lands, reaching 707 km^2 in 2016, which was the largest area of such cultivation in China [33]. The agricultural distribution pattern in Jingning County is divided into the south, mainly the apple orchard industry, and the north, which is primarily the grain growing industry [34]. In the southern part of the county, most farmed arable lands are used to plant apple trees, and it is difficult for the owners of livestock to breed livestock due to the lack of crop straw as fodder. There are almost no large livestock operations among the local fruit farming communities.

In 2015, we conducted a questionnaire survey of over 500 local fruit farmers. The results indicated that 66.5% of the households of the respondents did not breed any large livestock, such as pigs, cattle, and sheep; 32% of them had methane-generating pits; and 40% of the respondents thought that the main factor that hampered the improvement of biogas utilization efficiency was the lack of feedstock with a high heating value, such as cattle excrement. The survey results are in accordance with the research conclusions of local scholars. For example, Wang and Yan (2015) pointed out that, with an increase in the number of rural migrant farmers who had left for urban jobs, rural areas were dominated by the remaining elders and children, with a lack of skilled young and post-adolescent laborers. Meanwhile, with the promotion of industrial products, such as electric cookers in rural areas, the advantage of biogas in cooking has continuously weakened, and most people have begun to ignore the management and maintenance required for methane-generating pits. Thus, the utilization rate of biogas has decreased yearly [35].

In summary, there are two aspects to the problem of comprehensive utilization of agricultural by-products in the main apple tree production areas. On the one hand, influenced by many factors (such as insufficient manpower), small-scale livestock breeding and management are out of place, and the utilization rate of methane-generating pits is low. Enthusiasm for the building of new methane-generating pits is not high, and the circular economy model of "grass-livestock-biogas-fruit" does not work well in primary apple production areas. On the other hand, affected by traditional habits and development concepts, a large number of apple tree branches are treated as firewood and are burned directly for heating and cooking, which wastes resources and contaminates the environment.

This study explores the feasibility of generating power using apple tree branches as fuel inputs to replace biogas in the primary apple production areas using a case study in Jingning County, Gansu Province. We analyze the economic and environmental benefits, in order to provide theoretical guidance for the promotion of a circular economy model of "apple-biomass power generation-organic fertilizer" in the primary apple production areas and evaluate the reasonableness of the governmental support policy, based on an assessment of the positive externality of biomass power generation. Meanwhile, this study also builds a dynamic system model which can help to ascertain more accurately the subsidy intensity conducive to improving the efficiency of governmental support policies for the development of industries with positive externalities.

1.3. Basic Conditions of Jingning County

Located in the northern latitudes 35°01′~35°45′ and eastern longitudes 105°20′~105°05′, Jingning County belongs to a typical hilly-gully area in the Loess Plateau and is one of 18 arid counties in the central Gansu Province with an annual rainfall of only 423.6 mm. The total population of Jingning County is 48.78 million, and it has an arable land area of approximately 98,000 hm^2, where 93% of the ploughed land is located on the slope of a hill or ditch [36]. Jingning County has 22.4 million poor people in rural areas. The poverty rate, which is the proportion of poor people to all people in the rural areas, reached 59.6% in 1986 [37]. Based on the concept of poverty alleviation through agricultural industrialization, Jingning County's poverty alleviation plan prioritized the development of apple cultivation supplemented by grain planting from 1986 to 2002, and the county gradually formed a spatial production pattern of "north grain and south apple" [34]. As of 2015, the total area of apple orchards in the entire county reached 707 km^2 (of which, fruiting orchards accounted for nearly 50%), the total weight of apple products reached 0.6 million tons, and the total earnings reached US$0.3778 billion. The income from apple sales accounts for 80% of the rural per capita net income in Jingning County. By 2018, the poverty rate in Jingning County had dropped to 13.32% [38].

The samples of apple tree branches used in this study, which cover the main apple varieties in Jingning County, were collected from the towns of Chengchuan and Weirong. All samples were first dried and smashed, and then the heating values of the dry weights were measured through a microcomputer oxygen bomb calorimeter. The measurement process of every sample was repeated three times. The ash content was measured with a dry ash method. The ash-free heating value equals the heating value of dry weight/(1-ash content). The mathematical analysis adopted the *t*-test. Compared with crop straw and other high-energy plants, apple tree branches have obvious advantages with regards to the heating value of the dry weight and the ash-free heating value (see Table 1).

Table 1. Comparison of heating values of apple tree branches with other biomass fuels.

Index	Unit	Apple Tree Branches	Wheat Straw	Maize Straw	Reed
Dry weight heating value	MJ/kg	17.78	16.89	16.56	17.22
Ash content	%	2.47	4.21	6.07	7.89
Ash-free heating value	MJ/kg	18.23	17.63	17.63	18.72

Data source: The heating value of apple branches was sourced from the laboratory data and other values were sourced from the relevant research literature [39,40].

The ash-free heating value of apple tree branches in Jingning County reaches 18.23 MJ/kg, which is higher than the average value for terrestrial plants globally (17.79 MJ/kg) [41]. Compared with wheat, maize, and other crop straws, apple tree branches have the characteristics of higher heating values and lower ash contents. Research has shown that there is a significant correlation between the ash content and heating value of the dry weight [41]. Under the fixed conditions of the generating efficiency of a biomass power generation plant, apple tree branches are one of the most desired feedstocks.

The total area of apple cultivation in Jingning County was 707 km^2 in 2016. We calculated that the range of the annual total amount of apple branches is from 0.16 to 0.47 million tons. With an increase in the age of the apple tree, the annual total production of apple tree branches will reach its peak value, of more than 0.4 million tons, in approximately 10 years. Meanwhile, Jingning County also has many straw resources; their output reached 0.55 million tons in 2012, of which maize straw accounted for 66.30% and wheat straw accounted for 23.15% [42].

In the eastern developed regions of China, the biomass fuel cost of a biomass power plant is approximately US$0.068/kWh, which accounts for nearly 70% of the total cost [43]. In 2013, the biomass power generation capacity in China reached 7790.02 MW, of which the total scale of Eastern China, Central China, and South China accounted for 77.65% [44].

For economically developed regions in China, biomass resources are relatively rare, and the price of biomass feedstock is boosted when biomass power plants compete for the purchase of biomass fuel. At present, the purchase price of biomass fuel is basically between 30.22–52.89 US$ per ton, but the price in Jingning County was between 30.22–45.34 US$ per ton in 2012, which is obviously lower than that of China's eastern regions; additionally, the biomass feedstock sold in rural markets accounts for only 4% of the total production [23].

The results of the questionnaire given to fruit farmers in Jingning County in 2015 showed that over 95% of the respondent fruit farmers used apple tree branches—pruned down in the process of managing orchards—for heating and cooking; only very few fruit farmers directly burned them. In summary, the biomass feedstock market in Jingning County has not yet been exploited at present.

2. Methodology

2.1. Positive Externalities of Biomass Power Generation

An externality is an activity that imposes involuntary costs or benefits on others or an activity whose effects are not completely reflected in its market price [45]. According to this definition, the positive externality of a circular economy, which does not include the economic benefits of a circular economy because it can be reflected in the market price of the production, is embodied primarily in positive environmental and social benefits. We calculated the externality produced by biomass power generation as Equation (1):

$$E_t = E_{st} + E_{et},\qquad(1)$$

where E_t is the positive externality produced by biomass power generation in t year, E_{st} is the social benefits in t year, and E_{et} is the positive environmental externality in t year.

2.1.1. The Calculation of Social Benefits

Social benefits include two components. One component is the benefits of an employment increase, and another component is farmers' increased incomes from biomass sales. The other is increased income from biomass sales.

The construction and operation of biomass power plants can create jobs, thus transforming some farmers into workers. In China, the income level of industrial workers is higher than that of farmers. Compared with farmers' income levels, the fact that the project of building biomass power plants increases jobs is equal to the increase of social income.

In Jingning County, fruit farmers often use apple branches for heating or cooking. We assumed that this part of the energy consumption gap is supplemented by electricity after the peasants sell their apple branches, which means that the fruit farmers must purchase the same amount of energy (electricity) to satisfy their cooking and heating needs, and this part of the payment in the past has not been a requirement. Therefore, fruit farmers' incomes from biomass sales should be subtracted from this portion of the expenditure, leaving the remainder as a part of the positive externalities of biomass power generation.

Social benefits can be calculated by Equation (2):

$$E_{st} = E_{bst} + E_{eit},\qquad(2)$$

where E_{bst} is the increased income from biomass sales in t year, E_{eit} is the benefits of an employment increase in t year, and these two indicators can be calculated by Equations (3) and (4):

$$E_{bst} = C*P_1 - C*L*E/3.6*P_2,\qquad(3)$$

$$E_{eit} = N(W - (I*R + i*(1 - R))),\qquad(4)$$

where W is the average wage of employees working in a biomass power plant; N is the number of employees employed by the biomass power plant; R is the urbanization rate of the study area; I is the average urban income; and i is the rural average income; C is the weight of biomass sold by a fruit farmer; P_1 is the price of the biomass; L is the lower heating value of the biomass; E is the combustion efficiency of a farmer's kitchen range; and P_2 is the price of rural electricity.

2.1.2. The Calculation of Positive Environmental Externality

Compared with coal-fired electricity, biomass power is cleaner; emissions of the major pollutant and greenhouse gases, such as SO_2, NO_X, CO, and CO_2, can be decreased drastically. Therefore, it can be said that the positive environmental externality of biomass power generation is the reduction of negative environmental externality; its measurement can be converted into a market price accounting for the pollutants' emissions reduction. the positive environmental externality can be calculated by Equation (5):

$$E_{et} = \sum_{e=1}^{n} EM_e P_e,$$ (5)

where E_{et} is the positive environmental externality in t year, EM_e is the effective emission reduction of eth pollutants of biomass power generation compared with on-grid electricity, and P_e is the market price of the environmental loss generated by the eth pollutant. In this study, the main pollutants included four types: SO_2, NO_X, CO, and CO_2. The calculation of these pollutants' parameters is very complicated, and the process of these parameters' calculation is presented in detail in Sections 2.5.6–2.5.8.

2.2. Government's Fiscal Subsidy

China's feed-in tariff, similar to that of many nations in the world, has been enforced in the development of biomass power generation. At present, as far as biomass power generation is concerned, the unified purchase price of the State Grid Corporation of China is US$0.1133/kWh, while that of desulfurized coal-fired electricity ranges approximately from US$0.0408/kWh to US$0.0710/kWh. As far as desulfurized coal-fired electricity in Gansu Province is concerned, the purchase price of the State Grid Corporation of China is about US$0.0453/kWh, then the price subsidy intensity for biomass power generation is US$0.068/kWh.

Meanwhile, judging by the "Catalogue of Preferential Value-added Tax Policies for Products and Labor Services Generated from the Comprehensive Utilization of Resources" published by the Ministry of Finance of the People's Republic of China and State Administration of Taxation in 2015, the tax rebate ratio of biomass power generation is 100%. According to the "Enforcement Regulations of Enterprise Income Tax Law of the People 's Republic of China," published by the State Council of the People's Republic of China in 2007, the income tax on biomass power generation can be calculated as less than 10% because of the nature of the comprehensive utilization of resources; all corporate income taxes of corporations involved in biomass power generation are exempted in the first three years of their operational period, and 50% of corporate income taxes are exempted from the fourth to the sixth year. According to the support policy mentioned above and the gap between the prices in biomass power and desulfurized coal-fired electricity, the fiscal subsidy provided by the government can be calculated.

Therefore, besides the price subsidy, the subsidies obtained by biomass power generation plants because of the government's support policies is composed of two parts: Income tax exempted and value-added tax exempted. Because all the value-added tax has been exempted, the value-added tax exempted can be calculated by Equation (6):

$$V_t = T_o - T_i,$$ (6)

where V_t is the value-added tax exempted, T_o is the output tax, and T_i is the input tax.

Income tax exempted can be calculated by Equation (7):

$$I_t = a_t P_t * r * b_t + 0.1 * P_t *,$$ (7)

where I_t is the income tax exempted in t year, P_t is the current year's profit of a biomass power plant, and a_t is a coefficient of the current year's profit, which is equal to 90% from 2019 to 2024 and 0% after 2024. r is the income tax rate, and b_t is a coefficient of the income tax rate, which is equal to 100% from 2019 to 2021, 50% from 2022 to 2024, and 0% after 2024.

In conclusion, the government's fiscal subsidies for biomass power generation are composed of three parts: The price subsidy, income tax exempted, and value-added tax exempted. It can be calculated by Equation (8):

$$FS_t = B_t P_S + V_t + I_t,$$ (8)

where FS_t is the government's fiscal subsidy in t year, B_t is the electricity generated by a biomass power plant in t year, P_t is the price subsidy of biomass power, and V_t and I_t are the value-added tax exempted and income tax exempted in t year.

2.3. The Method to Judge Whether the Positive Externality Has Been Internalized

The economic theory of the externality tells us that industries with positive externalities will be underdeveloped. To promote the development of industries with positive externalities, it is necessary for the government to provide some support policies for the purpose of eliminating their positive externalities and achieving the promotion of the development of these industries. The essence of this kind of behavior of the government is called the internalization of externalities. Theoretically, the government's support should be equivalent to the positive externalities. In sum, whether the government's fiscal subsidy has matched the positive externality of biomass power generation can be judged by Equation (9):

$$Q = \sum_{t=2019}^{2039} (E_{st} + E_{et} - FS_t), \tag{9}$$

where Q is the indicator of externality internalization, E_{et} is the positive environmental externalities of biomass power generation in t year, E_{st} is the social benefits in t year, and FS_t is the government's fiscal subsidy in t year. We can say that the positive externality of biomass power generation has been internalized when the value of Q is equal to zero in 2039. The degree of externality internalization depends on the difference between Q and zero.

2.4. Six Scenarios of Analysis

As far as a planned biomass power plant is concerned, its operational circumstances are not stable, the market price of biomass fuels will fluctuate, and the governmental support policy can be variable. Moreover, it is not known whether the planned biomass power plant can be absorbed into the project of Clean Development Mechanism (CDM), which is one of the three flexible mechanisms to achieve partial emission reduction commitments abroad under Kyoto Protocol. Registered as a CDM project in the UN CDM Executive Board, the CO_2 emission reductions generated by the project can be sold in the regional carbon trading market. Thus, our analysis was based on several scenarios (see Table 2) to simulate possible practical situations. We analyzed the conditions that made the biomass power plant project feasible for every scenario.

Table 2. Several scenarios for the operation of a biomass power plant.

Index	Scenario					
	I	II	III	IV	V	VI
Price of biomass fuel (US$/ton)	variable	variable	variable	variable	variable	variable
Price of biomass power (US$/kWh)	0.1133	0.1133	0.0453	0.0453	0.0453	0.0453
CDM	no	yes	no	yes	yes	no
Support policy	yes	yes	yes	yes	no	no

Every scenario depicts the specific management environment of a biomass power plant. By comparing the conditions that can maintain the normal operation of a biomass power plant without a loss, the rationality of the government's support policy based on the idea of internalizing the positive externality of biomass power generation can be discussed.

2.5. Biomass Power Generation Technologies in the Designed Project and its Parameters

2.5.1. Biomass Power Generation Technologies

Many commercially proven power generation technologies which can use biomass as a fuel input are available. These technologies can generally be classified into three types: Biomass combustion, anaerobic digestion, and biomass gasification technologies [46]. Among these, the direct combustion of biomass for power generation is the most mature and common form of biomass power generation. Around the globe, over 90% of the biomass used for energy purposes goes through the combustion route. We did not consider anaerobic digestion as a feasible biomass generation technology, because the main biomass we used to generate power is composed of apple tree branches, which is not a feedstock used to produce biogas, in general. Although gasification technologies are commercially available, more work, such as R&D(Research and Development) and demonstrations, should be conducted to promote their widespread commercial use, as only approximately 373 MWth (megawatt-thermal) of the installed large-scale gasification capacity was in use in 2010, with only two additional projects totaling 29 MWth planned for the period up to 2016 [46]. The direct combustion of biomass is also a popular technology which has been adopted by China's biomass power generation plants. To discuss the economic and technological feasibility of biomass power generation using apple tree branches as a fuel input, a techno-economic analysis of a 30 MW biomass-fired power plant is provided.

2.5.2. Data Collection

Processing Equipment and Design Parameters of the 30 MW Biomass-Based Steam Power Plant

Currently, the main technologies for biomass power generation include water-cooled vibrating grate combustion technology and fluidized bed combustion technology. The former, which was developed by BWE—a famous corporation whose business is to produce industrial boiler in Denmark, adopts the grate-firing method in which the fuel can be diffused equally and burned with a vibrating grate, and the clinker is discharged at the terminal part of the boiler [47]. This method is an effective solution to inefficiency, and corrosive tail gas and clinker agglomeration. Thus, we assumed that the project of biomass power generation using apple tree branches as a fuel input would employ the water-cooled vibrating grate combustion technology.

To reduce the import cost of a water-cooled vibrating grate boiler, Chinese corporations have pursued domestically manufactured equipment by actively cooperating with BWE—a Danish boiler manufacturer. At present, many corporations, such as the Jinan Boiler Group Co., Ltd. and the Hangzhou Boiler Group Co., Ltd., manufacture their own water-cooled vibrating grate boilers and supply their products to domestic biomass power generation plants [48].

The operating principle of biomass direct combustion power generation technology can be depicted as follows: The biomass is sent directly into the boiler as a fuel input, after which high-temperature and high-pressure steam, produced by the boiler, pushes the steam turbine to drive a dynamo, which generates electricity. Table 3 lists the boiler and turbine data.

Table 3. Specifications of the boiler and turbine of the 30 MW plant.

Item	Value	Unit
Boiler capacity	130	t/h
Capacity of the turbine	30	MW
Steam requirement as per manufacturer	3.8	t/MWh
Steam required for full capacity (30 MW)	114	t/h
Pressure of steam	93	bar
Temperature of steam	540	°C
Heat required to produce 1 ton of steam at 93 bar and 540 °C	2.34	GJ/t
Boiler efficiency	92	%
Hours of operation	7000	h/y

Structure of the Total Investment and Financial Costs of the 30 MW Biomass Power Plant

The registered capital of the designed project, which accounts for 20% of the total investment, is approximately US$9.10 million. The remaining 80% of the investment will be loaned from a bank whose lending rate is 5.94%. The long-term loan amount is US$36.40 million with a repayment period of 13 years, of which the grace period is 2 years, and the repayment method is the average capital plus interest. The operation period of the biomass power generation project is from 2019 to 2039.

2.5.3. Mathematical Equations Used to Evaluate the Requirement of Biomass Fuel

In this study, the available biomass fuels include apple tree branches, wheat straw, and maize straw. The former is designated as the main biomass fuel, while the other two types are auxiliary fuels, used only when apple tree branches cannot be sufficiently supplied.

The calorific value of the dry weight cannot be used to evaluate the biomass requirement under boiler conditions in the process of biomass power generation. Cheng Xuyun et al. (2013) found that the lower heating value (LHV) has an intimate relationship with the structure of biomass fuel, which involves the mass fractions of moisture, ash, volatile matter, and fixed carbon. The LHV can be estimated from industrial analysis indices as follows [49]:

$$LHV = 6.22 - 5.38M + 11.68V - 8.42A + 10.89C, \tag{10}$$

where M, V, A, and C are the percentages of moisture, volatile matter, ash, and fixed carbon, respectively, accounting for straw biomass based on industrial analysis conditions; the unit of LHV is KJ/g.

The annual biomass requirement is also influenced by the structure of the biomass fuel. The requirements of the biomass fuel vary with the fraction of the moisture content in the biomass and the system efficiency of the equipment used for biomass power generation. We assumed that the biomass would be processed before consumption, and its mass fraction of moisture content is in accordance with the conditions under which the LHV is estimated. The following equation is used to estimate the annual biomass requirement (M_f) [50]:

$$M_f = \frac{E \times 3.6}{LHV \times \eta}, \tag{11}$$

where M_f, E, LHV, and η indicate the annual biomass requirement, energy demand, lower heating value, and combustion efficiency, respectively.

2.5.4. Net Present Value (NPV)

The method of net present value (NPV) is employed to judge whether a project is economically feasible. The net cash flow, which is the basis of an analysis of the net present value, is computed according to the item structure of the cash flow of the construction project, as shown in Figure 1.

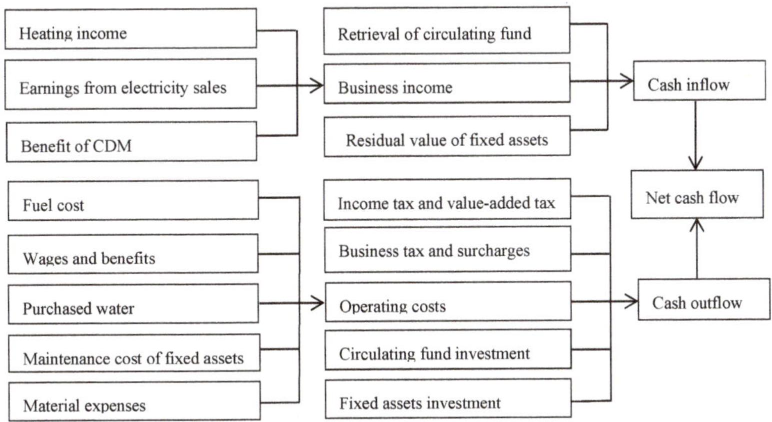

Figure 1. Item structure of the cash flow of a construction project [51].

The net earnings of each year in the lifetime of the project are discounted to year zero with the selected rate to satisfy the marginal average rate of return (MARR), and the investment is deducted from the sum of the present net earnings. This value of the *NPV* can be computed by the following equation [52]:

$$
\begin{aligned}
NPV &= -S + \frac{CF_1}{(1+r)^1} + \frac{CF_2}{(1+r)^2} + \cdots\cdots + \frac{CF_T}{(1+r)^T} \\
&= NPV = -S + \sum_{j=1}^{T} \frac{CF_j}{(1+r)^j},
\end{aligned}
\tag{12}
$$

where S is the initial investment, CF is the cash flow, r is the discount rate, and T is the economic life of the plant.

2.5.5. Internal Rate of Return (IRR) and Payback Period (PBP)

The definition the internal rate of return (IRR) is the discount rate that makes the NPV equal zero over the lifetime of the project, and it can be computed by Equation (13):

$$
NPV = -S + \sum_{j=1}^{T} \frac{CF_j}{(1+IRR)^j} = 0,
\tag{13}
$$

The payback period (PBP) is the number of years that the project takes to recover its total investment (TI) through earnings after interest and tax (*EAIT*). The PBP can be computed by Equation (14) [53]:

$$
PBP = \frac{TI}{EAIT}.
\tag{14}
$$

2.5.6. Emission Reductions by Displaced Energy

Compared with a coal-fired power plant, biomass power generation is considered to be a cleaner energy source, which reduces emissions, such as CO_2 and SO_2, as the biomass always releases these emissions during a biological degradation process, even though it is not used as biomass fuel to generate power. The amount of emissions that can be reduced by the displaced energy can be estimated as follows [54]:

$$
ER_d = E \times \left(PC \times EF_C + PP \times EF_P + PN \times EF_g \right),
\tag{15}
$$

where ER_d is theoretical emission reduction of pollutants of biomass power generation compared with on-grid electricity, in fact the value of ER_d is equal to the pollutants' emission amount generated by traditional electricity's production; E is the amount of electricity generated by the biomass power plant; PC is the share of coal in China's primary energy consumption; PP and PN indicate the shares of petroleum and natural gas in China's primary energy consumption, respectively; and EF_c, EF_p, and EF_g indicate the emission factors of coal, petroleum oil, and natural gas, respectively. These emission factors are shown in Table 4.

Table 4. Emission factors of several fuels [52].

Fuels	Emission Factors (Unit: t/MWh)			
	CO_2	SO_2	NO_X	CO
Coal	1.1800	0.0190	0.0052	0.0002
Petroleum oil	0.8500	0.0164	0.0025	0.0002
Natural gas	0.5300	0.0005	0.0009	0.0005
Renewable	0	0	0	0

In China, the consumption structure of its primary energy has varied rapidly in recent years. The 2015 Chinese energy consumption share is shown in Table 5.

Table 5. Chinese energy consumption share in 2015 (unit: %).

Country	Coal Power	Petroleum Power	Gas Power	Renewable Power
China	64	18.1	5.9	12

2.5.7. Emissions from the Transportation of Biomass Raw Materials (ET)

Emissions from the transportation of biomass raw materials can be calculated using the following Equation (16):

$$ET = \frac{M \times D \times T \times EF}{TT}, \tag{16}$$

where ET is the total emissions from the transportation of biomass raw materials; M is the weight of biomass fuel that must be transported; D is the distance that the biomass fuel needs to be transported; T is the tortuous coefficient of country roads; EF is the emission factor of diesel-based transportation; and TT is the truck load per trip.

We assume that the transportation vehicles from rural households to temporary storage stations are light-duty vehicles and that the transportation vehicles from temporary storage stations to biomass power plants are heavy-duty vehicles. The overall average emission factors are shown in Table 6.

Table 6. Emission factors (EF) of diesel vehicles.

Pollutant	Unit	Emissions Standards	EF_t	
			Diesel Light-Duty Vehicles (Average Mass Lower Than 3.5 t)	Diesel Heavy-Duty Vehicles (Average Mass Higher Than 12 t)
CO_2	t/km	EU III	0.356×10^{-3}	0.932×10^{-3}
NO_X	t/km	EU III	1.400×10^{-6}	8.400×10^{-6}
CO	t/km	EU III	0.700×10^{-6}	2.600×10^{-6}
SO_2	t/km	EU III	0.100×10^{-6}	0.300×10^{-6}

Data source: Determination of emission factors from motor vehicles under different emission standards in China [55].

2.5.8. Effective Emissions Reduction from Biomass Power Plants

In reality, the biomass power generation can also bring about pollutant emissions which should be subtracted from ER_d. Thus, the effective emissions reduction from a biomass power plant (EM_e) can be evaluated using Equation (17):

$$EM_e = ER_d - E \times E_b - ET, \tag{17}$$

where EM_e is the effective emission reduction of biomass power generation, ER_d is the theoretical emission reduction of pollutants of biomass power generation compared with on-grid electricity; E is the amount of electricity generated by the biomass power plant; E_b is the emission factors of pollutants; and ET is the emissions from biomass transportation.

The life cycle GHG (greenhouse gas) emission intensity of biomass power is 0.045 kg CO_2 e/kWh [56]; the emission factors of other emissions of a biomass power plant, including CO, SO_2, and NO_x, are 0.0080332 kg/kWh, 0.0000228 kg/kWh, and 0.0008626 kg/kWh, respectively [57]; ET were calculated using Equation (16).

2.6. System Dynamics and the BEP-SD Model

The system dynamics method, which was first founded in the 1950s [58], was applied to construct a circular economy systemic analysis model. This method has been widely used in the study of many complex systems of the economy, society, and ecology [59,60], and is characterized by quantifiability and controllability. It can reveal the dynamic changes, feedback, delays, and other processes of a system. Thus, this method has an obvious advantage in analyzing, improving, and managing systems characterized by a long developmental cycle and complex feedback effects [61].

The system dynamics model is generally composed of three types of variables: Level variable, rate variable, and auxiliary variable. The level variable is a cumulative variable which is like a water pool, it represents the status of the system and is always put in a rectangle. The rate variable makes the level variable change, which represents the rate at which the level variable changes. The auxiliary variable is the intermediate variable used to describe the information transfer and transformation process between the level variable and rate variable in the decision-making process. In addition, a constant is also often used in building a system dynamics model. The arrow direction indicates the causal relationship between variables.

To distinguish whether the construction of a biomass power generation plant is economically feasible and whether the China biomass power generation support policy is rational, the bioenergy-effect-policy system dynamics (BEP-SD) model, which is centered around biomass power generation using apple branches as a fuel, is constructed to quantitatively simulate various flows in the agricultural circular economy system of Jingning County, simulate their effects and long-run trends, and detect the system defects before providing some suggestions to improve the final system performance. According to the logical framework narrated above, the causal loop of the BEP-SD model was designed, as shown in Figure 2.

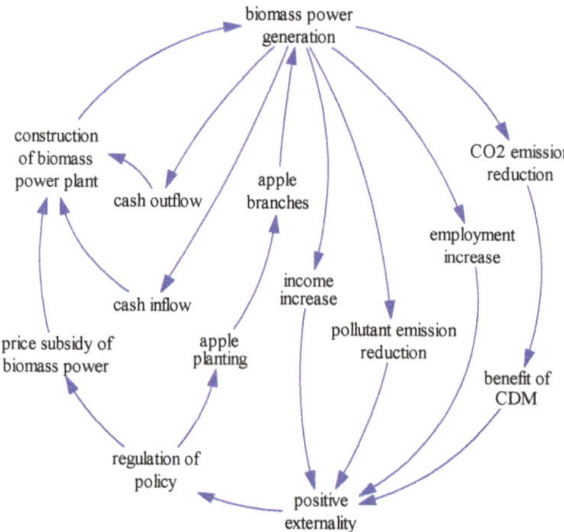

Figure 2. The causal loop of the bioenergy-effect-policy system dynamics (BEP-SD) model.

For the convenience of narration, we broke down the BEP-SD model into four subsystems: (1) Consumption of biomass fuel, (2) positive externalities, (3) fiscal subsidy, and (4) net present value. The consumption of a biomass fuel subsystem mainly includes one level variable, one rate variable, and 27 auxiliary variables. This subsystem simulates the growth process of the apple tree branches consumed by the biomass power generation plant and analyzes the cost of the biomass fuel (see Figure 3).

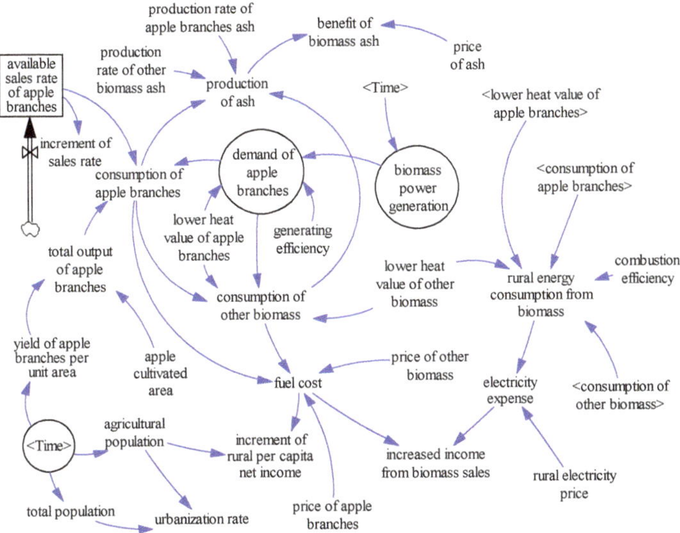

Figure 3. Subsystem of the consumption of biomass fuel.

The positive externalities subsystem involves one level variable, one rate variable, and 37 auxiliary variables. The operation of a biomass power generation plant generates some positive effects, such as economic growth, employment increases, and pollution reduction. This subsystem mainly analyses the positive environmental externalities relating to the environmental benefits generated by using biomass power to replace coal-fired power, such as emission reductions of CO_2, SO_2, NO_x, and CO. We mainly focus here on the environmental benefits generated by the emissions reductions of CO_2, SO_2, and NO_x (see Figure 4).

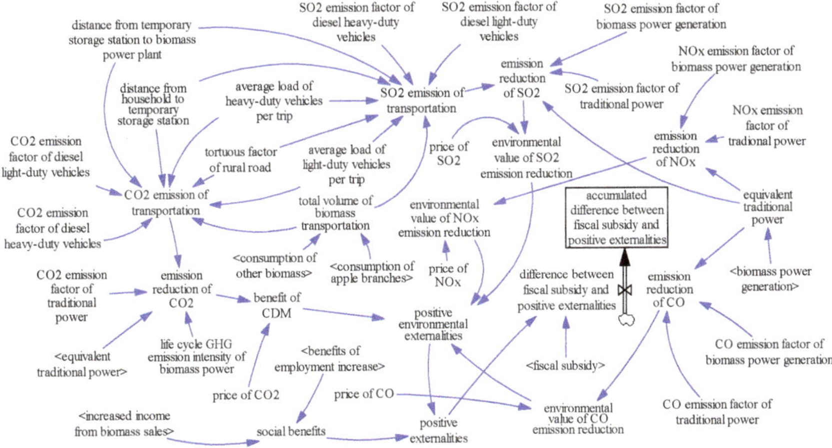

Figure 4. Subsystem of positive externalities.

The fiscal subsidy subsystem involves three level variables, four rate variables, and 45 auxiliary variables. This subsystem simulates the process of fiscal subsidy generation and depicts the gap between the fiscal subsidy and positive environmental externalities. It is used to judge the rationality of the governmental support policy for the development of biomass power (see Figure 5).

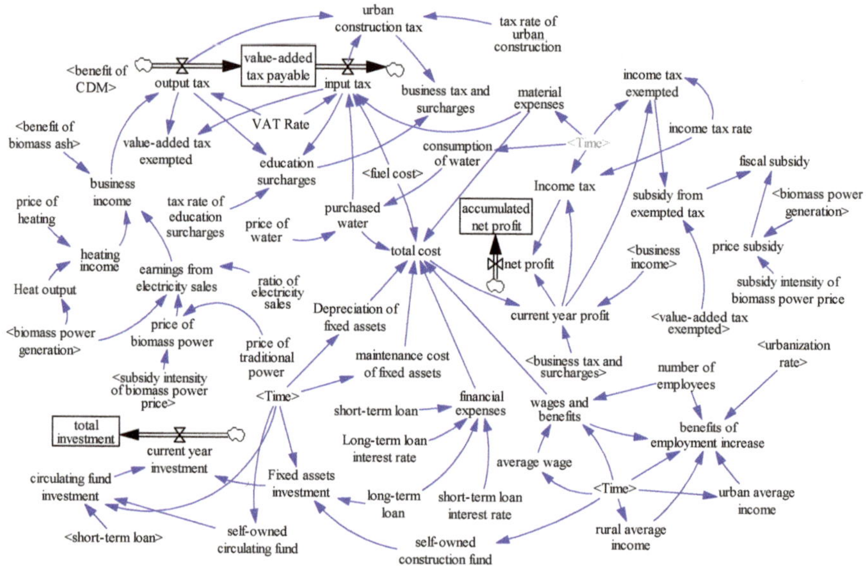

Figure 5. Subsystem of the fiscal subsidy.

The net present value subsystem involves two level variables, two rate variables and eight auxiliary variables. This subsystem is designed to compute the net present value of the project through which we judge whether the project of building a 30 MW biomass power plant using apple tree branches as biomass fuel is feasible (see Figure 6).

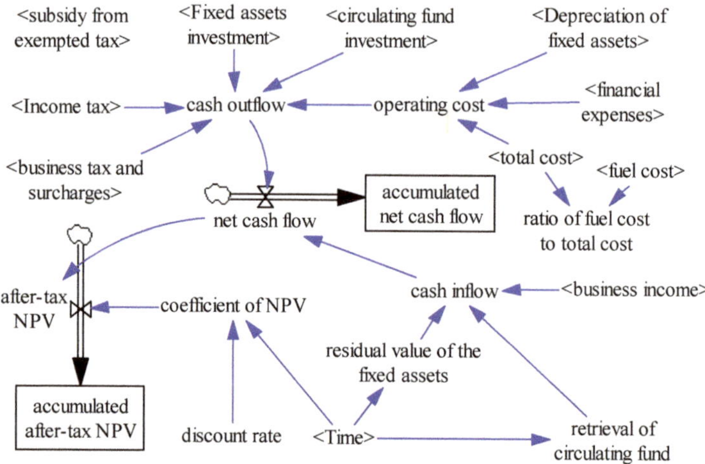

Figure 6. Subsystem of the net present value.

The BEP-SD is employed to analyze the interaction mechanisms among the different subsystems and the feedback relationships between the parameters of the model, thereby demonstrating and simulating the development of a biomass power plant quantitatively and dynamically.

In an allusion to the price fluctuations of biomass power and biomass fuel and the technological changes in biomass power generation, this study applied a scenario analysis to discuss the economic feasibility of a biomass power plant project. The project has positive externalities, and the internalization of these externalities is the theoretical basis of the governmental support policy for the development of biomass power generation. At present, China's unified on-grid price for biomass power is US$0.1133/kWh, which is approximately US$0.0529/kWh higher than the desulphurization benchmarking feed-in tariff; this higher part of the price can be regarded as an approximate compensation for the externalities. This study explores the rational space of the biomass power support policy and proposes policy recommendations by comparing the positive externalities generated by biomass power generation and the governmental subsidies for the development of biomass power.

3. Results and Discussion

3.1. Demand Analysis of Biomass Fuel

We assume that 70% of the total apple tree branches in Jingning County can be purchased ultimately in the rural market accompanied by an increase in the availability of apple tree branches; this amount will be nearly 0.11 million tons in the next few years. With an increase in the age of the apple trees, the available apple tree branches will increase yearly, and the highest available amount will reach 0.28 million tons, plus approximately 0.38 million tons of crop straw every year. These biomass resources can ensure the required biomass fuel inputs for the planned biomass power generation plant.

Based on the assumption that the planned biomass power plant will operate with a full load, we simulated the consumption trends of the biomass fuel in the planned biomass power plant (see Figure 7). The requirements of the apple tree branches and other biomass are within the scope of the supply capacity of Jingning County. To illustrate the relationship between the biomass price and the economic feasibility of the assumed project, the biomass price is marked in the figure legend; for example, scenario I 49 indicates that all data presented in the figure are calculated based on an assumption that the biomass price is US$49/ton in scenario I.

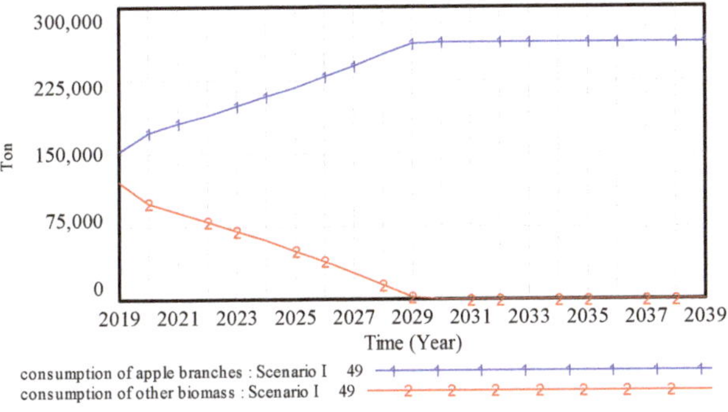

Figure 7. Consumption trends of biomass fuels in a biomass power plant.

Accompanied by an increase in the consumption of apple tree branches, the fuel cost of the project descends gradually as the LHV of apple branches is higher than for other biomasses, such as maize, wheat straw, and other crop straws. We made this conclusion under the assumption that the prices of apple branches and other biomass will not be raised with the building of the biomass power plant. As with biomass fuel, other crop straw biomass serve as substitutes for apple branches, and crop straw in the rural markets of Jingning County accounts for only 4% of the total scale; thus, this judgment is reliable.

3.2. Economic Feasibility Analysis of the Project of Building a 30 MW Biomass Power Plant

According to the NPV method, the project is economically feasible if the value of the NPV is larger than zero by the last year of its operation period. Thus, we can judge whether the planned project is economically feasible by analyzing which price for the biomass fuel makes the NPV larger than zero by 2039 when the indices of the IRR, the price of biomass power, the generating efficiency of the biomass power plant, and the other conditions of a plant's normal operations are fixed (see Table 2). The NPV values would be less than zero and the assumed project will not be economically feasible if the prices of biomass fuel are larger than the prices listed in Table 7.

Table 7. Conditions that make the project economically feasible.

Index	Scenario					
	I	**II**	**III**	**IV**	**V**	**VI**
Price of biomass (US$/ton)	49.2	54.7	1.3	6.8	−7.5	−13.0
PBP (year)	7.4	7.4	7.3	7.3	7.5	7.5
IRR (%)	14	14	14	14	14	14
NPV in 2039 (10^3 US$)	174.1	78.1	190.5	94.5	38.3	95.8

In the six scenarios, scenario I and scenario II are practical and feasible as the prices of crop straw (e.g., wheat and maize straw) were between US$30.22 and US$45.34 in 2012, which are less than those in scenario I and scenario II. The PBP is less than eight years, which is within the acceptable range, indicating that the planned biomass power plant is economically feasible (see Table 8).

Table 8. Two scenarios that can generate a higher economic efficiency.

Index	Scenario I			Scenario II		
Price of biomass (US dollars/ton)	30.2	37.8	45.3	30.2	45.3	52.9
PBP (year)	4.8	5.6	6.6	4.4	5.8	7.0
IRR (%)	14	14	14	14	14	14
NPV (10^3 US$)	29,505	17,773	6195	37,900	14,589	2857

As far as the cost of biomass power plant is concerned, the cost of biomass fuel accounts for about 60% of the total cost. So, to some extent, the price of biomass fuel seals the fate of biomass power plants. From scenario III to scenario VI, the prices of biomass fuels are far less than the prices that the market would accept (see Table 7), especially scenario V and scenario VI, whose prices are only US$0.0453/kWh, and there are no support policies for the investors in biomass power to help them decrease the cost of biomass fuel. We conclude that the biomass power generation project is not economically feasible without a support policy for biomass power generation and higher prices compared with fossil-based power.

3.3. Sensitivity Analysis of Profitability of Biomass Power Generation Project

3.3.1. The Influence of the Fluctuation of Biomass Price

Biomass fuel cost accounts for about 60% of the total cost of the biomass power generation project (Figure 8), and the fluctuation of the biomass price exerts an important impact on the profitability of the biomass power generation project.

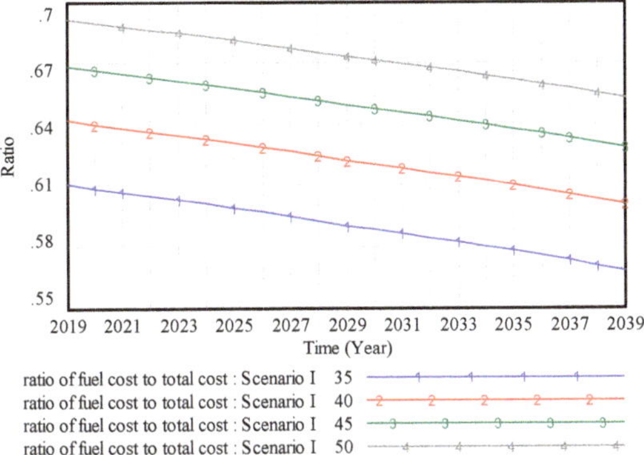

Figure 8. Ratio of fuel cost to total cost in scenario I.

According to the judgment standard of the net present value method, the project has investment value when the accumulated after-tax net present value is larger than zero at the end of the project. Obviously, in the project operation period, the earlier the accumulated after-tax NPV exceeds zero, the stronger the profitability of the project. In Figure 9, there is a trend that the profitability of the biomass power generation project declines together with the rise of the biomass fuel price under the backdrop of scenario I. When the price of biomass fuel is US\$35/ton, it needs less than nine years for the indicator value of the accumulated after-tax net present value to exceed zero; when the price of biomass fuel is US\$40/ton and US\$45/ton, this time will be shorter than 11 and 14 years, respectively; and when biomass fuel prices are above US\$50/ton, biomass power generation projects will lose investment attractiveness for industrial capital, as shown in Figure 9.

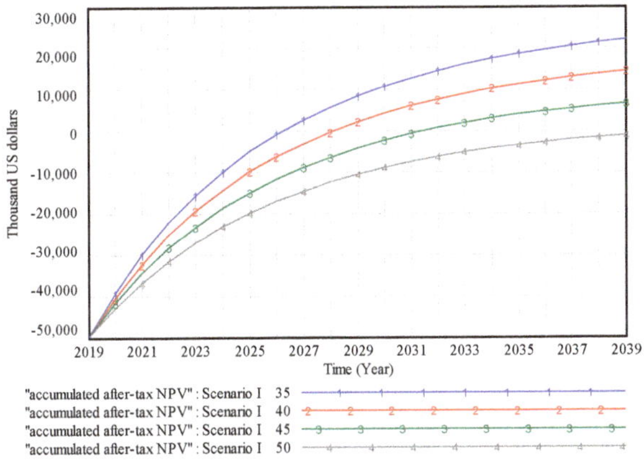

Figure 9. Sensitivity of profitability of the biomass power project to biomass price fluctuations.

3.3.2. The Influence of Governmental Price Subsidy Intensity

Among the policy tools that the government uses to support renewable electricity development, the price subsidy occupies a very important position, as shown in Figure 10.

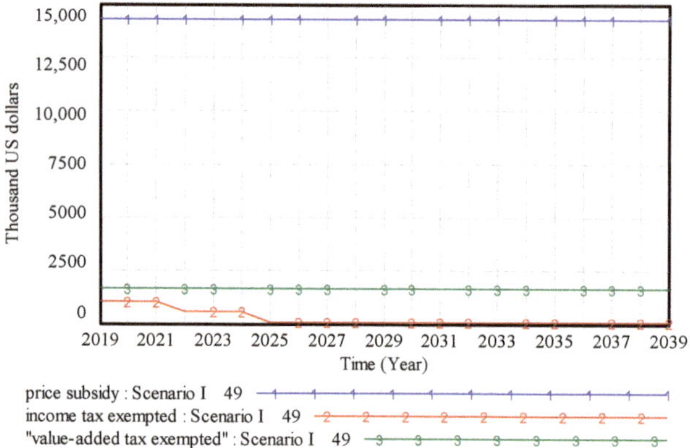

Figure 10. The structure of the governmental fiscal subsidy.

The governmental price subsidy intensity for the development of renewable electricity exerts an important impact on the profitability of the biomass power generation project. In Figure 11, scenarioI40-1 means that the price of biomass is US$40/ton, and governmental price subsidy intensity is US$0.068/kWh; from scenario I40-2 to scenario I40-4, the prices of all types of biomass are US$40/ton, but the governmental price subsidy intensity is US$0.0604/kWh, US$0.0529/kWh, and US$0.0453/kWh, respectively.

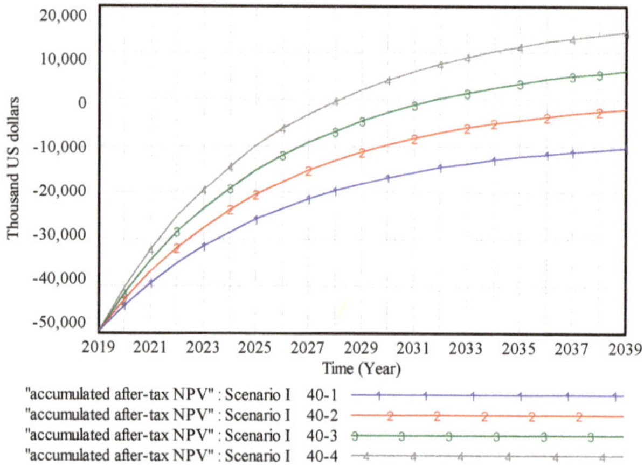

Figure 11. The influence of the governmental price subsidy intensity on the profitability of the project.

There is an obvious trend in Figure 11 that the profitability of the biomass power generation project declines rapidly together with the decline of the governmental price subsidy intensity. The indicator value of the accumulated after-tax net present value can be greater than zero in less than 11 years as the price subsidy intensity is US$0.068/kWh; as the price subsidy intensity is US$0.0604/kWh, it will take less than 14 years. However, as the subsidy intensity is US$0.0529/kWh and US$0.0453/kWh, the biomass power generation project loses its investment value.

3.4. Analysis of the Present Support Policy of Biomass Power Generation

3.4.1. To What Extent has the Positive Externality of Biomass Power Generation Been Internalized?

As mentioned before, the indicator value of Q (see Equation (9)) can help us judge whether the positive externality of biomass power generation has been internalized by the government's policy. For convenience, Q is replaced by another variable named the "accumulated difference between fiscal subsidy and positive externalities" in the system dynamics model, whose definition is the accumulated difference value that the fiscal subsidy is subtracted from the positive externality from 2019 to 2039.

The indicator value of the accumulated difference between the positive externalities and fiscal subsidy should be close to zero at the end of the project operation period if the positive externality was internalized completely. However, its values in scenario I49 and scenario II 54 in 2039 are US\$−140,729,000 and US\$−113,399,000, respectively, which are much lower than zero. Therefore, we can draw a conclusion that the subsidies provided by government policies far exceeds the positive externalities generated by the biomass power generation project.

The evolution trend of "difference between the fiscal subsidy and the positive externalities," which is far less than zero (see Figure 12), indicates that the fiscal subsidy far exceeds the positive externalities generated by biomass power generation. We can infer that the biomass power generation project is not economically feasible without the support of the government because the positive externalities of the biomass power generation project are not compensated.

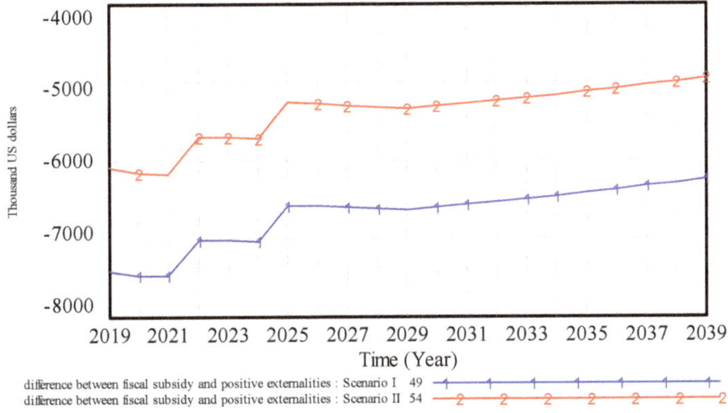

Figure 12. Difference between the fiscal subsidy and positive externalities in scenario I and scenario II.

The curves in Figure 12 present a similar developing trend, which indicates that the indicator's values are small from 2019 to 2024, and become bigger gradually after 2025. Because the value of positive externality of biomass power generation is relatively stable, the value of "difference between the fiscal subsidy and the positive externalities" mainly depends on the value of the fiscal subsidy. Therefore, the indicator curves in Figure 12 illustrate that the support intensity of the government's policy is larger in the first six years, and then its support intensity becomes weak, so the values of this indicator in the two scenarios will grow bigger and bigger.

3.4.2. Analysis of the Structure of Positive Externality and Fiscal Subsidy

The positive externality of biomass power generation mainly includes six items. According to their values by descending order, the list of these six items is the environmental value of SO_2 emission reduction, increased income from biomass sales, environmental value of NO_x emission reduction, benefit of CDM which is equal to the environmental value of CO_2 emission reduction, benefits of employment increase, and environmental value of CO emission reduction (see Figure 13).

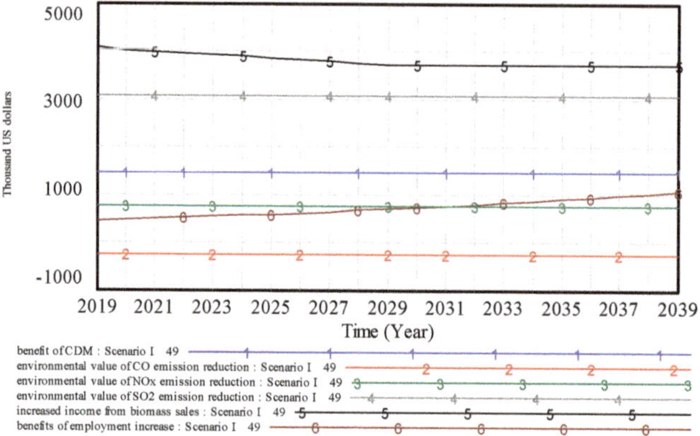

Figure 13. Items of positive externalities in scenario I.

Among them, the benefits of employment increase and increased income from biomass sales is a social benefit, which accounts for approximately 23% of the positive externality of biomass power generation.

In Figure 13, the environmental value of CO emission reduction is negative, which means that the emission level of CO for the biomass power plant is higher than traditional power.

Compared with the income tax exempted and value-added tax exempted, the price subsidy holds a large proportion of the government's fiscal subsidy, which can explain why the biomass power plant cannot operate economically without it from scenario IV to scenario VI (see Figure 10).

3.4.3. The Reasonable Governmental Price Subsidy Intensity and Prices of Pollutant Emission Reduction

From the previous analysis, we know that biomass power generation has a positive external effect. According to the theory of externality, the internalization of externality is an effective tool for optimizing resource allocation. Therefore, the reasonable government subsidy intensity should ensure that the net present value of biomass power generation projects is exactly equal to zero. The rational price subsidy intensity, which is that value that can make the indicator value of accumulated after-tax NPV equal to zero, is easy to calculate when the biomass price is set.

We calculated the reasonable price subsidy intensity when the biomass price is US$35, US$40, and US$45 per ton. The average value of the price subsidy intensity in these three scenarios is US$0.0548/kWh, which is US$0.0132 lower than the current price subsidy intensity. However, the values of accumulated after-tax NPV are still smaller than zero in these scenarios, as shown in Table 9.

Table 9. The reasonable governmental price subsidy intensity.

Indicators	Biomass Price (US$/Ton)		
	35	40	45
Subsidy intensity (US$/kWh)	0.0477	0.0548	0.0619
Accumulated after-tax NPV (10^3 US$)	25.297	25.649	25.978
Accumulated difference between fiscal subsidy and positive externalities (10^3 US$)	−161,026	−194,200	−227,373

What conclusion can we draw from this situation? Positive externality includes two parts, one is social benefits, and the other is positive environmental externalities. The indicator parameters, which are included in the calculation of social benefits, come from a market survey and are reliable. The quantities of the emission reduction of biomass power generation are reliable too. Thus, the only points that can arouse our suspicion are the prices of pollutant emission reduction. A reasonable inference is that the values of these indicators are smaller than their actual values. In order to obtain the actual values of these indicators, we promoted their values in the same proportion. These parameters can be calculated by Equation (9), where E_{et} can be replaced by E'_{et}:

$$E'_{et} = E_{et}(1 + a) = \sum_{e=1}^{n} EM_e P_e((1 + a)) \tag{18}$$

where E'_{et} is the positive environmental externality whose prices of the pollutant emission reduction are enhanced in the same proportion as a. The prices of the pollutant emission reduction that can make the value of the "accumulated difference between fiscal subsidy and positive externalities" close to zero are listed in Table 10 when the biomass price is US\$35, US\$40, and US\$45, respectively.

Table 10. Parameters that make the biomass power generation project economically feasible. (Equations and Parameters shows in Supplementary Materials).

Indicators	Biomass Price (US\$/Ton)		
	35	40	45
Price of CO_2 (US\$)	14.74	15.00	15.21
Price of CO (US\$)	335.84	340.90	345.72
Price of NO_x (US\$)	2687.99	2727.22	2765.74
Price of SO_2 (US\$)	2154.49	2181.78	2212.60
Subsidy intensity (US\$/kWh)	0.0477	0.0548	0.0619
Accumulated after-tax NPV (10^3 US\$)	25.297	25.649	25.978
Accumulated difference between fiscal subsidy and positive externalities in 2039 (10^3 US\$)	−15.8525	−7.2754	5.2422

The prices of the pollutant emission reduction listed in Table 6 are much larger than the corresponding values from scenario I to scenario VI. The latter is less than half of the former.

There is a huge controversy about the market value of pollutant emissions reduction. These indicators have a wide range of values. These parameters produced by our calculation may provide a useful reference for related research in the field of sustainability.

3.5. The Explanation of the Fiscal Burden Imposed by Renewable Electricity Policy

Renewable electricity subsidies are mainly derived from renewable energy surcharges for traditional power, which can be viewed as a type of Pigou tax. It was reported that the renewable electricity generated by biomass power plants reached US\$64.7 billion kWh. According to our research result that the current price subsidy intensity is US\$0.0132 higher than a reasonable level, among the governmental fiscal burden produced by its renewable electricity support policies, there is about US\$0.854 billion that can be attributed to an unreasonable price subsidy intensity in the field of biomass power generation, which accounts for 4.7% of the total amount [62], and is about two times the proportion that biomass power generation accounts for of the total installed capacity of renewable electricity. We can infer that the scale of the positive externality of renewable electricity we calculated is approximately accurate. As far as biomass power generation is concerned, the government's financial burden imposed by renewable electricity policy should be partly attributed to the fact that the tax rate of renewable energy surcharges imposed on traditional power is relatively low.

3.6. The Limitation of this Research

3.6.1. The Uncertainty of this Research

In general, the generation efficiency of biomass power generation equipment will improve with the development of technological innovation, and its price will decline. These factors will reduce the total cost of the biomass power generation project, and thus improve the feasibility of the project.

It will also affect the governmental support policies for the development of renewable electricity. At the same time, along with socioeconomic development, the gap between urban and rural labor prices tends to narrow, and the negative environmental externalities of traditional power will also decline due to technological progress; these factors will lead to a reduction in the positive externalities of biomass power generation, which in turn will affect the governmental subsidy intensity for the development of renewable electricity, thereby reducing the investment feasibility of a biomass power generation project.

3.6.2. The Issue of Monetization Assessment of the Positive Environmental Externality

A monetization assessment of the positive environmental externality of biomass power generation is a complicated process. The value of the positive environmental externality depends on many parameters. Among these parameters, the most significant parameter is the market value of the environmental loss per unit of pollutant emission, namely the market price of pollutants. The value of this parameter varies greatly from region to region. The same is true for the calculation of social benefits of the positive externalities of biomass power generation. In comparison, the calculation process of the government's financial subsidies is relatively simple, and there are few different viewpoints on the calculation results. The degree of internalization of the positive externality of biomass power generation mainly depends on the accuracy of the positive externality's calculation results, which is the reason why a monetization assessment of the scale of positive externality is one of the hot spots of academic debate.

4. Conclusions

Based on our research and analysis, we conclude the following:

(1) According to the established system dynamics model, the positive externalities of biomass power generation are influenced by many factors, such as the biomass market price, market price of pollutant emission reduction, labor wage, and so on. These factors have obvious regional differences. Subsidies for biomass power generation based on externalities' internalization should also reflect these differences, so as to adapt to local conditions. If the subsidy policy of biomass power generation is one-size-fits-all, it will inevitably lead to over-development of biomass power generation enterprises in areas where the externality scale is relatively small.

(2) From the current policy environment for biomass power generation, building a biomass power plant using apple branches as the primary biomass fuel is economically feasible, and the assumed biomass power plant has a larger profit margin. The price of biomass fuel, which accounts for 60% of the total cost of biomass power, has a significant impact on whether a biomass power generation project can be profitable. Because of the lower price of biomass fuel, projects of biomass power generation have broad market prospects in areas of western China with an abundance of biomass power fuel.

(3) It is difficult for a biomass power generation project to be profitable without governmental support policies. Comparing the scale of the positive externalities of biomass power generation with the scale of government subsidies, we believe that the scale of government subsidies exceeds the positive externalities that biomass power generation can generate. As far as biomass power generation is concerned, the current price subsidy intensity is about US\$0.0132 higher per kWh than the reasonable level. Furthermore, our study results indicate that the parameters frequently applied in the calculation of the prices of a pollutant emission reduction in Chinese research papers is relatively small, which would be only half of their actual values.

(4) The Chinese government should maintain their policy support intensity for the development of biomass power generation. China's support policy should be adjusted so that the subsidies that enterprises could receive from the government are as close as possible to the externalities that biomass power generation can generate. Whether renewable electricity policy brings about a

fiscal burden for the government is not a correct criterion to judge the rationality of this policy. The scientific criterion is whether this policy can internalize the positive externality of renewable electricity. Accurate monetization measurement of the positive externalities generated by biomass power generation is still a key question.

Supplementary Materials: The following are available online at http://www.mdpi.com/2071-1050/11/12/3426/s1, File 1: Explanation of equations and Parameters of Bioenergy-effect-policy System Dynamics model that make the biomass power generation project economically feasible.

Author Contributions: H.Y. and S.D., conceived the research idea and designed general framework of the research. Y.Y. and H.Y. built SD model, analyzed the data, and wrote the paper.

Funding: This study was funded by the Social Science Planning Project of Shandong Province (No. 18CSJJ14) andResearch Startup Fund of Ludong University (No. LB2017028), the National Natural Science Foundation of China (No. 41301642).

Conflicts of Interest: The authors declare no conflict of interest.

Nomenclature

a_t	coefficient of current year profit
b_t	coefficient of income tax rate
B_t	electricity generated by biomass power plant in t year
CDM	clean development mechanism
CF	cash flow
E_{bst}	increased income from biomass sales in t year
E_b	emission factors of pollutants
E_{et}	positive environmental externality in t year
E_{eit}	benefits of employment increase in t year
E_{st}	social benefits in t year
E_t	positive externality produced by biomass power generation in t year
$EAIT$	earnings before interest and tax
EFB	empty fruit bunch
EM_e	effective emission reduction of eth pollutants of biomass power generation compared with on-grid electricity
ER_d	theoretical emission reduction of pollutants of biomass power generation compared with on-grid electricity
ET	total emissions from the transportation of biomass raw materials
FIT	feed-in tariff
FS_t	government's fiscal subsidy in t year
GHG	greenhouse gas
I_t	income tax exempted in t year
IRR	Internal rate of return
LHV	lower heating value
M_f	annual biomass requirement
NPV	net present value
P_e	market price of the environmental loss generated by the eth pollutant
P_t	current year profit of biomass power plant
PBP	payback period
Q	indicator of externality internalization
RPS	renewable portfolio standards
S	initial investment
T_o	output tax
T_i	input tax
$TGCs$	tradable green certificates
TI	total investment
TT	truck load per trip
V_t	value-added tax exempted

References

1. Benato, A.; Macor, A. Italian Biogas Plants: Trend, Subsidies, Cost, Biogas Composition and Engine Emissions. *Energies* **2019**, *12*, 979. [CrossRef]
2. Kern, F.; Kivimaa, P.; Martiskainen, M. Policy packaging or policy patching? The development of complex energy efficiency policy mixes. *Energy Res. Soc. Sci.* **2017**, *23*, 11–25. [CrossRef]
3. Kwon, T.H. Is the renewable portfolio standard an effective energy policy? Early evidence from South Korea. *Util. Policy* **2015**, *36*, 46–51. [CrossRef]
4. Kwon, T.H. Rent and rent-seeking in renewable energy support policies: Feed-in tariff vs. renewable portfolio standard. Renew. Sustain. *Energy Rev.* **2015**, *44*, 676–681. [CrossRef]
5. García-Alvarez, M.T.; Cabeza-García, L.; Soares, I. Analysis of the promotion of onshore wind energy in the EU: feed-in tariff or renewable portfolio standard? *Renew. Energy* **2017**, *111*, 256–264. [CrossRef]
6. Zhang, Y.Z.; Zhao, X.G.; Ren, L.Z.; Zuo, Y. The development of the renewable energy power industry under feed-in tariff and renewable portfolio standard: A case study of China's wind power industry. *J. Clean. Prod.* **2017**, *168*, 1262–1276. [CrossRef]
7. Chinairn. China's Renewable Energy Generation Subsidy Gap has Exceeded 120 Billion CNY in 2018. Available online: http://www.chinairn.com/news/20180613/143748101.shtml (accessed on 13 June 2018).
8. National Energy Administration. Notice on the Implementation of Renewable Energy Electricity Portfolio Standards. Available online: www.nea.gov.cn/2018-11/15/c_137607356.htm (accessed on 15 November 2018).
9. Dong, S.C. Feed-in tariff vs. renewable portfolio standard: An empirical test of their relative effectiveness in promoting wind capacity development. *Energy Policy* **2012**, *42*, 476–485. [CrossRef]
10. Wang, B. Study on the Ways and Performance of Internalization of Positive Externalities. *Southeast Acad. Res.* **2002**, *2*, 158–165.
11. Wang, J.N. *Environmental Economics: Theory, Method and Policy*; Tsinghua University Press: Beijing, China, 1994.
12. Meng, C.; Gao, W. Establishing Support and Incentive Mechanism to Promote the Development of Agricultural Circular Economy. *Farmer's Daily*, 7 May 2013.
13. Ding, H.P.; He, M.F.; Deng, C. Lifecycle approach to assessing environmental friendly product project with internalizing environmental externality. *J. Clean. Prod.* **2014**, *66*, 128–138. [CrossRef]
14. Dong, S.C.; Yu, H.L.; Li, Y.; Li, Z.H.; Li, F.; Li, F.J. Analysis of Driving Factors of Development of China's Industrial Circular Economy. Chin. *J. Popul. Resour. Environ.* **2016**, *26*, 27–34.
15. Ci, F.Y. Study on Formation Mechanism, Structure and Optimization Countermeasures of Regional Cyclic Innovation System. *Morden Econ. Res.* **2016**, 7–8. [CrossRef]
16. Zhu, T.; Gao, S. Thinking on the key link of institutional system of ecological civilization construction. *Environ. Prot.* **2014**, *16*, 11–12.
17. EMF. Towards the Circular Economy: Opportunities for the Consumer Goods Sector. Available online: https://www.ellenmacarthurfoundation.org/publications/towards-the-circular-economy-vol-2-opportunities-for-the-consumer-goods-sector (accessed on 20 February 2019).
18. Murray, A.; Skene, K.; Haynes, K. The Circular Economy: An Interdisciplinary Exploration of the Concept and Application in a Global Context. *J. Bus. Eth.* **2017**, *1403*, 369–380. [CrossRef]
19. Picasso, E.; Escobar, M.B.; Harris, M.S.; Tanco, F. Measuring the externalities of urban traffic improvement programs. *Habitat Int.* **2016**, *55*, 10–16. [CrossRef]
20. Delucchi, M.A.; Murphy, J.J.; McCubbin, D.R. The health and visibility cost of air pollution: A comparison of estimation methods. *J. Environ. Manag.* **2002**, *64*, 139–152. [CrossRef]
21. Quah, E.; Boon, T.L. The economic cost of particulate air pollution on health in Singapore. *J. Asian. Econ.* **2003**, *14*, 73–90. [CrossRef]
22. Hao, J.M.; Li, J.; Duan, L.; He, K.B.; Dai, W.N. Valuation of forest damage cost from SO2 emission: A case study in Hunan province. *Environ. Sci.* **2002**, *6*, 1–5.
23. Yang, J.X.; Li, H.Q. Investigation and thinking on crop straws in Jingning County. *Gansu Agric.* **2013**, *13*, 56–57.
24. Nguyen, T.L.T.; Laratte, B.; Guillaume, B.; Hua, A. Quantifying environmental externalities with a view to internalizing them in the price of products, using different monetization models. *Res. Conserv. Recycl.* **2016**, *109*, 13–23. [CrossRef]

25. Zhao, X.L.; Cai, Q.; Ma, C.B.; Hu, Y.A.; Luo, K.Y.; Li, W. Economic evaluation of environmental externalities in China's coal-fired power generation. *Energy Policy* **2017**, *102*, 307–317. [CrossRef]
26. Ding, H.P.; Liu, L.; Fang, O.Y. Analysis of social cost internalization and investment valuation of reverse logistics. In Proceedings of the 2008 IEEE International Conference on Service Operations and Logistics, and Informatics, Beijing, China, 12–15 October 2008; pp. 2278–2283.
27. Li, S.Y. Study on Apple Pruning Branches as Biomass Source and Renewable Energy in Shaanxi Province. Master's Thesis, Northeast A&F University, Yangling, China, 2009.
28. Liang, L.Y. *Study and Promotion of Cultivating Shii—Take Using Branches of Fruit Tree as Substrate in Weibei Highland*; Northwest A&F University: Yangling, China, 2006.
29. Fan, H.X. Research and Popularization of Ecological Recycling Economy Model of "Fruits, Livestock and Methane" in Jingning County. *Xiandai Hortic.* **2013**, *18*, 22.
30. Lu, X.S.; Wang, G.H. Lingbao city develop circular economy of apple industry vigorously. *Farmer's Daily*, 20 December 2007.
31. An, K.C. Experience of Development of Circular Economy Centering on Fruit Industry in Luochuan County. In Proceedings of the 2006 Annual Meeting of China Association of Agricultural Science Societies, Beijing, China, 23–26 July 2006; pp. 270–271.
32. Lü, T.X. Reflections on the Development of Rural Biogas Transformation in Jingning County. *Mod. Landsc. Archit.* **2015**, *23*, 18–30.
33. Wang, X.H. Existing Problems and Measures of development of Apple Industry in Jingning County. *Developing* **2016**, *10*, 20.
34. Ma, F.P.; Han, X.F. Study on development mode of participatory and village-based poverty alleviation in Jingning County. *Anhui Agric. Sci. Bull.* **2013**, *19*, 5–6.
35. Wang, J.H.; Yan, W.G. Suggestions on Changing the Way of Rural Energy Development in Jingning County. *Agric. Sci.-Technol. Inf.* **2015**, *24*, 95–96.
36. Propaganda Department of Jingning County Party Committee. *One Industry One County in Order to Get Rid of Poverty and Achieve Rich*; Propaganda Department of Jingning County Party Committee: Lishui, China, 2013.
37. Compiling Committe of Jingning County Annals. *Jingning County Annals 1986–2002*; Zhonghua Book Company: Beijing, China, 2005.
38. Jingning County Party Committee Office. General situation of Jingning County. Available online: http://www.gsjn.gov.cn/wap/zjjn/jngk/ (accessed on 9 March 2019).
39. Jiang, W.Q.; Zhou, Z.Y.; Qin, Y.; Zou, L.N.; Yan, S.Y.; Li, X.Z.; Tian, F.Y. Study on heating value of forages and crop straws in Tibetan Autonomous Region. *Pratacultural Sci.* **2010**, *2707*, 147–153.
40. Ning, Z.L.; Chen, H.J.; Wang, Z.N.; Zhang, Z.W.; Qiu, Y.J. A study on the dynamic change of gross caloric value and ash content of the several tall grasses. *Acta Pratacul. Sin.* **2010**, *192*, 241–247.
41. Whittaker, R.H. *Communities and Ecosystems*; Science Press: Beijing, China, 1997.
42. Yang, D.H.; Li, H.L. Environmental pollution loss accounting based on damage and cost- the case of Shandong Province. *China Ind. Econ.* **2010**, *7*, 125–135.
43. Zhan, D.C.; Han, B. Investigation Report of Biomass Power Generation Plants and Generation Cost Research. 2011. Available online: https://wenku.baidu.com/view/aa5becc4aa00b52acfc7ca9d.html (accessed on 8 February 2019).
44. Fang, X.J. Grid-connected biomass power generation amounts to 35.6 billion kWh last year. *China Energy News*, 26 May 2014.
45. Samuelson, P.A.; Nordhaus, W.D. *Economics*, 16th ed.; Huaxia Press: Beijing, China, 1999.
46. IRENA. Biomass for Power Generation, Renewable Energy Technologies: Cost Analysis Series, 1. IRENA Secretariat: Abu Dhabi, UAE. Available online: https://www.irena.org/DocumentDownloads/Publications/RE_Technologies_Cost_Analysis-BIOMASS.pdf (accessed on 22 June 2016).
47. He, Z.C.; Yuan, Z.L. Analysis and research of various technical routes of crop straw power generation. *Energy Res. Util.* **2008**, *2*, 29–33.
48. Liu, Y. Design and Application of Biomass Boiler Burning Control Scheme. Master's Thesis, North China. Electric Power University, Beijing, China, 2012.
49. Cheng, X.Y.; Niu, Z.Y.; Yan, H.M.; Liu, M.Y. *Building of Heating Value Model of Straw Biomass Based on Industrial Analysis Indexes*; Transactions of the Chinese Society of Agricultural Engineering: Beijing, China, 2013.

50. Verma, V.; Bram, S.; Delattin, F.; Laha, P.; Vandendael, I.; Hubin, A.; De Ruyck, J. Agro-pellets for domestic heating boilers: Standard laboratory and real life performance. *Appl. Energy* **2012**, *90*, 17–23. [CrossRef]
51. Xie, Y.L.; Peng, D.S.; Xu, H.W. *Application of System Dynamics in Financial Evaluation of Construction Projects*; Metallurgical Industry Press: Beijing, China, 2010.
52. Malek, A.; Hasanuzzaman, M.; Rahim, N.A.; Al Turki, Y.A. Techno-economic analysis and environmental impact assessment of a 10 MW biomass-based power plant in Malaysia. *J. Clean. Prod.* **2017**, *141*, 502–513. [CrossRef]
53. Nishitani, K.; Kaneko, S.; Fujii, H.; Komatsu, S. Effects of the reduction of pollution emissions on the economic performance of firms: An empirical analysis focusing on demand and productivity. *J. Clean. Prod.* **2010**, *19*, 1956–1964. [CrossRef]
54. Dean, J.M. Does trade liberalization harm the environment? A new test. *Can. J. Econ.* **2002**, *35*, 819–842. [CrossRef]
55. Cai, H.; Xie, S.D. *Determination of Emission Factors from Motor Vehicles Under Different Emission Standards in China*; Acta Scientiarum Naturalium Universitatis Pekinensis: Beijing, China, 2010; Volume 46, pp. 319–326.
56. WNA.Comparison of Lifecycle GHG Emissions for Various Electricity Generation Sources. World Nuclear Association Report, London, UK. Available online: www.worldnuclear.org/World-Nuclear-Association/Publications/Reports/Lifecycle-GHGEmissions-of-Electricity-Generation (accessed on 10 October 2015).
57. Ye, X. Study on Characteristics of Pollutants Emission from Non-Road Mobile Source and Biomass Boilers on Real Work Conditions. Master's Thesis, South China University of Technology, Guangzhou, China, 2018.
58. Forrester, J.W. Industrial dynamics: A major breakthrough for decision makers. *Harv. Bus. Rev.* **1958**, *36*, 37–66.
59. Chang, Y.C.; Hong, F.W.; Lee, M.T. A system dynamic based DSS for sustainable coral reef management in Kenting coastal zone, Taiwan. *Ecol. Model.* **2008**, *211*, 153–168. [CrossRef]
60. Wang, Y.Q.; Zhang, X.S. A dynamic modeling approach to simulating socioeconomic effects on landscape changes. *Ecol. Model.* **2001**, *140*, 141–162. [CrossRef]
61. Tao, Z.P. Scenarios of China's oil consumption per capita (OCPC) using a hybrid Factor Decomposition-System Dynamics (SD) simulation. *Energy* **2010**, *35*, 168–180. [CrossRef]
62. Wang, J. The Installed Capacity of Biomass Power Generation reached 12.14 Million kWh. *China Electric Power News*, 15 July 2017.

© 2019 by the authors. Licensee MDPI, Basel, Switzerland. This article is an open access article distributed under the terms and conditions of the Creative Commons Attribution (CC BY) license (http://creativecommons.org/licenses/by/4.0/).

Article

Selection of the Most Sustainable Renewable Energy System for Bozcaada Island: Wind vs. Photovoltaic

Elif Oğuz [1,*] **and Ayşe Eylül Şentürk** [2]

[1]   METUWIND, Hydromechanics Laboratory, Civil Engineering Department,
      Middle East Technical University, 06800 Ankara, Turkey
[2]   Hydromechanics Laboratory, Civil Engineering Department, Middle East Technical University,
      06800 Ankara, Turkey
*   Correspondence: elifoguz@metu.edu.tr

Received: 24 June 2019; Accepted: 23 July 2019; Published: 29 July 2019

Abstract: Energy production without destroying the environment has been one of the most crucial issues for people living in today's world. In order to analyze whole environmental and/or economic impacts of the energy production process, life cycle assessment (LCA) and life cycle cost (LCC) are widely used. In this study, two distinct renewable energy systems are assessed. First, a land-based wind farm, which has been operating in Bozcaada Island since 2000, is compared to a proposed solar photovoltaic power plant in terms of Energy Pay-Back Time (EPBT) periods and greenhouse gas (GHG) emissions and life cycle cost. The energy production process including the recycling phase evaluated "from cradle to grave" using GaBi software for both cases. All scenarios are compared by considering different impact categories such as global warming potential (GWP), acidification potential (AP), and eutrophication potential (EP). Following this, levelized unit cost to produce 1 MWh electricity (LUCE) is calculated for both systems. This study revealed that LCA and LCCA are useful and practical tools that help to determine drawbacks and benefits of different renewable energy systems considering their long-term environmental and economic impacts. Our findings show that onshore wind farms have a number of benefits than proposed photovoltaic power plants in terms of environmental and cost aspects.

Keywords: LCA; LCC; photovoltaic; onshore wind; renewable energy

---

1. Introduction

Climate change is one of the most critical issues for the future of the world. One of the major reasons for climate change is greenhouse gas (GHG) emissions. Therefore, reduction in greenhouse gas emissions is recognized as an important step [1] to avoid the devastating effects of climate change. As the focus on reduction in GHGs increases, researchers are seeking ways to mitigate climate change by applying environmentally-friendly solutions such as green technologies, hybrid systems, and renewable energy systems. In this context, a significant amount of research and development activities have been carried out to determine feasible renewable energy systems for a wide range of engineering fields. The International Maritime Organization (IMO) introduced a number of regulations such as maritime pollution (MARPOL) Annex VI (2014). According to MARPOL Annex VI, oil companies and engine manufacturers have to follow stringent rules and guidelines in order to decrease SOx and NOx emissions [2]. Since the regulatory legislation for ship emissions in the maritime industry has brought the need for improving energy efficiency for environmental benefits, organizations, and governments that started to give funds for research and development studies in this field. Recently, three HORIZON2020 Projects (namely Ship Life cycle software solutions (SHIPLYS, HOLISHIP and LINCOLN) are funded by the European Union (EU) [3], which aims to develop the software integration process for ships considering environmental benefits and their outcomes published in References [4–9]. Integration of renewable energy technologies to short-route ferries [10] are also carried out in order to emphasize the importance of such systems in reducing the emissions for the countries that currently have no stringent regulation. Similarly, renewable energy sources (RES) are suggested by a number of researchers [11–15] to reduce air pollution and climate change instead of traditional energy production systems where fossil fuels utilize. The integration of renewable systems with the idea of local co-production is considered an excellent way for mitigating climate change [16]. In recent years, sustainable building with the idea of local co-production has been investigated all around the world [17–19] in order to follow International Partnership for Energy Efficiency Cooperation (IPEEC) regulations [20] for zero energy building with the aim to decrease in emissions. After successful applications in the construction sector, some governments started to give funds for the integration of renewables for larger-scale construction areas, namely, new cities. In the 21st century, smart city concepts have been developing to integrate renewable technologies [21] into the production of self-electricity requirements of the new cities. One of the main reasons of enormous greenhouse gas emissions are considered as the emissions from the maritime application as well as the self-electricity need of an island, known as hydrogen islands [22], can be one of the solutions for decarbonizing maritime transportation by means of hydrogen refueling stations. In this study, Bozcaada Island is chosen not only to be the potential of a smart city but also to be the first hydrogen island of Turkey, according to the United Nations Industrial Development Organization (UNIDO) project [23]. Bozcaada island has significant solar energy potential, 308.0 cal/cm^2 sunshine radiation per day, and 7.5 h sunshine duration per day, as well as wind energy potential with 8.4 m/s average wind speed at 50 m [24]. Although there is an operating wind farm in Bozcaada island, investigation of the photovoltaic power plant should be carried out due to solar energy potential in this region. Two different configurations, namely, existing wind farm and proposed photovoltaic (PV) plant, are compared in terms of environmental and economic aspects by using life cycle assessment (LCA) and life cycle cost analysis (LCCA). Both previously mentioned renewable energy technologies are evaluated in terms of "cradle to grave perspective". The production processes of raw materials are included throughout the analysis even though transportation of raw materials such as silica in the LCA of the PV system and steel in the LCA of onshore wind farm are excluded.

The introduction starts with research motivation of the study and a review of LCA applications in various industries to determine the applicability of the method. Following this, limited application of solar energy and LCA of the renewable energy systems are summarized. The model structures and assumptions related with renewable energy technologies and lifecycle inventory analyses are presented in Sections 3 and 4. A sensitivity analysis is carried out for both systems in order to determine the most feasible recycling strategy, which is presented in Section 5.3. However, the end of life approach is considered only as a suggestion in the case of a PV system due to the fact that there is still no strict way to recycle the procedure of photovoltaic technology. Both technologies are modelled and evaluated using GaBi software. Following the evaluation of environmental impacts, life cycle cost analyses are carried out with both renewable systems. LCA and LCCA results are presented in Section 5. Lastly, discussions are made in Section 6 and future directions are mentioned in Section 7.

1.1. Research Motivation

One of the key objectives of this study is carrying out LCA and LCCA of the photovoltaic power plant since there is no specific study published in Turkey. In this study, photovoltaic power is evaluated for the first time in terms of environmental impacts in Turkey.

1.1.1. Review of Life Cycle Assessment (LCA) and Life Cycle Cost (LCC)

LCA methodology can be defined as an evaluation procedure of a product, a process, or a system in terms of environmental characteristics with a "cradle to grave perspective." By means of this method, the systems, containing extraction or acquisition of raw materials and manufacturing, transportation of raw materials, and construction procedures starting with infrastructure and finishing with the installation of design structure, use process and reuse, recycling, or disposal of waste materials that can be evaluated for an entire life. LCA methodology has a widespread utilization area around the world. For example, LCA in the infrastructure structure [25] is applied for capturing all changes due to the decisions instead of utilization of embodied carbon calculator tools in order to manage emissions of infrastructure projects. As another example from the infrastructure sector, the requirement of integration of disassembly and deconstruction phase into building information systems [26] are shown by means of LCA methodology. Apart from infrastructure, Australian read meat supply chains focusing on Australian beef and lamb exported to USA [27] is evaluated in terms of environmental impacts with the application of LCA. In the case of the energy sector, Atilgan and Azapagic [28] state that the expansion of renewable technologies is essential for Turkey's electricity from the medium to long term and Atilgan and Azapagic [29] focus not on the sun but on renewable electricity generated with the utilization of hydro, wind, and geothermal sources in Turkey. As another example of the energy sector, the electricity mix of French territories are studied with the cradle-gate electricity production model and Guyana is reported as the lowest GHG emissions than other islands by Rakotoson and Praene [30].

As an economic evaluation procedure, the life cycle cost analysis, which aimed to predict the total cost of a system, a product, or a process throughout the lifespan, is becoming popular around the world. For instance, the analysis of life cycle cost is carried out by Utne [31] for Norwegian fishing fleet whether LCC can be used as a tool to improve sustainability. In automotive manufacturing, the life cycle cost analysis is conducted [32] for a distinct type of composite materials in order to select an appropriate one for the design of a lightweight automotive. Jeong et al. [9] tried to draw a framework for selecting an optimal propulsion system by combining the life cycle and cost assessment for the shipping industry. Life cycle cost analysis of a defense electronic system [33] is utilized to be able to determine end-of-life cost during the early design step. LCA and LCC for the electricity mix of Turkey are carried out as a first time by Yılan [34]. She selected a levelized cost of electricity as an indicator for the future electricity mix.

The leading software products, which are GaBi and SimaPro, are utilized in the market more than 20 years [35] as life cycle assessments tool. Although there are few studies related with the compassion of LCA software tools, the comparison between GaBi and SimaPro was conducted for the packaging system [36] and continued with simplified systems [37] for the creation and disposal of four basic materials, which are aluminum, polyethylene terephthalate, glass, and corrugated board. All of these are 1 kilogram. Despite the fact that the comparison of the simplified systems concluded that the differences of impact analysis, which is greater than 20%, are caused by the differences in the characterization factors used by the two programs, it is reported that [38] there is no difference between using SimaPro and GaBi for ordinary and even skilled LCA software users in terms of the capability to detect potential errors. GaBi software is presented as the best one [39] in terms of user-friendliness, service, and functionality whereas the best specification for SimaPro is determined to be the cost. Due to a user-friendly specification, GaBi is utilized for the modelling and evaluation procedure of this study.

1.1.2. LCA Applications of Renewable Technologies

Since two distinct renewable systems to generate electricity in Bozcaada Island are investigated in this study. Previous LCA studies of solar panels and onshore wind farm applications are presented in the following.

LCA of Solar Panels

Application of life cycle assessments methodology for the open ground mounting is very rare in the literature since the grid connection of photovoltaic technology and its application is relatively newer than the grid connection of wind power. For example, the installation of photovoltaic power plant composed of multi-Si solar panels is recommended Fu et al. [40] without taking into account therecycling phase of plants even though multi-Si production leads the highest contribution to environmental impacts. Although there are a lot of LCA studies in the literature, they differ from each other by either one aspect or more since there are a lot of classifications of photovoltaic electricity. Classification type of photovoltaic technology can be summarized as the type selection of solar modules and installation styles of solar panels basically. Solar cells produced a multi-Si wafer has the most widespread application than other type of solar cells, including mono-crystalline, thin-film layer, CdTe, and CIS. In the case of installation styles, ground mount installation, namely land-based application, and roof-top installation, which can be categorized as slanted-roof and flat-roof installation, can be regarded. In addition to this, photovoltaic systems are divided into two basic groups as stand-alone (independent, off-grid) PV systems and network connected (grid-connected, on-grid, grid-tied) systems. In order to make a meaningful classification, LCA studies about photovoltaic electricity especially based on crystalline-cell type tabulated in Table 1 by noting the results of energy pay-back time and GHGs. In Table 1, the first row of phases shows the onset of the life cycle analysis as the second row demonstrates end phases of the research studies listed.

Since LCA of PV systems considering the "cradle to grave" perspective are limited, as indicated by Nugent and Sovacool [41].

Table 1 presents publicly available studies for different locations in the world.

LCA of the Onshore Wind Farm

Applications of life cycle assessment methodology for onshore wind farm is widespread around the world as opposed to a land-based photovoltaic power plant. For example, in the European region, the climate change impacts during the life cycle and energy payback time (EPBT) of the onshore wind plant are reported as less than 7 g CO_2-eq./kWh and 7 years, respectively [42]. In the case of Texas [43], GHG emissions are classified for different turbine sizes. Global warming potential of 1 MW is 7.35 g CO_2-eq. GWPs of 2MW and 2.3 MW turbines are found as 7.09 g CO_2-eq. and 5.84 g CO_2-eq., respectively. The comparison between the model type of wind turbines is conducted [44] and EPBTs are calculated as 0.43 and 0.53 years for the selected turbine models. As a result, the LCA application of wind energy is more common than the photovoltaic system since wind power technology is older than photovoltaic technology.

LCA of Photovoltaic vs. Wind Systems

To the author's best knowledge, research on the selection of appropriate systems for the selected region is limited. Schmidt, et al. [45] used a functional unit in order to adopt the amount of electricity required for 10,000 citizens in Toronto. The existing photovoltaic system and proposed wind farm are compared in terms of environmental impacts and damage assessment. Although the PV plant demonstrates less damage to human health than the wind system, the wind farm shows lower impact to ecosystems than the PV plant for the Canadian region.

Table 1. Summary of LCA studies about photovoltaic electricity.

Phases		Installation		Solar Cells		PV System		GHGs [kg/MWh]	EPBT [years]	Location	References
Onset	Final	Roof	Ground	Module Efficiency	Cell Type	On-Grid	Off-Grid				
Material choices	Recycling	√	√	-	All	√	√	-	-	Indonesia	[46]
Extraction of raw materials	Recycling	-	√	14.4%	mc-Si	√	-	106	4.17	Perugia, Italy	[47]
Production	O&M	-	√	12.8%	multi-Si	-	-	12.0	1.9	Gobi Desert	[48]
Production	O&M	√	-	-	multi-si	-	-	20.9–30.2	1.01–1.08	Singapore	[49]
Production	Recycling (BOS)	√	-	14.2%	mono-si	-	√	0.053	2.5	Copenhagen, Denmark	[50]
Production	O&M	√	-	14.7%	multi-si	√	-	-	2.33	Australia	[51]
Production	Recycling or disposal	-	√	-	multi-Si	√	-	-	2.3	China	[52]
Production	O&M	-	√	12.8%	multi-si	√	-	13.9–14.9	2.2–2.3	Gobi Desert	[53]
Production	Disposal	-	√	-	multi-si	√	-	with ReCiPe method		Toronto	[45]
-	Disposal or recycling	-	-	-	multi-si	-	√	-	-	-	[54]
Production	Disposal	-	√	12.2%	BOS (multi-si)	√	-	$29/m^2$	0.21–0.37	Springerville, USA	[55]

2. System Boundaries and Methodology

This section introduces the approach of LCA including the framework, assumptions, and related activities. The formulas associated with the LCC are given to demonstrate the calculation part of the life cycle cost section of all technologies.

2.1. System Boundaries

In this study, two different configurations of power plants have been selected. As a general assumption, transportations of primary raw materials such as silica and steel are not considered, since the wafer is produced in Taiwan in the case of a photovoltaic power plant and nacelle, rotor, and tower are produced in Germany for the wind farm case. Relevant processes for production of primary raw materials are included, but transportations of related materials, which is necessary for their production, are excluded throughout this study. GaBi is used to establish LCA modelling and assessments. Furthermore, the circular economy approach is applied for scrapping materials for both systems. Other assumptions related to the technologies are listed in Sections 3.1 and 4.1 which are model structure parts for the PV system and wind farm, respectively.

A normalization procedure is applied in order to present the results as a single emission type. In other words, different emissions like acidification potential and eutrophication potential convert into an equivalent quantity of CO_2 by means of the normalization process. In this study, CML2001-Jan 2016 database [56] is used from GaBi. For example, the emissions caused by global warming potential are represented as kg CO_2-equivalent. The functional unit of LCA is taken as the unit of power, MWh. Additionally, kg CO_2-eq./MWh is the measure parameter for global warming potential, MJ/MWh is the parameter for cumulative energy demand (CED), year for energy pay-back time, and money for the lifecycle cost.

The main purpose of the study is the comparison of different renewable systems for Bozcaada Island in order to select the most sustainable option. Therefore, the selection of energy mixes is based on the location where the materials are produced, bought, and transported, which is carried out separately for each system and summarized in related assumption tables in the model structure and assumption sections. Section 3.1 for the photovoltaic power plant and Section 4.1 for the onshore wind farm.

2.2. Life Cycle Assessment (LCA) and Life Cycle Cost (LCC)

Life cycle assessment can be defined as a method to evaluate environmental characteristics of a system, product or processes [57] and [58].

2.2.1. Life Cycle Assessment (LCA)

Life cycle stages are divided into four main phases (namely production, construction, operation and maintenance, and decommissioning and recycling) following the statement in the International Agency Report Methodology Guidelines on Life Cycle Assessment of Photovoltaic Electricity [59].

Production Phase

It starts with raw material extraction, which includes manufacturing all the components including infrastructure materials and transmission materials utilized to the grid.

Construction Phase

It starts with transportation of all materials to the operation site. It includes the test procedure for initialization as well as commissioning of the system together with the foundation and supporting structures.

Operation and Maintenance Phase

It starts with electricity production and it consists of dusting and cleaning and periodic controls of the power plant as a maintenance procedure.

Decommissioning and Recycling (or Disposal) Phase

It includes disassembly of plant parts and decomposition and separation of them for recycling or disposal.

2.2.2. Life Cycle Cost (LCC)

The life cycle cost (LCC) is a technique to evaluate all costs during the life cycle [60]. For the case of life cycle costs, design costs containing feasibility and improvements of projects are not considered. In terms of feasibility costs, both systems are renewable and the selected region has high potential for each case, according to the UNIDO project [23]. Project costs might be considered approximately the same for both technologies since the same land is selected for their applications. Since the purpose of the study is to create the framework of the selection between two distinct renewable technologies in terms of life cycle specifications, life cycle costs are divided into three categories. During the procedure, labor costs is not included in LCC calculations since labor costs are expected to be the same due to the requirement of qualified workers for both systems. In this study, the costs of material flows are only considered and are divided into three parts as follows.

Initial Investment Costs

All costs like materials of the systems and construction period are included. Construction costs such as infrastructure for the investments are evaluated in this section for both cases. Costs of material transportation to construction site are included in the initial investment costs.

Operation and Maintenance Costs

Materials required during the operation procedure like lubricants for wind farm and maintenance procedure like tap water for the cleaning of the PV system are considered. The cost of the replacements of components is another item that is included in operation and maintenance costs. The cost of transportations of either the replacements of components or transportation of necessary materials for the maintenance procedure are considered in the operation and maintenance costs even though they have minor effects as opposed to costs of transportation in the initial investment process.

Disposal or Recycling Costs

Costs for decomposition of the plants and transportation costs for the decomposed materials are included. Decomposed materials are transported to either the recycling facility or to the landfill area for treatment.

In part of the life cycle cost, there is an assumption related to the cost calculation that is the difference between the costs of the projects of two renewable systems, which can be neglected due to the fact that the same land area is selected for both cases. Therefore, Equation (1) is developed for the life cycle cost of renewable energy technologies in the light of the explanations for three cost categories in order to make the comparison possible between two configurations.

$$LCC = C_{inv} + C_{O\&M} + C_{DorR}, \tag{1}$$

where $C_{O\&M}$ indicates operation and maintenance costs, $C_{inv.}$ and C_{DorR} show initial investment costs, and costs of the phase of disposal or recycling, respectively. $C_{inv.}$ can be utilized as the initial investment cost with the shortage of IIC shown, especially in tables in the rest of the study.

Details of the costs for the configurations are classified in the associated sections for both the photovoltaic power plant and the wind farm.

3. Ground-Mounted Photovoltaic Plant

3.1. Model Structure and Assumptions

In this section, a ground mounted photovoltaic system is proposed on the site that is currently the location of Bozcaada wind farm (see Figure 1). The total area of wind farm is taken as 20,560 m^2 based on discussions with experts from the operating company. Since this study aims to compare two configurations, the total area of the PV system should be considered close to this area. Moreover, land-use requirement of fixed-tilt PV system is estimated by Denholm and Margolis [61] as 3.8 acres/MWac in the United States. The land use requirement is calculated as 17,391 m^2/MW for the Konya Plain Region [62] even though there is no information for the Bozcaada region. The PV system capacity can be estimated to be approximately 1.2 MW for the same land area of wind farm since land occupation for 1 MW photovoltaic plant is around 16,400 m^2, which was calculated by taking the average of land use requirements stated for the U.S. and Turkey.

Figure 1. Selected site on Bozcaada Island [63].

Due to long exposure time of environmental conditions, the efficiency of solar panels decreases. This is called *degradation*. The main factors of degradation are regarded as temperature differences and humidity conditions and soiling of solar panels. Total energy production should be estimated correctly with appropriate degradation ratio. Therefore, the overall system degradation ratio for the photovoltaic system based on polycrystalline cells are assumed to be 0.6% per year for this study [64].

The lifetime of the photovoltaic power plant is assumed to be 30 years, according to the list, which is in the part of lifetime [65].

The performance ratio is another important factor in order to predict total energy production of the proposed photovoltaic system throughout its life. The performance ratio can be defined as the ratio of actual output and theoretical output for an ideal case. In other words, in real life, actual electricity production of a PV system is less than the calculated one for an ideal case. For this purpose, the appropriate performance ratio is substantial for predicting total energy production. According to Karadogan et al. [66], on-grid applications of the photovoltaic system was started in 2012 in Turkey. Therefore, actual data for electricity production of these systems is not sufficient to estimate a reliable performance ratio in Turkey. In the previously mentioned study [66], it has been found that the deviation between the real energy production and the estimated values for electricity production for seven PV power plants in different sites of Turkey is around 5.54% and, in most of the cases in the study, the Photovoltaic Geographical Information System (PVGIS) database demonstrates closer results. Therefore, the PVGIS calculator is utilized for the expected electricity production of the proposed PV power plant for this study. The performance ratio of the photovoltaic system is determined to be 0.80 by the PVGIS calculator. This performance ratio is also recommended for a ground mounted system [67]. In the PVGIS calculator, the optimum design is selected for the array of solar panels with 32° slope and 3°azimuth angles for Bozcaada Island.

Assumptions based on the photovoltaic system are listed in Table 2.

Table 2. Assumptions for the proposed PV system.

PV system Assumptions and Specifications		Comments
Nominal power of photovoltaic plant	1.2 MW	Extrapolation with land–use for 1 MW
Area of plant for infrastructure	20,560 m^2	Established wind farm area
Plant area	16,400 m^2	Land requirement for 1MW
Performance ratio	0.80	PVGIS database and Reference [67]
Degradation ratio	0.6%	[64]
Lifetime of the plant	30 years	[65]
Lifetime of the inverters	15 years	[65]
Lifecycle inventory assumptions		
Production up to wafer	In Taiwan	Including processes excluding transportation
Electricity mix	Chinese	No Taiwanese grid mix in GaBi
Transportation of wafers	By ocean-going ship	8689 nautical miles as shortest route
Production of solar cell	In Turkey	Assembly with metallization pastes and wafers
Production of other parts	In Turkey	Ground mounting structures (aluminum frames)
Electricity mix	Bulgarian mix	No Turkish grid mix and production in Tekirdag
Transportation to site	By truck and ferry	By ferry to island
Initialization	Greek mix	No Turkish grid mix (for installation)
Operation and Maintenance	By truck	Carrying waterand replacements of broken panels and inverters

In this study, the Eco-Invent database is used for components and processes for the life cycle inventory of the photovoltaic power plant.

3.2. Life Cycle Inventory (LCI) of the Photovoltaic System

All components in order to establish the 1.2 MW photovoltaic power plant are listed in Table 3.

Table 3. Configurations of the proposed PV system.

List of Materials for Solar Cells	
Nominal power of the solar module	265 Wp
Number of solar module	4615 + 15
Number of solar cell in a module	60
Number of solar cell	277,800
Area of one solar cell	243 cm^2
Area of photovoltaic modules	6855 m^2
List of materials used for the support structure (open ground mounting structure)	
Steel, zinc coated	3909 kg
Aluminum	3111 kg

They include solar panels consisting of multi-crystalline cells, inverters, support structures, containing foundation and fence equipment, and are named as an open ground mounting structure in the rest of the study, and electrical components are comprised of cables, low and medium voltage switchboards.

Multi-crystalline solar cells can be produced in Turkey after raw material extraction, which is silicon wafer. Solar cells are made from metallurgical grade silicon. The Bulgarian grid mix is used to produce solar cells since there is no available data for the Turkish grid mix. In addition to this, raw materials are obtained from Taiwan for multi-Si technologies by the selected local company. The Chinese grid mix is used until solar cell production due to the fact that there is no data available for the Taiwanese grid mix. In the production phase, silicon wafers from Taiwan transported by an ocean-going ship and transportation distance is assumed as 8689 nautical miles. Inverter and open ground mounting structure are transported to the site in the construction phase.

The infrastructure requires cleaning of the area and construction of the building, which is necessary for the operation stage. Additionally, it requires roads between arrays due to the requirement of the maintenance period including cleaning solar cells with acetone and changing broken ones. Therefore, in the production stage, necessary materials for infrastructure are added to the open ground mounting structure process to be transported to the site area. The open ground mounting structure process is created as a unit process. Material and energy flows are added to the related process in GaBi. Fence and foundation parts are considered by means of relevant flows inside this process.

For construction, first, the raw material extraction phase excluding transportation is completed. Following this, solar modules, open ground mounting structure, and inverters are transported to the site for assembling the power plant. The transportation distance for solar modules produced in Tekirdag is 291 km. The distance between the city and Kilitbahir ferry dock is 228 km. After 2 km by ferry, the distance between Geyikli and Çanakkale ferry docks is 54 km by truck. Lastly, the distance along the Bozcaada ferry dock up to the construction site is 9 km. Distances for transportation are shown in Table 4 in the section of the life cycle cost.

As the initialization procedure, a unit process is first generated in GaBi and named as electric installation with the extrapolation from 570 kWp photovoltaic plant in the database [68] in order to observe mass and energy flows related lightning protection, cabling in the module area, cabling from the module to the inverter, cabling from the inverter to the electric meter, and the weight of the fuse box. In order to finalize the construction phase, a unit process named as the PV plant installation is created. Additionally, the Greek electricity mix is used to initiate electricity production on this process.

Operation and maintenance stages are considered in the third phase. The utilization of tap water for the cleaning of the solar cell is assumed as the main maintenance procedure of solar modules. In the case of cleaning procedure (throughout the whole life), the transportation distance of truck carrying tap water is assumed to be 80 km. Another procedure for the maintenance is the change of 15 solar modules, which are assumed to break during the lifespan of the plant. In addition, all inverters are changed once since their lifetime is accepted as 15 years, according to the list for the PV system component [65] during the operation phase. Total transportation distance for square parts containing the inverters and 15 broken solar cells is assumed to be 80 km. Total electricity production throughout the whole life of a power plant is estimated to be 52.31 GWh with a 0.6% annual degradation rate and an 80% performance ratio, which are shown in Figure 2.

Lastly, as the fourth phase, which is the deconstruction and recycling stage, two distinct recycling procedures are investigated. For the first case, with the end of life approach, on-site basic deconstruction is assumed. A unit process is created in the software for the recycling procedure. Aluminum, copper, and steel scrap from frames and inverters coming from the balance of the system components are evaluated as recycling materials, and their weights are calculated separately, and solar panels are added to the previously mentioned unit process, which is named as the total weight of solar panels. The rest of the material are added as municipal waste to the structural scrap materials process. Scrap materials and solar modules called structural scrap materials are transported to İzmir, which is 300 km far away from the construction site for the recycling procedure. In the Izmir region, aluminum, copper, and steel scraps from the balance of system can be recycled. However, there is no recycling technology for solar panels in Turkey. The end-of-life approach is applied for solar panels by creating a takeback and recycling unit process and avoid the burden form solar module unit process. According to Frischknect et al. [59], the end-of-life approach is based on the recycling of glass from solar modules. In this approach, there is no exact information related to the recycling process for other materials such as silicon wafer and metallization pastes. The unit process is defined based on the recycling technology for glass in the end-of life approach.

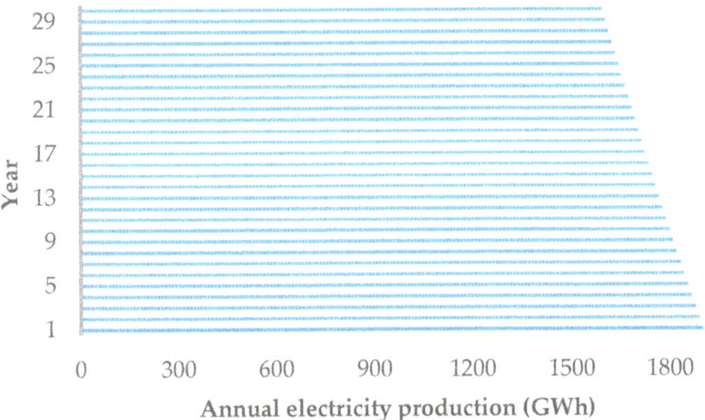

Figure 2. Prediction of electricity production from the PV plant during its life cycle.

As the second case, named as the real recycling plant case, like in the similar case study [47] deconstruction and recycling or disposal stage, disposed solar panels are transferred to the Deutsche Solar AG recycling plant [69] by a cargo plane and only recycled solar modules, with 3.48% mass fraction of total solar panels, are transferred back to the construction site by a truck since there is no recycling plant for solar cells in Turkey. During the first transfer part of the second case, transportation by ferry and truck are neglected due to the fact that a cargo plane is considered for the main transportation vehicle.

Open loop recycling is applied for other parts except solar panels.

In the rest of the study, the first case is mentioned as recycling with the end-of life approach and the second case is named as the real recycling plant case.

In the comparison part of the technologies, which is Section 5, a real recycling plant case is utilized for the photovoltaic power plant in order to make a realistic comparison between the technologies since recycling with the end-of-life approach is an imaginary scenario based on the imaginary recycling plant in İzmir.

3.3. Life Cycle Cost (LCC) of the Photovoltaic System

The LCC model of the PV plant is applied, according to Equation (2), which is adopted with the aid of the LCC equation in the article of Abu-Rumman et al. [70] and the LCC equation, Equation (1), which is developed for this study. Therefore, Equation (2) is utilized for the calculation of the life cycle cost of PV configurations.

$$LCC_{PV} = C_{panels} + C_{P_{elec}} + C_{P_{inf}} + C_{P_{O\&M}} + C_{P_{tr}}, \tag{2}$$

where

$$C_{P_{inv}} = C_{panels} + C_{P_{inf}} + C_{P_{elec}} + C_{P_{t1}}, \tag{3}$$

and

$$C_{P_{tr}} = C_{P_{t1}} + C_{P_{t2}} + C_{P_{t3}}, \tag{4}$$

In the above equations, $C_{P_{tr}}$ is the total transportation costs from cradle to grave. In addition to this, $C_{P_{t1}}$, $C_{P_{t2}}$, and $C_{P_{t3}}$ are the transportation costs of three categories. In other words, transportation costs for the initial investment is $C_{P_{t1}}$, $C_{P_{t2}}$ is transportation costs for the operation and maintenance procedure, and, lastly, $C_{P_{t3}}$ is the transportation cost for disposal or recycling. Therefore,

$$C_{P_{O\&M}} = C_{tap\ water} + C_{spare\ panels} + C_{P_{t2}}, \tag{5}$$

and

$$C_{P_{DorR}} = C_{P_{t3}}, \cdots \tag{6}$$

Details for transportation costs are summarized in Table 4.

The costs of solar panels are calculated with the price of solar panels for the selected company in Tekirdag.

Infrastructure costs for a photovoltaic system requires site preparation initially. However, there is no need for the selected area, as seen in Figure 1 since the selected region has no meaningful slope and there is no vegetation that prevents the application. During site clearance for the infrastructure of the PV system, general cleaning will be sufficient. Therefore, no material flows are considered for general cleaning of the infrastructure. Furthermore, there is the assumption of the elimination of labor costs to make the comparison easier between the renewable configurations with the estimation of approximately equal labor costs for both systems. Therefore, the open ground mounting structure cost can be regarded as one of the sources of the costs for the infrastructure of the photovoltaic power plant. For the costs of open ground mounting structure, the costs of settings, wiring etc. [71] for 1 kW is extrapolated with the inflation rate of Turkey [72] since the production of all other materials except the wafer is assumed to be produced in Turkey. Another source of the infrastructure is the costs of the building and landscape. The working area, which will be used for the operation procedure, is assumed to be 100 m^2 and its costs is assumed to be \$34,180 as in the wind farm case. $C_{P_{elec}}$ consists of the costs of inverters and 9 km-cables either between solar panels or the transmission line to the grid. The Turkish inflation rate is applied to the cost of the grid tie inverter price [71] to obtain the costs of an electrical apparatus containing cables for this study.

Table 4. Transportation costs for the materials of the photovoltaic power plant.

Materials	Weight (kg)	By Truck (km)	By Ferry (km)	Transportation Costs	Cost Categories
Solar panels	77,840	291	10	\$171.56	IIC
Open ground mounting structure	8990	390	8	\$135.50	IIC
Inverters	9849	451	8	\$878.01	IIC
Spare inverters	9849	451 + 80	8	\$1032.60	O&M
Spare solar panels	252	291 + 80	10	\$218.72	O&M
Tap water	46,593	80	-	\$372.20	O&M
Solar panels for recycling	77,776	Without ferry and truck (with cargo plane)		\$8525.77	DorR
Aluminum scrap	3111	300	8	\$71.70	DorR
Copper scrap	1100	300	8	\$33.04	DorR
Steel scrap	8636	300	8	\$111.44	DorR

The costs of operation and maintenance are based on costs of the replacements of spare parts and cost of cleaning the solar panels. Throughout the life of the photovoltaic plant, 46.6 tons tap water is utilized and its transportation distance is assumed to be 80 km similar to the replacement distance for the inverters and spare parts, which are *broken solar panels*, as seen in Table 4. Spare solar panels and spare inverters are not allocated initially due to the difficulty of their protection. Hence, their transportation distances are added to the transportation distances of the original parts.

Transportation costs as seen in Table 4 are related to the weight of materials and distances of the materials and are calculated based on the consumption of diesel. Diesel consumptions are taken from GaBi and the price of diesel is found in the archived list of the BP company [73]. Ultimate diesel price for Istanbul is utilized for the calculation procedure. Solar panels for recycling are transferred with a cargo plane. Kerosene price [74] is taken from the Alibaba website as \$300/tons.

Results of the life cycle costs of not only the PV system but also the wind farm are compared in the life cycle cost analysis part.

4. Onshore Wind Farm

4.1. Model Structure and Assumptions

As seen in Figure 1, there is a wind farm consisting of 17 wind turbines with a linear arrangement. The wind farm has 10.2 MW installed capacity with Enercon E-40 (600 kW) wind turbines [75]. It covers 20,560 m^2 and it is located on the west-side of Bozcaada Island. Tower height is taken as 44 m based on discussions with the experts from the operating company. Considering environmental impacts, the company decided to paint the wind farm using an earth color and to connect turbines by underground wiring which is 9 km long to the central transformer of the island in order to preserve the natural appearance of the island.

Assumptions for the wind farm and the basic characteristics of the wind farm are listed in Tables 5 and 6, respectively.

Enercon E-40 type wind turbines were used in Bozcaada Island. However, there are no detail data for the case of specific tower height on the manufacturer's website. Therefore, tower weight is utilized from the work [76] since tower height is 46 m in that work and it is 44 m in Table 5.

There is no production line for this kind of turbine in Turkey when this wind farm was installed to operate. Therefore, wind turbines are transported from Enercon Company in Germany by truck in the modelling procedure. Transportation distances between Bozcaada Island and Enercon Company in Germany are measured using google maps as 2640 km by truck and 8 km by inland ship to reach Bozcaada Island.

Cables and an inverter for electric installation of the farm were brought from the company's own cable factory in Bilecik. Hence, transportation distance for them was 441 km by truck and 8 km by ferry.

There is no information about acquisition of concrete materials. Hence, transportation distance of concrete is assumed to be 305 km.

Table 5. Assumptions for the wind farm.

Wind Farm Assumptions and Specifications		Comments
Nominal power of photovoltaic plant	10.2 MW	Established wind farm capacity
Area of plant for infrastructure	20,560 m^2	Established wind farm area
Average produced electricity per year	34 GWh	From the discussion with operating company
Lifetime of the plant	20 years	[43]
Lifecycle inventory assumptions		
Production up to wind turbine	In Germany	Including processes excluding transportation
Electricity mix	Deutch	Production in Germany
Transportation of wind turbines	By truck	2640 km
Production of other parts	In Turkey	Concrete, cables, and inverters
Electricity mix	Bulgarian mix	No Turkish grid mix and production in Turkey
Transportation of other parts	By truck	305 and441 km for concrete and cables and inverters
Transportation to site	By truck and ferry	By ferry to the island
Initialization	Greek mix	No Turkish grid mix (for installation)
Operation and Maintenance	By truck	3400 kg lubricant [77]

In GaBi, cables and inverter, nacelle, foundations and roads, tower, and rotor are created as a unit process. However, transportation of primary raw material such as cast iron is not considered since the production line for basic components of wind turbine is in Germany. Although other components such as foundations, roads, cables and inverters are produced in Turkey. Transportation distance for their raw materials are neglected in this study.

Table 6. Basic characteristics of wind farm.

Onshore Wind Farm on Bozcaada Island	
Location of the wind farm	West side of the island
Number of turbines	17
Nominal power of turbine	600 kW
Underground wiring between turbines	9 km
Rotor diameter *	43.7 m
Tower height	44 m

Note: *: It is taken from [78].

4.2. Life Cycle Inventory (LCI) of Onshore Wind Farm

Specifications of Enercon E-40 (600 kW) wind turbines are demonstrated in Table 7. Tower weight is reduced by means of linear interpolation technique to correct material flows. Furthermore, the total weight is decreased for material requirement calculations.

Table 7. Enercon E-40 (600 kW) specifications [78].

Enercon E-40 Wind Turbine	
Power capacity	0.60 MW
Rotor diameter	43.70 m
Tower height	44.00 m
Rotor weight	8.27 tons
Nacelle weight	19.77 tons
Tower weight	29.91 tons
Base weight	220.00 tons
Total weight	277.95 tons

As a beginning, the wind farm is divided into two parts as a fixed part of the wind turbine and moving parts of the wind turbine, which are compatible with GaBi flows. Moving parts of the wind power plant compromises of cables, inverters, nacelle, and rotor.

4.2.1. Moving Parts of the Wind Turbine

Moving parts of the wind turbine contains nacelle, rotor, cable, and an inverter for this study. The transportation distance assumed as 441 km for cables. In GaBi, a unit process is created to mimic cables and the inverter.

Nacelle

It was the main parts of the wind turbine for housing the generator, gearbox, and brakes. It was produced in Germany. It contains steel and a cast-iron part. As a process, the metal roll forming process [79] was applied for its production in GaBi. Nacelle is transported to the construction site by ferry.

Rotor

The rotor contains a hub and blades. Glass fiber, epoxy resin, and cast iron are the materials for its production. It is another component from the moving parts of the turbine.

4.2.2. Moving Parts of the Wind Turbine

It includes tower and foundations as well as access roads requiring to not only construct but also to maintain the wind farm.

Tower

It is a painted steel tube. The main function of it is carrying the rotor and nacelle. As a process, metal roll forming [79] applied for its production in GaBi is similar to nacelle production. The energy requirement of the process and cast-iron parts Deutsch grid mix is selected due to the production of it in Germany when the plant was established.

Foundations and Roads

The foundation is necessary for assembling a wind turbine on it. Access roads are constructed to carry out maintenance in order to unite whole turbine components. Concrete and steel are the main materials to model its production. In the disposal procedure, foundations and roads are left on the plant site in order to construct a new plant in the future.

The lifespan of the wind farm is assumed to be 20 years [43] for this study.

When extraction of the raw materials phase is finalized, in the construction phase, which is the second phase for the life cycle assessment, moving parts of the wind turbine and the fixed part of wind turbine are connected to turbine assembly by means of the excavator for construction since there is no crane or lifter in GaBi. This is one of the limitations of this study.

As a third phase, operation and maintenance procedures are defined as the use phase above. Energy production from the wind farm is calculated as 680 GWh for the entire life cycle. Furthermore, the wind farm requires four types of periodic controls based on discussions with the operator of the plant. Visual controls are one of the periodic controls, but there is no material flow for it. Second type of periodic control is oiling of the parts. Therefore, lubricants are required for the maintenance stage. The lubricant requirement for the power plant calculated as 3400 kg throughout the 20-year life of the plant, according to the report [77], and its transportation distance is assumed to be 300 km. According to a discussion with the operator, there are failures of moving parts especially electronic devices. For spare parts, replacement of 1% of moving parts of the wind turbine is assumed and added to material flow as spare parts. In addition to information taken from the operators, wind turbines require mechanical maintenance twice a year [80]. Site maintenance is neglected since spare parts allocated initially and there is no requirement for access roads due to the fact that the area has no traffic except a maintenance procedure.

In the fourth phase, which is deconstruction and disposal or recycling phase, onshore wind plant decomposed into the main production components including tower, nacelle, rotor, foundation, and decomposition of electronic parts coming from cables and inverters. The end of-life treatment for foundation is 100% landfill as in the DTU International Energy Report [81] and Haapala and Prempreeda [44] advised. Decomposition of electronic parts is sorted as waste for disposal and aluminum scrap. The landfill process is applied for a decomposed rotor since recycling of the composite is not an easy task for the current technology [81]. Nacelle and tower are decomposed as decomposition for iron sorting and its end-of life treatment is applied as 90% recycling materials and 10% landfill. In the recycling phase, open loop recycling is applied since recycling strategies are beyond the aim of the study. Recycling of electronic parts is assumed to be 95% recycling and 5% landfill.

Treatment of materials and material quantities is demonstrated in Table 8. For open loop recycling, transportation distances utilized during the end of-life treatment are seen in Table 9 in the life cycle cost of the wind farm.

Table 8. End of-life summary.

Name of the Components	Treatment	Ratio	Materials Treated	Mass of Components
Nacelle + tower	Landfill Recycling	10% 90%	Iron	845 tons
Rotor	Landfill	100%	Composite	142 tons
Foundation	Landfill	100%	Concrete	3740 tons
Decomposition of electronic parts	Landfill Recycling	5% 95%	Aluminum	132 tons

4.3. Life Cycle Cost (LCC) of Onshore Wind Farm

The LCC model of the wind farm is applied according to Equation (7), which is adopted with the aid of the LCC equation in the article [82] and developed the LCC equation for this study, which is Equation (1). Therefore, for this study, Equation (7) is utilized for calculating the life cycle cost of the wind farm.

$$LCC_{ONW} = C_{turbines} + C_{W_{elec}} + C_{W_{inf}} + C_{W_{O\&M}} + C_{W_{tr}}, \tag{7}$$

where

$$C_{W_{inv}} = C_{turbines} + C_{W_{inf}} + C_{W_{elec}} + C_{W_{t1}}, \tag{8}$$

and

$$C_{W_{tr}} = C_{W_{t1}} + C_{W_{t2}} + C_{W_{t3}}, \tag{9}$$

In the above equations, $C_{W_{tr}}$ is the total transportation costs from cradle to grave. In addition to this, $C_{W_{t1}}$, $C_{W_{t2}}$ and $C_{W_{t3}}$ are transportation costs of three categories. In other words, transportation costs for the initial investment is $C_{W_{t1}}$, $C_{W_{t2}}$ is transportation costs for the operation and maintenance procedure, and, lastly, $C_{W_{t3}}$ is the transportation cost for disposal or recycling. Therefore,

$$C_{W_{O\&M}} = C_{W_M} + C_{W_{t2}}, \tag{10}$$

and

$$C_{W_{DorR}} = C_{W_{t3}}, \tag{11}$$

where C_{W_M} is the material costs in the operation and maintenance phase.

During the calculation of costs for infrastructure and costs of electrical apparatus are extrapolated from the cost [83] and [84]. For example, the costs of foundation and roads is calculated to be $136,724 by means of the previously mentioned study [84] for the year 2015. The building and landscape area is assumed to be 100 m^2 and its cost is taken as $34,180. $C_{W_{inf}}$ is determined as $170,904, which is the sum of the costs of foundation and roads and the costs of the building and landscape area. Following that, the Turkish inflation rate is applied to the costs to reach the infrastructure costs for 2019 with the aid of the inflation calculator [72] for Turkish Republic in order to be able to make a comparison between two renewable configurations.

The same procedure is conducted for $C_{W_{elec}}$ since cables and inverters are also assumed to be produced in Turkey. In the case of $C_{turbines}$, first, one of the Enercon E-40 turbine cost is found from the article [85] in the year 2006.

Total turbine costs are calculated and extrapolated with the producer price indices [86] by using Germany's price indicator.

The operation and maintenance costs consist of the costs of lubricants, transportation costs of spare parts for the necessary replacements, and transportation cost of lubricants when labor costs are excluded from operation and maintenance costs based on the assumptions made in the beginning of the life cycle cost section. The costs of spare parts are considered in the initial investment costs since the allocation of them is conducted before the initialization of the wind farm. The cost of the maintenance procedure, which is basically independent of size [87] and includes replacements of spare parts and lubricants is found in the article [88] as $5770.77 for the Enercon E-40 turbine. It is extrapolated with the quarterly producer price indices [86] by using Germany's price indicator since the prices of lubricants and other consumables are increased between 2008 and 2019.

Costs of transportation due to scrap materials are calculated by means of the weights and hauls of scrap materials. They are calculated based on the consumption of diesel. Diesel consumptions of the transportation process for all scrap materials are taken from GaBi and the price of diesel is calculated by means of the archived list of the BP company like in the life cycle cost of the photovoltaic plant. The ultimate diesel price for Istanbul is utilized for the calculation procedure. The procedure mentioned above and results are listed in Table 9.

Table 9. Transportation costs for the materials of the wind farm.

Materials	Weight (t)	By Truck (km)	By Ferry (km)	Transportation Costs	Cost Categories
Nacelle	336	2640	8	$24,009	IIC
Rotor	142	2640	8	$15,925	IIC
Cables and inverter	132	441	8	$2275	IIC
Foundation and roads	3740	305	8	$28,187	IIC
Tower	508	2640	8	$31,735	IIC
Spare parts	6.1	300	Allocated	$141	O and M
Lubricants	3.4	300	8	$34	O and M
Iron	845	100	15	$2712	DorR
Composite	142	100	15	$681	DorR
Concrete	3740	-	-	Landfill	DorR
Aluminum	132	100	15	$424	DorR

Results of life cycle costs of the wind farm are compared with the cost of the PV plant in a life cycle cost analysis.

5. Results

In this section, environmental quantities of both the onshore wind farm and the photovoltaic power plant are described and compared.

Average annual electricity production is 34 GWh/year for the wind farm and 1.74 GWh/year for the PV plant, respectively.

5.1. Life Cycle Impact Assessment (LCIA)

Primary energy demands from renewable and non-renewable resources (net calorific value) for onshore wind farm and photovoltaic power plant are 71,160,356.64 MJ and 12,268,381.46 MJ, respectively.

Cumulative energy demand values are calculated for each system by using the primary energy requirement values as in the study [89]. The ratio of the total energy embedded in the system as the primary energy [90] and average annual electricity production is defined as energy payback time for this study. Energy pay-back time values for not only onshore wind farm but also the photovoltaic power plant can be seen in Table 10.

The CML2001-Jan 2016 [56] method is utilized for all environmental quantities including global warming potential, acidification potential and eutrophication potential.

Air pollution is caused by acid rain and it leads to air pollution [91]. Air pollution and eutrophication are the major reasons for water pollution. Both acidification and eutrophication are considered as other environmental impact categories for this study due to the selection of the location, which is an island to be established as the renewable energy generation systems.

Acidification potential of the PV system is shown in Figure 3 and acidification potential of onshore wind farm is shown in Figure 4 whereas eutrophication potential of PV system is shown in Figure 5 and eutrophication potential of the onshore wind farm is shown in Figure 6. In the case of the PV system, the production phase leads to the highest acidification level as expected due to extensive energy requirements from different sources such as thermal energy and electricity. Similar to the production phase of the PV plant, the disposal or recycling phase causes a high acidification level due to the need of fuel for transportation of scrap materials. As seen in Figures 3 and 4, the construction phase for the onshore wind farm demonstrates a higher acidification level than the photovoltaic power plant. During the production process for the wind turbine, the unit process of aluminum ingot mix is the main contributor of acidification in the case of onshore wind farm. In the production phase of the photovoltaic power plant, the energy requirement processes either thermal energy or electricity and the unit process of float flat glass can be regarded as the major cause for the acidification. As seen in Figure 4, the acidification potential of the disposal or recycling phase of the PV system is another highest share due to the airline transport of the scrap materials to the real recycling plant.

In terms of eutrophication level, the disposal or recycling phase of onshore wind farm demonstrates the highest one due to a disposal or recycling phase. The unit process of municipal solid waste on landfill which derived from the disposal of foundation and roads, is the major reason of high eutrophication. Furthermore, the result of the production phase of onshore wind farm does not indicate the specific unit process like in the case of disposal or recycling phase of the wind farm when the results are examined in detail. In the case of eutrophication, the production of multi-Si wafer can be regarded as another cause as well as the energy requirement of the processes and the unit process of float flat glass for the first phase of the PV system. The unit process of the cargo plane has the biggest share in the disposal or recycling phase of the photovoltaic plant in terms of the eutrophication level like in the acidification potential.

Global warming potential of the photovoltaic power plant and the onshore wind farm are shown in Figures 7 and 8, respectively. While total GWP of the PV system is 958,858.26 kg CO_2-eq., total GWP of the onshore wind farm is 7,194,780.48 kg CO_2-eq.

As seen in Figure 9, the ratio of energy demand of the production phase for the PV power plant is 91.738% whereas the second most energy required phase is disposal or recycling with the ratio of 7.923%. The least energy requirement phase is operation and maintenance. The construction phase needs 0.333% of total energy demand in the case of the photovoltaic power plant.

In the case of the onshore wind farm (Figure 10), the most energy required, which is, 92.208%, is the production phase like in the case of the PV system. However, the construction phase of the wind farm, which is, 4.104%, needs more than the disposal or recycling of the onshore wind farm, which is, 3.493%, unlike in the photovoltaic power plant. The energy requirement of the operation and maintenance phase for the onshore wind farm, which is, 0.195%, shows a similar trend like in the case of the PV system.

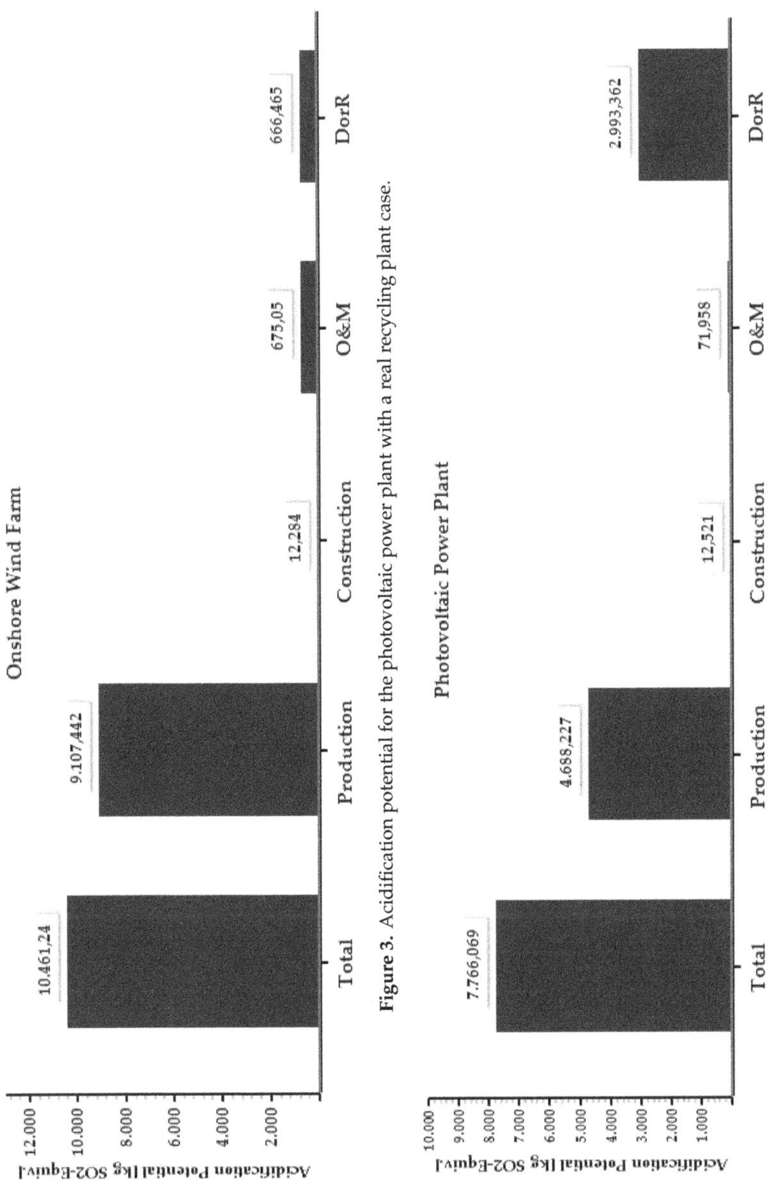

Figure 3. Acidification potential for the photovoltaic power plant with a real recycling plant case.

Figure 4. Acidification potential for the onshore wind farm.

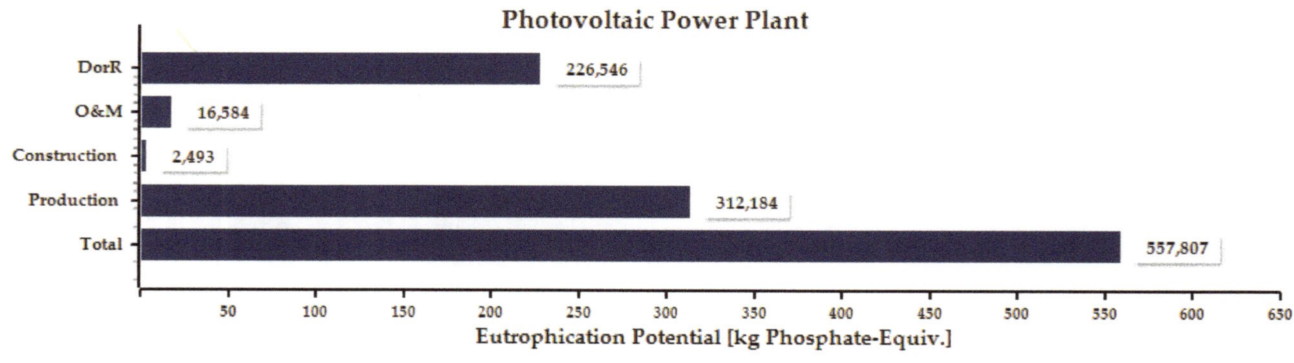

Figure 5. Eutrophication potential for the photovoltaic power plant with a real recycling plant case.

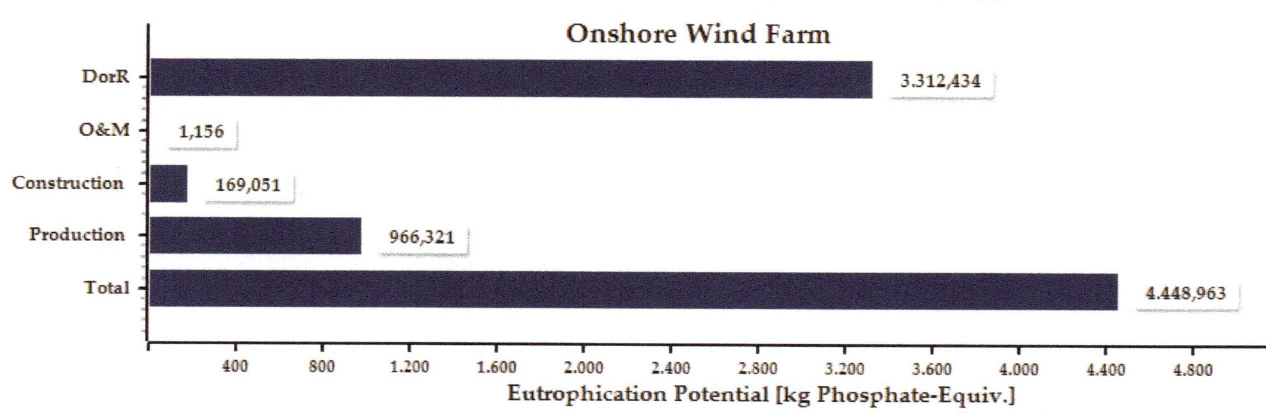

Figure 6. Eutrophication potential for the onshore wind farm.

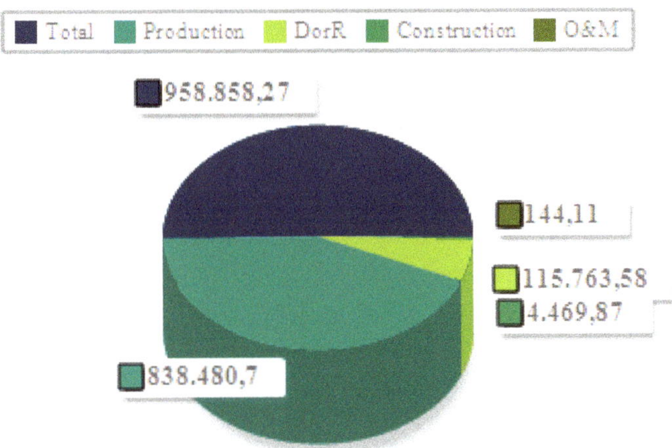

Figure 7. Global warming potential of photovoltaic power plant with real recycling plant case.

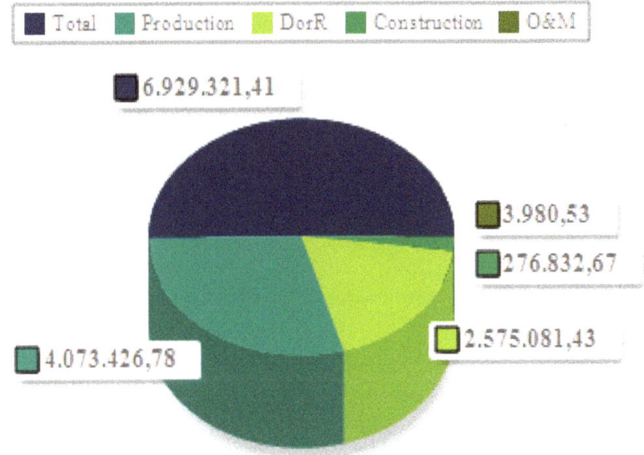

Figure 8. Global warming potential of onshore wind farm.

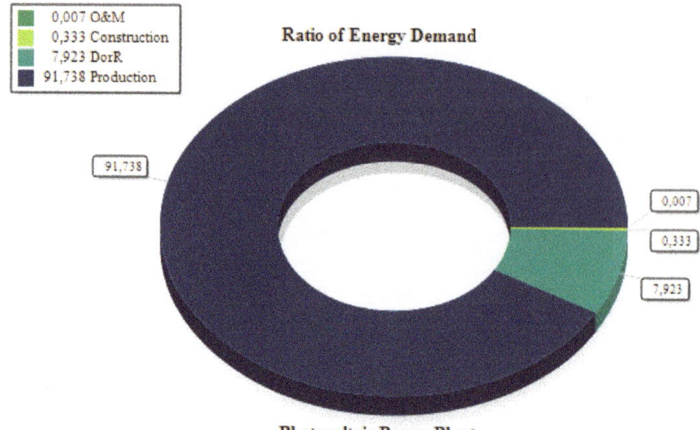

Figure 9. Energy requirement ratios for the phases of the photovoltaic power plant.

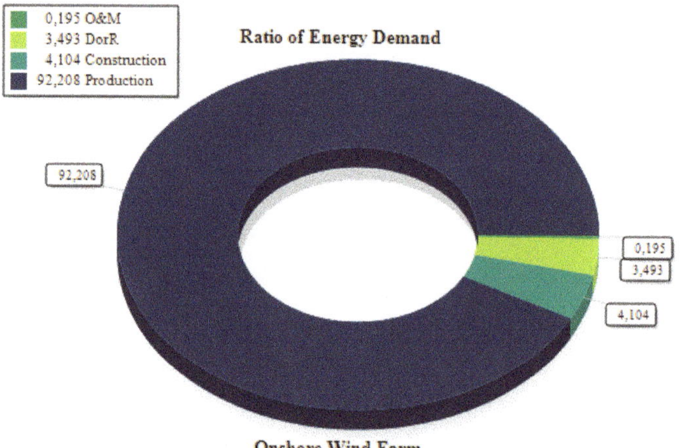

Figure 10. The energy requirement ratios for the phases of the onshore wind farm.

Table 10. Selected LCIA results.

	Onshore Wind Farm	Photovoltaic Power Plant
Global Warming Potential [kg CO$_2$-eq./MWh]	10.58	18.33
Acidification Potential [kg SO$_2$-eq./MWh]	0.01538	0.09816
Eutrophication Potential [kg Phosphate-eq./MWh]	0.00654	0.00794
Energy Pay Back Time [year]	0.62	2.06
Cumulative Energy Demand [MJ/MWh]	104.65	234.53

5.2. Life Cycle Cost Analysis (LCCA)

The results of life cycle calculations based on the prices of 2019 are summarized in Table 11. As seen in Table 11, LCC$_{ONW}$ is $23,949,194.42 and LCC$_{PV}$ is $2,826,759.22. Ratios of transportation cost during the phases are shown in Table 11 for each configuration. The ratio of the transportation cost of the third phase for each system is equal to 1 since disposal or recycling phase costs for each configuration are based on the transportation costs from Equation (6) and Equation (11). The transportation costs are not crucial since the material costs for each phase are more dominant than the transportation costs for each case, as seen in Table 11. However, whether LCC is a useful tool or not for decision-making between distinct renewable configurations for a selected region, Bozcaada Island, cannot be understood from these cost results. Hence, levelized unit costs for producing 1 MWh electricity are calculated to make a comparison. In other words, the required cost during their lifecycle is measured with a levelized unit cost to produce 1MWh electricity. The levelized unit cost for electricity is shown as LUCE in the rest of the study and Equation (12) shows its calculation methodology.

$$LUCE = \frac{\text{Lifecycle cost of the system [\$]}}{\text{Expected electricty generation from the system [MWh]}} \tag{12}$$

Table 11. Life cycle costs of the configurations.

	Onshore Wind Farm		Photovoltaic Power Plant	
Cost specifications	Total cost for the phases [$]	Ratios of the transportation cost	Total cost for the phases [$]	Ratios of the transportation cost
IIC	23,838,415.88	0.0043	2,104,369.07	0.0006
O and M	106,961.61	0.0016	713,648.20	0.0023
DorR	3816.94	1	8741.95	1
LCC	23,949,194.42	0.0044	2,826,759.22	0.0041

5.3. Sensitivity Analysis

Sensitivity analysis of each system are conducted separately for the photovoltaic power plant and the onshore wind farm.

5.3.1. Sensitivity Analysis for the PV System

Sensitivity analysis of the PV system include three cases. For the first and second case, recycling strategies of the photovoltaic power plant are changed, as seen in Table 12. A detailed description of recycling with the end of-life approach and real recycling plant case are conducted in the disposal or recycling phase of the photovoltaic system. In addition to this, the transportation distance for all scrap materials is indicated in Table 4 by a 300 km truck and an 8 km ferry. The difference between the second and the third case is applied for the recycling ratios for aluminum scraps in the open loop recycling. In the second case, the ratio is taken as 0.7 and 0.9 is the selected value for the third case. The results are also presented in Table 12.

Table 12. Sensitivity case specifications for the PV system.

	Photovoltaic Power Plant		
Cases	First Case	Second Case (Base Case)	Third Case
Phase	Recycling with the end-of life approach	Real recycling plant case	Real recycling plant case
Differences	Without scrap materials transportation	Aluminum recycling ratio = 0.7	Aluminum recycling ratio = 0.9
GHGs [kg CO_2-eq./MWh]	16.07	18.33	18.33
EPBT [years]	1.82	2.06	2.06
CED [MJ/MWh]	207.76	234.53	234.58

The results of them are compared as the sensitivity analysis of the PV plant. The LCIA results for environmental characteristics of the recycling strategies are demonstrated in Figures 11 and 12, respectively.

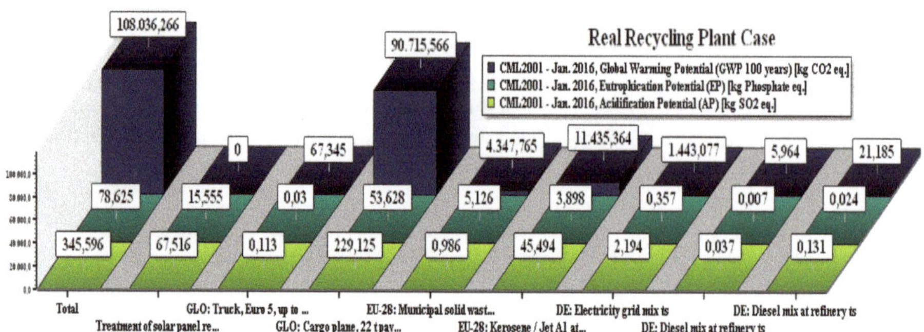

Figure 11. Quantities of recycling in a real recycling plant.

As seen in Figures 11 and 12, greenhouse gas emissions for the real recycling plant case are higher than greenhouse gas emissions for recycling with the end-of-life approach. The major reason is the difference between transportation distances in order to reach recycling plants. In this respect, with the recycling with end-of-life approach, which is based on glass recycling, is more meaningful for the recycling of the photovoltaic power plant system. Recycling of glass from solar panels in İzmir prevents greenhouse gas emission required to produce glass from sand.

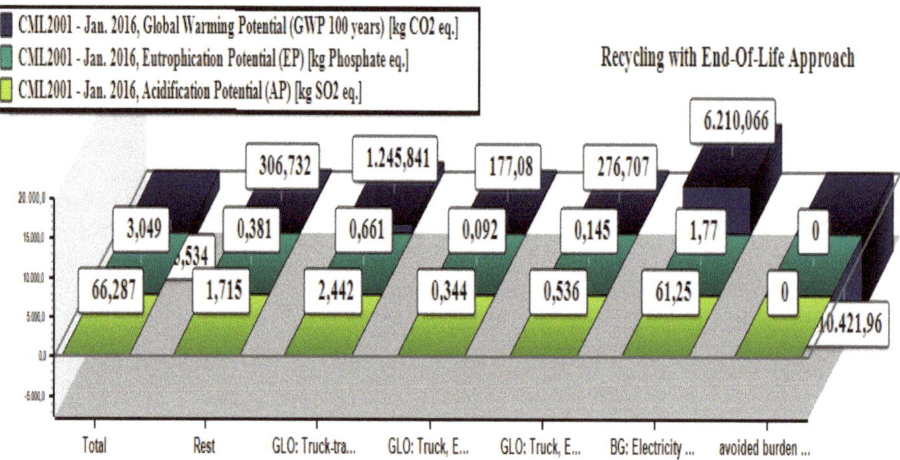

Figure 12. Environmental quantities of recycling with end-of-life approach.

5.3.2. Sensitivity Analysis for Wind Farm

In the case of the wind farm, scraps containing iron from the decompositions of nacelle and tower, composite from the decomposition of rotor blades, and aluminum from the decomposition of cables and inverters, are focused materials for the sensitivity analysis of the onshore wind farm. Sensitivity analyses are divided into two parts and named as Case A and Case B. Transfer procedures of scrap materials to the distribution center are excluded from the life cycle analysis of the wind plant in Case A. In other words, open loop recycling is applied to scrap materials without considering transportations of them for Case A. On the other hand, scrap materials are transferred to the distribution center for the application of open loop recycling procedure by a 15km ferry and a 100km truck in the second option of sensitivity analysis of the wind plant and it is called Case B, as summarized in Table 13. Results of the cases are also demonstrated in the table.

Transfer procedure of scrap materials to the distribution center increased global warming potential, energy pay-back time of the plant, and cumulative energy demand as expected. However, a difference between the cases are not significant. For example, transportation of scrap materials causes 1.6% (approximately four days) increase in the energy pay-back time for the onshore wind farm. As concrete from foundations and tower are left in the construction site for future investments, other scraps, which are iron, aluminum, and composite, should be transferred the distribution center for a cleaner environment after disposal of the plant.

Table 13. Sensitivity case specifications for the wind farm.

	Onshore Wind Farm	
Cases	Case A	Case B (Base case)
Phase	Disposal or recycling	Disposal or recycling
Differences	without scrap materials transportation	with scrap materials transportation (100 km by truck and 15 km by ferry)
GHGs [kg CO$_2$ eq./MWh]	10.57	10.58
EPBT [years]	0.61	0.62
CED [MJ/MWh]	104.46	104.65

6. Discussion

In this section, results presented in Section 5 are evaluated for two distinct renewable systems.

It can be seen in Figures 3 and 4 that a total acidification potential of the PV system and total acidification potential of the onshore wind farm throughout the lifecycle are higher than the photovoltaic power plant.

Acidification potential of the onshore wind farm is 523.1 kg SO$_2$-eq./year while acidification potential of the photovoltaic power plant is 171.2 kg SO$_2$-eq./year, annually. However, acidification potential of the onshore wind farm is lower than the photovoltaic power plant in terms of levelized characteristics, which are found by dividing with a functional unit of LCA and are shown in Table 10. As a result, the photovoltaic power plant is less advantageous than the onshore wind farm in terms of acidification potential.

Although eutrophication potential of the wind farm, which is annually 222.4 kg phosphate-eq./year, is higher than the eutrophication potential of the photovoltaic power plant, which is annually 13.8 kg phosphate-eq./year. The onshore wind farm shows a better trend than the photovoltaic power plant in terms of levelized characteristics of eutrophication potential based on a functional unit, as indicated in Table 10, in the case of eutrophication potential throughout life cycle analyses of different technologies.

In the case of energy pay-back time, photovoltaic technology requires approximately 25 months to produce primary energy which is initially embedded, whereas the onshore wind farm needs 7.5 months. It can be explained by the primary energy requirements of each technology and the power production capacity for the technologies. The energy requirement of production technologies can be clarified by a cumulative energy demand. As seen in Table 10, cumulative energy requirement of onshore wind farm is less than cumulative energy demand of photovoltaic power plant to be invested in Bozcaada. Onshore wind farm with 34 GWh average annual electricity production has higher potential than photovoltaic power plant in terms of power production capacity on the selected area which is approximately the same for both technologies.

When annual greenhouse gas emissions of both configurations are examined, onshore wind farm emits 359,739.0 kg CO$_2$-eq. per a year and PV system emits 31,961.9 kg CO$_2$-eq. per a year. However, onshore wind farm emits less greenhouse gases than photovoltaic power plant to produce 1 MWh electricity as seen in Table 10 by means of levelized characteristics of global warming potential.

Figures 9 and 10 indicate that operation and maintenance phase shows the least primary energy demand for both renewable systems as expected. Production phase leads to the highest value of greenhouse gas emissions not only for onshore wind farm but also for photovoltaic power plant. Decommissioning and recycling phase is the second most energy required phases for both technologies.

The construction phase of power plant demonstrates a lower ratio than the construction phase of onshore wind farm as expected since there is no requirement of utilization of construction machines in order to establish a power plant unlike in the case of the establishment of wind farm. Moreover, the construction of infrastructure is simpler for the power plant than the onshore wind farm.

In terms of levelized unit cost to produce 1 MWh electricity, onshore wind farm is more logical investment than photovoltaic technology for Bozcaada Island although the initial investment cost of onshore wind plant configuration is approximately 11 times of the initial investment cost of photovoltaic power plant configurations. When the life cycle costs of both technologies are considered, transportation costs of the material are insignificant not only PV system but also onshore wind farm as seen Table 11. However, the transportation cost in initial investment is higher for onshore wind farm than photovoltaic technology due the fact that the total weight of wind turbines is higher than the weight of solar panels.

In the case of sensitivity analysis of PV plant, the increase in the recycling ratio for aluminum scraps leads to 0.02% change in the cumulative energy demand, change in the EPBT and GWP are insignificant as seen Table 12 while change in the recycling strategies causes approximately 11.7 % decrease in the global warming potential. In terms of environment, recycling with end-of life approach is more logical for the PV plant. For the sensitivity analysis of wind farm, the changes between Case A and Case B are insignificant.

As seen in Table 14, phases up to disposal or recycling phases of each technology require less energy and less costs than the entire life cycles of each configurations as expected. In fact, the levelized unit cost of electricity should be lower with open loop recycling procedure in the real life. In other words, in real life, it should be less than the costs calculated in this study when scraps are sold. However, it can be utilized for the comparison between the systems. In addition to this, operation and maintenance procedure of the PV system is approximately 25% of total cost of photovoltaic power plant due to mainly replacement of inverters even though it is almost insignificant for the case of onshore wind farm disposal or recycling cost is less dominant in the case of wind technology than in the PV system. It can be explained by both developed recycling in the wind sector and large amount of landfill treatment which has no cost.

Table 14. Results for the different life cycle boundaries.

Phases Results	Production	Construction	O&M	DorR	System
GHGs [kg CO_2-eq./MWh]	16.0291	0.0854	0.0028	2.2130	PV
	6.0102	0.4077	0.0059	4.1567	Onshore
EPBT [years]	1.7881	0.0102	0.0002	0.2649	PV
	0.5562	0.0347	0.0016	0.0229	Onshore
CED [MJ/MWh]	203.7367	1.1390	0.0249	29.6316	PV
	94.7149	5.8463	0.2716	3.8147	Onshore
LUCE [$/MWh]	40.2288		13.6427	0.1671	PV
	35.0565		0.1573	0.0056	Onshore

The production phase of both configurations requires extensive energy than other phases. Although the PV system causes enormous GHGs in the production phase, both production and disposal or the recycling phase are the main cause in the wind energy. It can be said that both configurations lead to almost zero global warming potential during their operation.

In terms of energy pay-back time, recycling of the wind farm is more efficient than recycling of the photovoltaic power plant. Furthermore, the disposal or recycling phase of the onshore wind farm requires less primary energy than disposal or recycling phase of the PV system by means of the comparison between the values of cumulative energy demand in Table 14 found. However, disposal or the recycling procedure of onshore wind farm leads to higher greenhouse emissions than the procedure of disposal or recycling for the photovoltaic power plant. It can be explained that recycling strategies of wind farm is cleaner than the photovoltaic power plant in terms of process chain in the production phase of wind turbine.

In the case of the life cycle cost calculations, $LUCE_{ONW}$, 35.2194 \$/MWh, and $LUCE_{PV}$, 54.0386 \$/MWh are validated. The costs calculated in this study is lower than costs in Reference [34], as expected, because the main assumption of this study was not including insurance costs, labor costs and project costs.

7. Conclusions

This paper investigated the environmental and economic impacts of two renewable energy systems for a selected region, which is Bozcaada Island. Our findings indicate that the onshore wind farm is more appropriate for the selected region. All of the environmental specifications show a better trend for the onshore wind farm than the photovoltaic power plant. In terms of costs, establishment of wind farm is more meaningful for the generation of electricity than the establishment of the PV system even though annual parameters of global warming potential indicate that the onshore wind farm emits more greenhouse emissions than the land-based photovoltaic plant. In other words, the wind farm is cleaner to generate 1 MWh electricity than PV technology when the lifespans of systems are considered in terms of selected environmental quantities. In the comparison of levelized unit cost, the onshore wind farm is more economic than the PV system for Bozcaada Island to generate electricity. This study shows that the LCA and LCCA of wind and solar energy systems will contribute to investment decision-making by considering environmental impacts and economic analysis.

This study indicates that LCA and LCCA should be used to determine the most feasible option for a selected region (i.e., Bozcaada Island), which has many renewable energy potentials like wind and solar. As is mentioned in Section 1, Bozcaada Island has 308.0 cal/cm^2 sunshine radiation per day and 7.5 hours sunshine duration per day, as well as wind energy potential with 8.4 m/s average wind speed at 50 m [24].

During this study, the life spans of the proposed the PV system and existing wind farm are considered like in the other studies suggested for each system. In other words, the life of the photovoltaic power plant is accepted 30 years and the useful life of the existing wind farm is determined to be 20 years. However, there is a still a requirement to investigate about real lifespans of the systems by means of the sensitivity analysis including either change of the life spans or a fixed economic life. Issues like risk have a significant importance for solar panel system applications as well as wind turbines. Therefore, a future study should also include life cycle impacts of risks to determine a more comprehensive LCA study for evaluating renewable energy systems.

Decommissioning and the recycling phase with the current technology causes enormous greenhouse gas emissions not only in PV plant configurations but also the onshore wind farm. The main reasons can be decommissioning and disposal of the plant as well as the transportation of waste and recycling materials. Undeveloped recycling technologies especially in the case of photovoltaic power plant can be another reason for the extensive global warming potential of the fourth phase of photovoltaic technology. There is an urgent requirement of recycling strategies of solar panels.

In the case of land-based photovoltaic technology, selection of system boundaries affects results significantly. For example, up to the disposal and recycling phase, the PV system emits 16.12 kg CO_2-eq. for the generation of 1 MWh electricity. The disposal and recycling phase of the PV plant leads to approximately 13.7% increase in the greenhouse emissions, according to the boundaries defined as from the production to the disposal phase. However, disposal and recycling of the onshore wind farm causes almost a 65% increase in GWP, according to its assessment from production to the disposal phase. With the aid of EPBT and CED in Table 14, the reason a 65% increase for the wind plant can be claimed as landfill processes for concrete and composite materials since there is no significant increase in energy requirements which is required for the transportation of waste materials for the fourth stage of the wind farm. For the photovoltaic power plant, its fourth phase needs extensive primary energy.

This can be related with utilization of cargo plane for the transfer of solar panels waste since there is no other energy embedded procedure during the modelling part of the fourth phase of the photovoltaic power plant. As mentioned in fourth phase of the PV system in Section 3.2, recycling of solar panels investigations including energy flows are not well-defined in the literature so far. It can be suggested that recycling technologies of the PV system should be improved in the near future.

Results of this study indicate that wind farm is cleaner than photovoltaic power plant in terms of greenhouse gas emission, acidification potential and eutrophication potential for Bozcaada island similar to Canadian case study without the adoption procedure which includes normalization, weighting and single score [45]. It is critical to be able to determine whether the wind system is cleaner than PV system for any location. However, there is still a need to carry out further research to demonstrate cleanness of wind technology compared to photovoltaic technology.

Author Contributions: The authors contributed equally to this paper.

Funding: This research received no external funding.

Acknowledgments: It should be reported that short part of this paper will be presented at 18th International Congress of the International Maritime Association of the Mediterranean in 2019.

Conflicts of Interest: The authors declare no conflict of interest.

Nomenclature

IMO	International Maritime Organisation
MARPOL	Maritime Pollution
LCA	Life cycle assessment
LCC	Life cycle cost
LCCA	Life cycle cost analysis
LCI	Life cycle inventory
LCIA	Life cycle impact analysis
EPBT	Energy-payback time
GHGs	Greenhouse emissions
GWP	Global warming potential
AP	Acidification potential
EP	Eutrophication potential
RES	Renewable energy sources
CO_2	Carbon dioxide
CO_2 eq	Carbon dioxide equivalent
PV	Photovoltaic
O&M	Operation and maintenance
DorR	Disposal or recycling
C_{inv}	Investment cost
$C_{O\&M}$	Operation and maintenance costs
C_{DorR}	Disposal or recycling costs
PVGIS	Solar radiation database
LCC_{PV}	Life cycle cost of photovoltaic power plant
LCC_{ONW}	Life cycle cost of wind farm
C_{panels}	Cost of solar panels
$C_{P_{elec}}$	Cost of electrical apparatus for PV plant
$C_{P_{inf}}$	Cost of infrastructure of PV plant
$C_{P_{O\&M}}$	Cost of operation and maintenance procedure of PV system
$C_{P_{tr}}$	Total transportation cost of PV plant throughout lifespan
$C_{P_{t1}}$	Transportation costs of PV plant during initial investment phase
$C_{P_{t2}}$	Transportation costs of PV plant for operation and maintenance phase
$C_{P_{t3}}$	Transportation costs of PV plant during disposal or recycling phase

$C_{tap\ water}$	Cost of tap water
$C_{spare\ panels}$	Cost of spare solar panels
$C_{P_{DorR}}$	Cost of disposal or recycling of photovoltaic power plant
BP	British Petrol
$C_{turbines}$	Cost of turbines
$C_{W_{elec}}$	Cost of electrical apparatus for wind farm
$C_{W_{inf}}$	Cost of infrastructure of wind farm
$C_{W_{O\&M}}$	Cost of operation and maintenance procedure of wind farm
$C_{W_{tr}}$	Total transportation cost of wind farm throughout lifespan
$C_{W_{t1}}$	Transportation costs of wind farm during initial investment phase
$C_{W_{t2}}$	Transportation costs of wind farm for operation and maintenance phase
$C_{W_{t3}}$	Transportation costs of wind farm during disposal or recycling phase
C_{W_M}	Cost of maintenance procedure of wind farm in terms of material costs
$C_{W_{DorR}}$	Cost of disposal or recycling of wind farm
OECD	Organization for Economic Co-operation and Development
IIC	Initial investment cost
LUCE	Levelized unit cost to produce 1 MWh electricity
$LUCE_{ONW}$	Levelized unit cost to produce 1 MWh electricity for wind farm
$LUCE_{PV}$	Levelized unit cost to produce 1 MWh electricity for photovoltaic power plant

References

1. Schanes, K.; Giljum, S.; Hertwich, E. Low carbon lifestyles: A framework to structure consumption strategies and options to reduce carbon footprints. *J. Clean. Prod.* **2016**, *139*, 1033–1043. [CrossRef]
2. Kavli, H.P.; Oguz, E.; Tezdogan, T. A comparative study on the design of an environmentally friendly RoPax ferry using CFD. *Ocean. Eng.* **2017**, *137*, 22–37. [CrossRef]
3. European Union. *Energy Road Map 2050*; Publications Office of the European Union: Brussels, Belgium, 2012.
4. Jeong, B.; Oğuz, E.; Wang, H.; Zhou, P. Multi-criteria decision-making for marine propulsion: Hybrid, diesel electric and diesel mechanical systems from cost-environment-risk perspectives. *Appl. Energy* **2018**, *230*, 1065–1081. [CrossRef]
5. Aysar, M.; Lorenzo, S.; Ponci, F.; Monti, A. Multi-agent based intelligent frequency control in multi-terminal dc grid-based hybrid ac/dc networks. *IET Renew. Power Gener.* **2018**, *12*, 1434–1443.
6. Adrees, A.; Milanovic, J.V.; Mancarella, P. The Influence of Location of Distributed Energy Storage Systems on Primary Frequency Response of Low Inertia Power Systems. In Proceedings of the IEEE Power & Energy Society General Meeting (PESGM), Portland, OR, USA, 5–10 August 2018.
7. Wang, H.; Oguz, E.; Jeong, B.; Zhou, P. Life cycle cost and environmental impact analysis of ship hull maintenance strategies for a short route hybrid ferry. *Ocean. Eng.* **2018**, *161*, 20–28. [CrossRef]
8. Marzi, J.; Papanikolaou, A.; Corrignan, P.; Zaraphonitisc, G.; Harries, S. HOLISTIC Ship Design for Future Waterborne Transport. In Proceedings of the 7th Transport Research Arena TRA 2018, Austria, Vienna, 16–19 April 2018.
9. Jeong, B.; Wang, H.; Oguz, E.; Zhou, P. An effective framework for life cycle and cost assesment for marine vessels aiming to select optimal propulsion systems. *J. Clean. Prod.* **2018**, *187*, 111–130. [CrossRef]
10. Wang, H.; Oguz, E.; Jeong, B.; Zhou, P. Life cycle and economic assesment pf solar panel array applied to a short route ferry. *J. Clean. Prod.* **2019**, *219*, 471–484. [CrossRef]
11. Hernández, C.V.; González, J.S.; Blanco, R.F. New method to assess the long-term role of wind energy generation in reduction of CO_2 emissions- Case Study of european Union. *J. Clean. Prod.* **2019**, *207*, 1099–1111. [CrossRef]
12. Özkale, C.; Celik, C.; Turkmen, A.C.; Cakmaz, E.S. Decision analysis application intended for selection of a power plant running on renewable energy sources. *Renew. Sustain. Energy Rev.* **2017**, *70*, 1011–1021. [CrossRef]
13. Santoyo-Castelazo, E.; Azapagic, A. Sustainability assessment of energy systems: Integrating environmental, economic and social aspects. *J. Clean. Prod.* **2014**, *80*, 119–138. [CrossRef]
14. Hong, J.H.; Kim, J.; Son, W.; Shin, H.; Kim, N.; Lee, W.K.; Kim, J. Long-term energy strategy scenarios for South Korea: Transition to a sustainable energy system. *Energy Policy* **2019**, *127*, 425–437. [CrossRef]

15. Keleş, S.; Bilgen, S. Renewable energy sources in Turkey for climate change mitigation and energy sustainability. *Renew. Sustain. Energy Rev.* **2012**, *16*, 5199–5206. [CrossRef]

16. Panwar, N.; Kaushik, S.; Kothari, S. Role of renewable energy sources in environmental protection: A review. *Renew. Sustain. Energy Rev.* **2011**, *15*, 1513–1524. [CrossRef]

17. Li, Q.S.; Chen, F.; Li, Y.; Lee, Y. Implementing wind turbines in a tall building for power generation: A study of wind loads and wind speed amplifications. *J. Wind Eng. Ind. Aerodyn.* **2013**, *116*, 70–82. [CrossRef]

18. Vourdoubas, J. Review of sustainable energy technologies used in buildings in the Mediterranean basin. *J. Build. Sustain.* **2018**, *1*, 2.

19. Yuan, X.; Wang, X.; Zuo, J. Renewable energy in buildings in China—A review. *Renew. Sustain. Energy Rev.* **2013**, *24*, 1–8. [CrossRef]

20. IPEEC Building Energy Efficiency Taskgroup. *Zero Energy Building Definitions and Policy Activity-An International Review;* International Partnership for Energy Efficiency Cooperation: Paris, France, 2018.

21. Eremia, M.; Toma, L.; Sanduleac, M. The Smart City Concept in the 21st century. *Procedia Eng.* **2017**, *181*, 12–19. [CrossRef]

22. Nistor, S.; Carr, S.; Sooriyabandara, M. The Island Hydrogen Project: Electrolytic Generated Hydrogen for Automotive and Maritime Applications. *IEEE Electrif. Mag.* **2018**, *6*, 55–60. [CrossRef]

23. Yazici, M.S.; Hatipoğlu, M. Hydrogen and fuel cell demonstrations in Turkey. *Energy Procedia* **2012**, *29*, 683–689. [CrossRef]

24. Oğulata, R.T. Energy sector and wind energy potential in Turkey. *Renew. Sustain. Energy Rev.* **2003**, *7*, 469–484. [CrossRef]

25. Jackson, D.; Brander, M. The risk of burden shifting from embodied carbon calculation tools for the infrastructure sector. *J. Clean. Prod.* **2019**, *223*, 739–746. [CrossRef]

26. Akanbi, L.A.; Oyedele, L.O.; Omoteso, K.; Bilal, M.; Akinade, O.O.; Ajayi, A.O.; Delgado, J.M.D.; Owolabi, H.A. Disassembly and deconstruction analytics system (D-DAS) for construction in a circular economy. *J. Clean. Prod.* **2019**, *223*, 386–396. [CrossRef]

27. Wiedemann, S.; McGahan, E.; Murphy, C.; Yan, M.-J.; Henry, B.; Thoma, G.; Ledgard, S. Enviromental impacts and resource use of Australian beef and lamb exported to the USA determined using life cycle assessment. *J. Clean. Prod.* **2015**, *94*, 67–75. [CrossRef]

28. Atilgan, B.; Azapagic, A. Life cycle enviromental impacts of electricity from fossil fuels in Turkey. *J. Clean. Prod.* **2014**, *106*, 555–564. [CrossRef]

29. Atilgan, B.; Azapagic, A. Renewable electricity in Turkey: Life cycle enviromental impacts. *Renew. Energy* **2016**, *89*, 649–657. [CrossRef]

30. Rakotoson, V.; Praene, J.P. A life cycle assessment approach to the electicity generation of French overseas territories. *J. Clean. Prod.* **2017**, *168*, 755–763. [CrossRef]

31. Utne, I.B. Life cycle cost (LCC) as a tool for improving sustainability in Norvewgian fishing fleet. *J. Clean. Prod.* **2009**, *17*, 335–344. [CrossRef]

32. Delugo, M.; Zanchi, L.; Maltese, S.; Bonoli, A.; Pierini, M. Enviromental and economic life cycle assessment of a lightweight solution for an automotive component: A comparison between talc-filled and hollow glass microspheres-reinforced polymer composites. *J. Clean. Prod.* **2016**, *139*, 548–560. [CrossRef]

33. Cheung, W.M.; Marsh, R.; Griffin, P.W.; Griffin, P.W.; Newnes, L.B.; Mileham, A.R.; Lanham, J.D. Towards cleaner production: A roadmap for predicting product end-of-life costs at early design concept. *J. Clean. Prod.* **2015**, *87*, 431–441. [CrossRef]

34. Yılan, G. *Comparison of Life Cycle Assessment of Electricty Production Mix in Turkey with Future Electricty Production Scenarios;* Marmara University Institute for Graduate Studies in Pure and Applied Sciences: Istanbul, Turkey, 2018.

35. Ormazabal, M.; Jaca, C.; Puga-Leal, R. Analysis and Comparison of Life Cycle Assessment and Carbon Footprint Software. *Adv. Intell. Syst. Comput.* **2014**, *281*, 1521–1530.

36. Speck, R.; Selke, S.; Auras, R.; Fitzsimmons, J. Choice of Life Cycle Assessment Software Can Impact Packaging System Decisions. *Packag. Technol. Sci.* **2015**, *28*, 579–588. [CrossRef]

37. Speck, R.; Selke, S.; Auras, R.; Fitzsimmons, J. Life Cycle Assessment Software: Selection Can Impact Results. *J. Ind. Ecol.* **2016**, *20*, 18–28. [CrossRef]

38. Herrmann, I.T.; Moltesen, A. Does it matter which Life Cycle Assessment (LCA) tool you choose? A comparative assessment of SimaPro and GaBi. *J. Clean. Prod.* **2015**, *86*, 163–169. [CrossRef]

39. Silva, D.; Nunes, A.O.; da Silva Moris, A.; Moro, C.; Piekarski, T.O.R. *How Important Is the LCA Software Tool You Choose Comparative Results from GaBi*; openLCA, SimaPro and Umberto: Medellin, DC, USA, 2017.

40. Fu, Y.; Liu, X.; Yuan, Z. Life-cycle assessment of multi-crystalline photovoltaic (PV) systems in China. *J. Clean. Prod.* **2015**, *86*, 180–190. [CrossRef]

41. Nugent, D.; Sovacool, B.K. Assessing the lifecycle greenhouse gas emissions from solar PV and wind energy: A critical meta-survey. *Energy Policy* **2014**, *65*, 229–244. [CrossRef]

42. Bonou, A.; Laurent, A.; Olsen, S.I. Life cycle assessment of onshore and offshore wind energy—From theory to application. *Appl. Energy* **2016**, *180*, 327–337. [CrossRef]

43. Chipindula, J.; Botlaguduru, V.S.V.; Du, H.; Kommalapati, R.R.; Huque, Z. Life Cycle Environmental Impact of Onshore and Offshore Wind Farms in Texas. *Sustainability* **2018**, *10*, 2022. [CrossRef]

44. Haapala, K.R.; Prempreeda, P. Comparative life cycle assessment of 2.0 MW wind turbines. *Int. J. Sustain. Manuf.* **2014**, *3*, 170–185. [CrossRef]

45. Schmidt, K.; Alvarez, L.; Arevalo, J.; Abbassi, B. Life Cycle Impact Assessment of Renewable Energy Systems: Wind vs. Photovoltaic Systems. *Int. J. Curr. Res.* **2017**, *9*, 59140–59147.

46. Yudha, H.M.; Dewi, T.; Risma, P.; Oktarina, Y. Life Cycle Analysis for the Feasibility of Photovoltaic System Application in Indonesia. *IOP Conf. Ser. Earth Environ. Sci.* **2018**, *124*, 012005. [CrossRef]

47. Desideri, U.; Proietti, S.; Zepparelli, F.; Sdringola, P.; Bini, S. Life Cycle Assessment of a ground-mounted 1778 kWp photovoltaic plant and comparison with traditional energy production systems. *Appl. Energy* **2012**, *97*, 930–943. [CrossRef]

48. Ito, M.; Kato, K.; Sugihara, H.; Kichimi, T.; Song, J.; Kurokawa, K. A Preliminary Study on Potential for Very Large-Scale Photovoltaic Power Generation (VLS-PV) System on the Gobi Desert from Economic and Environmental Viewpoints. *Sol. Energy Mater. Sol. Cells* **2003**, *75*, 507–517. [CrossRef]

49. Luo, W.; Khoo, Y.S.; Kumar, A.; Low, J.S.C.; Li, Y.; Tan, Y.S.; Wang, Y.; Aberle, A.G.; Ramakrishna, S. A comparative life-cycle assessment of photovoltaic electricity generation in Singapore by multicrystalline silicon technologies. *Sol. Energy Mater. Sol. Cells* **2018**, *174*, 157–162. [CrossRef]

50. Palanov, N. *Life-Cycle Assessment of Photovaltaic Systems*; Lund University: Lund, Sweden, 2014.

51. Yu, M.; Halog, A. Solar Photovoltaic Development in Australia—A Life Cycle Sustainability Assessment Study. *Sustainability* **2015**, *7*, 1213–1247. [CrossRef]

52. Wu, P.; Ma, X.; Ji, J.; Ma, Y. Review on life cycle assessment of energy payback of solar photovoltaic systems and a case study. *Energy Procedia* **2017**, *105*, 68–74. [CrossRef]

53. Ito, M.; Kato, K.; Komoto, K.; Kichimi, T.; Kurokawa, K. A comparative study on cost and life-cycle analysis for 100 MW very large-scale PV (VLS-PV) systems in deserts using m-Si, a-Si, CdTe, and CIS modules. *Prog. Photovolt. Res. Appl.* **2008**, *16*, 17–30. [CrossRef]

54. Zhong, Z.-W.; Song, B.; Loh, P.E. LCAs of a polycrystalline photovoltaic module and a wind turbine. *Renew. Energy* **2011**, *36*, 2227–2237. [CrossRef]

55. Mason, J.; Fthenakis, V.M.; Hansen, T.; Kim, H.C. Energy Pay-Back and Life Cycle CO_2 Emissions of the BOS in an Optimized 3.5 MW PV Installation. *Prog. Photovolt. Res. Appl.* **2006**, *14*, 179–190. [CrossRef]

56. Institute of Environmental Sciences. *CML, 2016 CML-IA Characterisation Factors*; Leiden University: Leiden, The Netherlands, 2016.

57. ISO. *ISO 14040:2006—Environmental Management–Life Cycle Assessment–Principles and Framework*; International Organization for Standardization: Geneva, Switzerland, 2006.

58. ISO. *ISO 14044:2006—Environmental Management–Life Cycle Assessment–Requirements and Guidelines*; International Organization for Standardization: Geneva, Switzerland, 2006.

59. Frischknecht, R.; Heath, G.; Raugei, M.; Sinha, P.; de Wild-Scholten, M. *Methodology Guidelines on Life Cycle Assessment of Photovoltaic Electricity*, 3rd ed.; Brookhaven National Laboratory: New York, NY, USA, 2016.

60. Lee, D. Fundamentals of Life-Cycle Cost Analysis. *Transp. Res. Rec. J. Transp. Res. Board* **2002**, *1812*, 203–210. [CrossRef]

61. Denholm, P.; Margolis, R. Land-Use Requirements and the Per-Capita Solar Footprint for Photovoltaic Generation in the United States. *Energy Policy* **2008**, *36*, 3531–3543. [CrossRef]

62. Ministry of Development, Regional Development Administration for Konya Plain Project. *KOP Bölgesinde Arazi ve Enerji Üretimi Planlaması*; Regional Development Administration for Konya Plain Project: Konya, Turkey, 2012.

63. Bores Bozcaada Santrali. Available online: http://www.demirer.com.tr/santral/bores/index.html (accessed on 13 July 2019).

64. Jordan, D.C.; Kurtz, S.R. *Photovoltaic Degradation Rates—An Analytical Review*; NREL: Golden, CO, USA, 2012; Volume 5200, p. 18.

65. Ito, M. Life Cycle Assessment of PV systems. In *Crystalline Silicon-Properties and Uses*; Basu, S., Ed.; IntechOpen: London, UK, 2011.

66. Karadogan, O.; Kilicarslan, T.; ve Celiktas, M.S. *The Actual Performance Value of Photovoltaic Solar System In Turkey and Compare with Software Result*; SOLARTR: İzmir, Turkey, 2014.

67. Erik, A.; Fraile, D.; Frischknecht, R.; Fthenakis, V.; Held, M.; Kim, H.C.; Pölz, W.; Raugei, M.; de Wild Scholten, M. *Methodology Guidelines on Life Cycle Assessment of Photovoltaic Electricity*; International Energy Agency, Photovoltaic Power Systems Programme, Rolf Frischknecht, ESU-services Ltd.: Uster, Switzerland, 2009.

68. Jungbluth, N.; Stucki, M.; Frischknecht, R.; Büsser, S. *Photovoltaics*; ESU-services Ltd. & Swiss Centre for Life Cycle Inventories: Uster, Switzerland, 2010.

69. Appleyard, D. Light Cycle: Recycling PV Materials. *Renew. Energy World Mag.* **2009**, *109*, 28–35.

70. Abu-Rumman, A.K.; Muslih, I.; Barghash, M.A. Life Cycle Costing of PV Generation System. *J. Appl. Res. Ind. Eng.* **2017**, *4*, 252–258.

71. Batman, A.; Bagriyanik, F.G.; Aygen, Z.E.; Gul, O.; Bagriyanik, M. A feasibility study of grid-connected photovoltaic systems in Istanbul, Turkey. *Renew. Sustain. Energy Rev.* **2012**, *16*, 5678–5686. [CrossRef]

72. Inflation Calculator. Available online: http://www3.tcmb.gov.tr/inflationcalc2/inflationcalc.php (accessed on 13 July 2019).

73. BP Türkiye-Ürünler ve Servisler-Akaryakıt-Akaryakıt Pompa Satış Fiyatları. Available online: http://www.bppompafiyatlari.com/ (accessed on 13 July 2019).

74. Kerosen-Kerosen Manufacturers, Suppliers and Exporters on Alibaba.comJetFuel. Available online: https://www.alibaba.com/trade/search?fsb=y&IndexArea=product_en&CatId=&SearchText=kerosen (accessed on 13 July 2019).

75. Turkish Wind Energy Association. *Turkish Wind Energy Statistic Report*; Turkish Wind Energy Association: Ankara, Turkey, 2018.

76. Lee, Y.; Tzeng, Y.; Su, C. Life Cycle Assessment of Wind Power Utilization in Taiwan. In Proceedings of the 7th International Conference on Eco Balance, Tsukuba, Japan, 14–16 November 2006.

77. Razdan, P.; Garrett, P. Life Cycle Assessment of Electricity Production from an onshore V110-2.0 MW Wind Plant; Vestas Wind Systems A/S, Hedeager 42, Aarhus N, 8200, Denmark, 2015. Available online: https://www.vestas.com/~{|/media/vestas/about/sustainability/pdfs/lcav11020mw181215.pdf (accessed on 13 July 2019).

78. Enercon E-40/6.44-600,00 kW-Wind Turbine. Available online: https://en.wind-turbine-models.com/turbines/68-enercon-e-40-6.44 (accessed on 13 July 2019).

79. Ghenai, C. Life Cycle Analysis of Wind Turbine. In *Sustainable Development—Energy, Engineering and Technologies—Manufacturing and Environment*; Ghenai, C., Ed.; InTech: Melbourne, FL, USA, 2012; p. 27.

80. Chan, D.; Mo, J. Life cycle reliability and maintenance analyses of wind turbines. *Energy Procedia* **2017**, *110*, 328–333. [CrossRef]

81. Andersen, P.D.; Bonou, A.; Beauson, J.; Brøndsted, P. *Recycling of Wind Turbines*; Technical University of Denmark (DTU): Lyngby, Denmark, 2014.

82. Abu-Rumman, A.K.; Muslih, I.; Barghash, M.A. Life Cycle Costing of Wind Genaration System. *J. Appl. Res. Ind. Eng.* **2017**, *4*, 185–191.

83. Erdem, O. Bozcaada'da CFD Programı Kullanarak Bir Rüzgar Enerji Santralinin Potansiyelinin Belirlenmesi. *Tesisat Mühendisliği* **2015**, *147*, 20–26.

84. Erdem, O.; Batur, B.; Bilge, Z.D.; Temir, G. Bozcaada'da Kurulacak Olan Bir Rüzgar Enerjisi Santralinin Ekonomik Analizi. *Tesisat Mühendisliği* **2015**, *148*, 22–27.

85. Ozerdem, B.; Ozer, S.; Tosun, M. Feasibility study of wind farms: A case study for Izmir, Turkey. *J. Wind Eng. Ind. Aerodyn.* **2006**, *94*, 725–743. [CrossRef]

86. OECD. Producer Price Indices (PPI). 2019. Available online: https://data.oecd.org/price/producer-price-indices-ppi.htm#indicator-chart (accessed on 2 May 2019).

87. Henderson, G. Potential for Reducing Cost of Energy by Scaling Up a Low-Mass wind turbine design. In Proceedings of the 16th International Workshop on Large-Scale Integration of Wind Power, Berlin, Germany, 25–27 October 2017.
88. Fathiyah, R.; Mellott, R.; Panagoda, M.; Lane, M. *Windmill Design Optimization through Component Costing*; IEEE Seminar: London, UK, 2000.
89. Merta, G.; Linkeb, B.; Auricha, J. Analysing the Cumulative Energy Demand of Product-Service Systems for wind turbines. *Procedia CIRP* **2017**, *59*, 214–219. [CrossRef]
90. Gkantou, M.; ve Baniotopoulos, C.C. Life Cycle Analysis of Onshore Wind Turbine Towers. In Proceedings of the 2nd International TU1304 WINERCOST, Catanzaro, Italy, 21–23 March 2018.
91. Kim, T.H.; Chae, C.U. Environmental Impact Analysis of Acidification and Eutrophication Due to Emissions from the Production of Concrete. *Sustainability* **2016**, *8*, 1. [CrossRef]

© 2019 by the authors. Licensee MDPI, Basel, Switzerland. This article is an open access article distributed under the terms and conditions of the Creative Commons Attribution (CC BY) license (http://creativecommons.org/licenses/by/4.0/).

 *sustainability*

Article

Renewable Energy Prosumers in Mediterranean Viticulture Social–Ecological Systems

Ines Campos [1,*], Esther Marín-González [1], Guilherme Luz [1], João Barroso [2] and Nuno Oliveira [3]

[1] Centre for Ecology, Evolution and Environmental Changes (CE3C), Faculty of Sciences, Lisbon University, 1649-004 Lisboa, Portugal; emgonzalez@fc.ul.pt (E.M.-G.); gpluz@fc.ul.pt (G.L.)

[2] Sustainability Plan for Wines of Alentejo, Comissão Vitivinícola Regional Alentejana, 7005-485 Évora, Portugal; joao.barroso@vinhosdoalentejo.pt

[3] Ecosystem management, Esporão SA, 1400-315 Lisboa, Portugal; nuno.oliveira@esporao.com

* Correspondence: iscampos@fc.ul.pt

Received: 8 November 2019; Accepted: 27 November 2019; Published: 29 November 2019

Abstract: The significant energy demands of wine production pose both a challenge and an opportunity for adopting a low-carbon, more sustainable and potentially less expensive energy model. Nevertheless, the (dis)incentives for the wider adoption of local production and self-consumption of energy (also known as "prosumerism") from renewable energy sources (RESs) are still not sufficiently addressed, nor are the broader social–ecological benefits of introducing RES as part of a sustainable viticulture strategy. Drawing on the social–ecological systems (SESs) resilience framework, this article presents the results of a Living Lab (an action-research approach) implemented in Alentejo (South of Portugal), which is an important wine-producing Mediterranean region. The triangulation of results from the application of a multi-method approach, including quantitative and qualitative methods, provided an understanding of the constraining and enabling factors for individual and collective RES prosumer initiatives. Top enablers are related to society's expectation for a greener wine production, but also the responsibility to contribute to reducing carbon emissions and energy costs; meanwhile, the top constraints are financial, legal and technological. The conclusions offer some policy implications and avenues for future research.

Keywords: prosumers; renewable energy sources; Mediterranean wineries; constraints and enablers; social–ecological system; resilience

---

1. Introduction

Energy producers and consumers using renewable energy sources (RESs), often referred to as RES prosumers [1–3], may play an important role in the transformation of the energy system [4,5], which could evolve from a centralized and fossil-fuel-based model to a decentralized low-carbon system [6,7].

Notwithstanding the relevance of adopting RESs across most economic and industrial sectors [8,9], viticulture and winemaking are particularly relevant. According to Smyth and Russel [10], between 0.41 kg to 2 kg of CO_2 per wine bottle is released into the atmosphere, which could result in a total carbon footprint for the wine industry of 76.3 million tonnes of CO_2 (p. 1990). The same study concludes that if the global wine industry replaces its electricity needs with solar photovoltaics (PV), this renewable alone would eliminate 10.9 million tonnes of CO_2. Thus, adopting a renewables-based energy model is an important option in dealing with carbon-intensive agriculture practices, such as wine production [11,12].

Mediterranean wine-producing regions are vulnerable to climate change impacts, such as extreme weather events (e.g., irregular precipitation patterns, severe droughts, and heat waves), loss of biodiversity and soil degradation [13,14]. These regions require a long-term strategy to both mitigate and adapt to climate change impacts [13,15–17]. In this context, the high energy demands of wine production pose both a challenge and an opportunity for adopting a low-carbon, more sustainable and potentially less expensive energy model [18,19]. Furthermore, as vine-growers and winemakers become self-consumers of the energy they produce (i.e., become prosumers), new forms of collaboration may emerge that can increase this sector's adaptability and resilience to external challenges (e.g., climate change).

Alentejo in Portugal is a wine-producing region with a warm temperate climate and Mediterranean and Continental characteristics. It is, therefore, a case study for examining the interactions between viticulture and winemaking and the use of renewable energy sources.

Despite recent advances in the study of prosumers as key players in the energy transition [2], including prosumers in the wine market [20], a better understanding of the constraining and enabling factors that may act as (dis)incentives for local production and self-consumption of renewables in the wine sector is still needed. Some studies focused on the barriers faced by winemaking companies when adopting RESs [11,20]. Yet, in order to understand the transformative potential of prosumerism, it is critical to go beyond the individual adoption of RES systems and understand the potential for collaboration between viticulturists and winemakers with other social actors (municipalities, residents or other farming industries). New business models and governance arrangements may emerge through new collective self-consumption schemes involving these different actors.

This paper draws on the social–ecological systems (SESs) resilience framework [21] to analyse factors leading to a wider adoption of RES prosumerism in Alentejo's winemaking region. As an SES, winemaking is dependent on interrelated and co-evolving ecological, social and economic aspects. The sector is particularly dependent on a high (fossil fuel) energy expenditure [10]. Decentralised renewable energy production could reduce energy costs, as well as carbon emissions [22]. Thus, a wider adoption of a novel energy production and consumption model may increase SES resilience in winemaking.

The goal of this paper is to understand what the enabling and constraining factors are for individual and collective RES prosumers in the wine sector. The collective aspect is of importance, since we hypothesize that some factors acting as disincentives for individual prosumers may become incentives for collective initiatives. The paper also aims to capture how viticulturists and winemakers perceive the adoption of renewables, namely as a specific answer to a problem (e.g., the need to reduce energy costs) or as part of a strategy that aims to increase the resilience of the wine SES. This distinction is relevant since it can shape the way RESs are adopted through the establishment of interactions with other sub-systems (e.g., water management, land management, consumer behaviours and community engagement), and through collaborations with local communities and other stakeholders (e.g., tourism officers and local administrations). To understand this, the study draws on a co-creation approach (Living Labs).

The Living Labs (LLs) approach [23] is a form of participatory action-research [24,25], characterised by involving a multi-stakeholder community who shares a common goal. Setting up a LL provides an innovation space where new ideas, concepts, services or prototypes are co-created by the stakeholders to respond to a need and/or to find integrated solutions to problems [26,27]. For instance, Niitamo and colleagues [27] referred to LLs as an "emerging Public Private Partnership concept in which firms, public authorities and citizens work together to create, prototype, validate and test new services, businesses, markets and technologies in real-life contexts" (p. 45). Thus, the key characteristics of LLs are an active user involvement, real-life setting, involving multi-stakeholders, drawing on multiple methods (qualitative and quantitative) for data collection and analysis and following a co-creation approach [23].

Likewise, as a constellation of actors, including those that interact daily with the SES (e.g., vine growers and winery managers, often over generations of winemakers), a LL can replicate the functioning of an SES, integrating resilience thinking, increasing the capacity for learning and innovation and providing a space for adaptive co-management [28].

This study draws on the results of the first research cycle of a Living Lab named "RES for Sustainable Alentejo Viticulture". The LL brought together researchers, vine growers, managers and owners of wineries, entrepreneurs, companies working in the RESs sector, and the Alentejo Regional Winegrowing Commission (Comissão Vitivinícola Regional Alentejana (CVRA)), which is a private institution responsible for certifying, controlling and protecting Alentejo's wine production. The key goal set by the LL stakeholders was to develop a new energy model for the wine SES and contribute to CVRA's sustainability programme by promoting and facilitating the adoption of RES. Following a multi-method approach, qualitative and quantitative methods were applied to develop baseline knowledge on the potential of RES prosumerism in the context of viticulture.

We proceed as follows: the SES resilience framework and the Alentejo wine SES are explained in Section 1.1. The multi-method approach is described in Section 2 and the results are given and discussed in the following Sections 3 and 4, respectively. Finally, the conclusion distils our main findings and provides some policy implications and avenues for future research.

1.1. Social–Ecological Systems Resilience Framework and the Alentejo Wine SES

Natural and human systems are interdependent and require an interdisciplinary and integrated approach to understand the co-evolution of social and ecological elements [29]. The SES framework results from the meeting of socio-cultural and ecological sciences in an effort to achieve an integrated analysis of complex problems (e.g., resource depletion and climate change) [30,31].

SESs are dynamic systems, their elements are anything but stable; instead, they are constantly co-evolving, adapting or transforming through ongoing interactions within the bio-physical and socio-cultural environment. SESs are also complex, non-linear systems, which leads to unforeseeable outcomes emerging from the capacity of the system to adapt, cope with, re-arrange or renew itself [32,33]. As an SES cannot be controlled, command-and-control approaches to ecosystem management need to give way to an adaptive co-management approach [28]. The approach relies on the building-up of social networks and on the collaboration of multiple stakeholders (e.g., researchers, practitioners, local communities and policymakers), who embody different forms of knowledge, thus harnessing the potential of a system to deal with change.

According to Carl Folke, resilience is both a quality of an SES and a way of thinking since it refers to the ability of a system to absorb shocks, but also to its capacity for "renewal, re-organization, and development" [21] (p. 254) and to "the degree to which the system can build and increase the capacity for learning and adaptation" [21] (p. 260). Innovation is a scale within the SES, which may increase the adaptability of the system and/or help manage its resilience. Thus, innovation is central to resilience thinking as it may help change dominant patterns in order to harness the SES potential for transformative change.

Adaptability is a quality of resilience and losing one implies losing the other, leaving the system more vulnerable to environmental, social, economic or political shocks. Within the SES framework, "adaptability is referred as the capacity of people in a social-ecological system to build resilience through collective action, whereas transformability is the capacity of people to create a fundamentally new social-ecological system when ecological, political social or economic conditions make the existent system untenable" [21] (p. 262). Collective action is therefore a critical element for enabling either the adaptability or transformability of a system.

Alentejo's wine culture, a set of viticulture and vine growing practices leading to wine production and consumption, is an example of a social–ecological system. The region has traditionally been characterized by agro–silvo–pastoral systems [34,35]. The use of the land, adapted to its Mediterranean climate, included regions of evergreen forest (olive and oak trees), grazing areas and cultivation. Since the late twentieth century, Alentejo has witnessed a growth in intensive and extensive cultivation, as well as socioeconomic and policy changes, resulting in increased land abandonment, land degradation and high soil losses [36]. Sustainable agriculture and the capacity to attract population back to rural areas are considered essential strategies to tackle these problems [37,38]. Furthermore, Mediterranean wine regions are extremely vulnerable to the impacts of climate change due to their already high climate variability and propensity for extreme events (such as heat waves) [39] and soil degradation [13].

The availability of water for irrigation is becoming a persistent problem that can lead to serious water shortages in the future depending on different climate change scenarios [12,14]. However, irrigation is increasingly used in viticulture and winemaking, and consequently, energy needs are also increasing as a direct function of water use in the different processes carried out (i.e., irrigation, grape harvesting, cleaning of equipment, etc.) [10].

Renewable energy sources (RESs) have multiple benefits across the full cycle of viticulture and winemaking processes. This is well described in a study by Smyth and Russel [10], who show how RES can be integrated from the initial process of site preparation—using electric vehicles and equipment—to soil and disease control—using soil solarisation techniques to help retain heat in the soil and as a form of pest control and frost protection—as well as the use of PV-powered pumping systems for irrigation. For oenology, the needs of heating, cooling and maintenance of a controlled environment can be partially satisfied by both PV and thermal solar energy. Smyth and Russel equally mention the destemming, crushing and pressing processes of winemaking, arguing for the use of different solar solutions at each step.

Regarding solar photovoltaics, as well as other renewable energies (solar thermal, hydro from private dams and biomass), recent in-depth studies of eight Alentejo wineries [40–42] indicate that investments in renewables will allow for reductions in carbon emissions of between 19% and 35%. Thus, in the context of local energy production and self-consumption (i.e., prosumerism), the Alentejo wine SES shows a high potential for utilising both solar and biomass. Additionally, as they become RES prosumers, winemakers could potentially reduce energy bills and increase energy efficiency, which would create an upsurge in profits and reduce dependency on energy price fluctuations [43,44]. Alentejo may also be considered an example of a "wine tourism cluster" [45]; the growth of sustainable wine tourism may be a driver for attracting people to the region, creating new green jobs and helping to fight land abandonment [46].

2. Materials and Methods

In an SES, knowledge procurement results from the meeting of various knowledge systems and is gained through an ongoing learning process [47]. In this context, the wine SES benefits from the interaction between scientific knowledge (i.e., oenology) and traditional and local ecological knowledge [48,49]. Therefore, the methodology applied aimed to capture this transdisciplinarity [35].

2.1. Multi-Method Approach

The multi-method approach included a workshop, field diary notes and a questionnaire. The triangulation of the methods provided a good understanding of the constraining and enabling factors for increasing the adoption of RES technologies in the wine SES.

First, a "needs-assessment" workshop was conducted with the participation of LL stakeholders, including representatives of 10 wineries in the region. The participating wineries were among the largest companies in the region, including representatives of a cooperative of wineries. About a third of the participants were already using RESs, with installed capacities between 60 kW to 250 kW (photovoltaic). Another group had small installations and sought to increase their adoption of RESs, and finally, the last group had not yet installed any RES system. The workshop provided an overview of the challenges and opportunities regarding becoming RES prosumers from the perspective of local viticulture and winemaking employees and managers. A world café, i.e., a participatory method that facilitates debate involving a group of people working on different interconnected topics [37], was used to discuss barriers and opportunities for winemaking companies to adopt RESs.

Second, two case studies were explored, which we refer to as Company E and Company S. These companies were chosen because both come from a tradition of sustainable viticulture and winemaking in the region. Company E has been producing wine in the region for 40 years and has been innovating through its experimental land use management approach. Company S comes from a tradition of winemakers (which has been the family business for about 350 years), and its owners aim to continue increasing their knowledge through harmonising their past legacy with innovation and experimentation. Both companies are using RESs in their wineries and aim to continue increasing their installed capacity. Company E has an installation of 250 kW next to its grape vines, while Company S has a PV installation of 90 kW on the winery's rooftop. Despite being in the same region, there are key differences between the two in terms of energy needs due to the difference in the volume of wine produced and irrigation needs (e.g., Company E irrigates, while Company S does not).

Lastly, an online questionnaire was given (in Portuguese). The questionnaire included seven closed questions (see Table 1) and one open question. The response options of the questions were informed by the results of the first "needs-assessment" workshop and by the notes collected during the field trips to wineries in the region. The questionnaire was produced using the Google forms app and shared via email (between February and April 2019). It was given to a total of 1800 registered viticulturists in the region, of which 280 were also winemakers. Data regarding the collection of registered viticulturists was provided by CVRA, who also sent out the questionnaire to its members. With 59 valid responses (see Table S1), the sample was not representative of the entire sector; however, with a confidence level of 90% and a sample error of 10%, the sample size was statistically significant for the collection of wine producers in the region ($N = 55$), which corresponded to the main target population.

Table 1. Challenges and opportunities for a renewable energy source (RES)-based energy system in Alentejo's Wine social–ecological system (SES) online questionnaire (translated from Portuguese).

Question	Response Options
1. Does your wine company or cooperative have installed RES systems?	No, we do not have systems installed, nor intend to No, but we intend to have RES systems installed in the future Yes, we have installed RES systems
1.1. Why?	Open question (optional)
2. What renewable energies does your company/cooperative produce or plan to produce in the near future?	We do not intend to produce renewable energies Solar photovoltaics (PV) Solar Thermal Wind Energy Biogas Other
3. What is the regulatory regime of your current and future RES installation(s)?	We do not intend to produce renewable energies Self-consumption unit Small-production unit Former micro-generation scheme Do not know Other
4. What other technologies are in use in your wineries' energy system?	Thermal storage Electricity storage Smart meters Demand-side management systems Not applicable Other
5. If you produce or intend to locally produce energy from renewable sources, would you be interested in selling excess energy to other companies or local entities?	On a scale from 1 (not interested) to 6 (very interested)

Table 1. *Cont.*

Question	Response Options
6. In your opinion, what are the top five challenges constraining your local production and self-consumption of renewable energies?	Systems adapted to the seasonality of viticulture Cost of storage technologies Capacity of current storage technologies Cost of electric vehicles Initial cost of installation of photovoltaic system Cost of renewable technologies Low remuneration of surplus electricity sold to the grid Limitation of 250 kW of installed capacity in small production units Capacity of renewables to ensure a continuous and safe supply Adapting viticulture to the impacts of climate change in Alentejo Changes in landscape due to renewables Difficulty in accessing relevant information
7. In your opinion, what are the top five opportunities that motivate your local production and self-consumption of renewable energies	Capacity of installing and using renewable technologies Transition to electric mobility (e.g., introducing electric vehicles in grape harvesting) Rises in oil prices Decrease in the prices of renewables Decreased dependence on future price fluctuations in the energy market Possibility of selling excess energy to the grid New green jobs Good company image (sustainability) Expectations of society for a cleaner environment Decrease carbon emissions from wine production in Alentejo Adapt the wine industry to the impacts of climate change in Alentejo Other

3. Results

The following sections present the viticulturists and winemakers' perspectives regarding the constraining and enabling factors for adopting RESs, but also concerning the way RES adoption is perceived in the context of a wider sustainability approach.

Although this study has focussed on RESs, including different technologies, such as solar thermal, photovoltaic, wind, biomass, hydro or biogas, Living Lab stakeholders have addressed mainly solar photovoltaics and expressed little interest in discussing other technologies. This is possibly due to wind energy not having a high potential in the region, hydro energy being confined to private dams, and biomass and biogas appearing to be less appealing than solar energy (although throughout the LL interactions, researchers frequently mentioned the potential of biomass).

3.1. Needs Assessment Workshop: Constraints and Enablers

Factors acting as constraints and enablers are related to financial, regulatory and technological aspects, as well as environmental and socioeconomic aspects.

Concerning constraining financial factors, viticulturists and winemakers highlighted the initial cost of an installation of photovoltaic systems, which is the most commonly used renewable in the region. Additionally, the high cost of storage technologies (batteries) is also a problem. The payback period of the initial investment for a solar panel installation would take up to between 7 to 12 years (depending on the installed capacity). However, wineries were not yet able to profit from selling surplus energy to another consumer. Until October 2019, Portuguese law (i.e., Decree-Law 153/2014, Self-Consumption Law) allowed the sale of surplus energy to the grid (in a system of self-consumption) for a low price (90% of the average wholesale market price). A new law for collective self-consumption (Decree-Law 162/2019) was issued in October 2019, yet it does not include any additional support schemes (such as feed-in-tariffs) for the compensation for surplus energy. Given the high environmental and economic costs of using batteries, prosumerism would be more attractive if financial compensation for locally produced electricity sold to the grid was higher.

On the other hand, the progressively lower prices and increase in productivity of solar panels appeared as opportunities. Workshop participants also agreed they would like to reduce their dependency on fossil fuels and on the fluctuations of energy market prices. It was found that should the Portuguese self-consumption law change to allow selling or exchanging surplus energy directly with other consumers (e.g., village residents, municipality or other farmers), this would enable new investments since the return on investment could be higher.

Regarding constraining technological factors, participants found that storage technologies are not yet sufficiently developed (and have high environmental costs, linked to lithium extraction), but a major challenge for wineries is that any energy model must be adapted to the seasonality of production activities. This poses a problem for the installed capacity and the contracted capacity. If wineries opt for an installation that covers the peak energy needs of the harvest season, which corresponds to the months when solar energy is abundant (i.e., between July/August and October/November), there will be a significant surplus for most of the year. Should the Portuguese law allow seasonal contracted capacity for farmers, this would provide a good incentive. On the upside, new RES technologies tend to be cheaper and increasingly easier to install and use. Also, the development of electric vehicles, including electric tractors, could offer an important incentive since surplus energy could be used to charge these vehicles' batteries.

Regarding environmental issues, producers mostly pointed to enabling factors and incentives to act. Alentejo is perceived as a region with an enormous solar potential due to its warm Mediterranean climate and its geographic characteristics. All agreed that wine production is a high energy intensity activity and the need to reduce carbon emissions is an important driver for adopting RES. Not surprisingly, viticulturists and winemakers are concerned with the need to adapt viticulture to the impacts of climate change and its consequences (e.g., soil degradation, alteration of grape properties and wine quality and increased fire risk), and to increase the resilience to external shocks, such as eventual energy supply disruptions.

Regarding sociocultural factors, workshop participants referred to some constraints, namely a cultural resistance to change, including among those who work in viticulture. There is a prevailing notion that renewables are not yet capable of ensuring a continuous and safe supply. Nevertheless, integrating RES was found to be good marketing for wine companies, who would be meeting society's expectations for a cleaner and more sustainable environment. The possibility of creating new "green jobs" was also mentioned.

3.2. Field Notes: A Holistic Perspective

The owners and managers of the two wineries visited (i.e., Company E and Company S) showed a holistic perspective about winemaking. In these wineries, adopting renewables was understood as part of a broader ecological vision, wherein the environmental, economic and social domains are intrinsically connected. These winemakers were aware of the impacts of climate change and the need for an energy transition in the sector.

In both wineries, caring for the soils by reducing or eliminating chemical pesticides and fertilizers was considered a crucial part of a more sustainable business. While in Company E, all production is already organic, in Company S, the owners opted for not irrigating, gradually moving toward an organic production and adopting soil regeneration techniques using organic compost. One of the owners of Company S explains that: "Vines can have small roots, but if you irrigate, they get addicted to water. If they have no water, roots can grow up to five metres to find water, and the vines become more resilient to weather changes."

In Company E, climate change adaptation is well-integrated into wine production. The company had a testbed of nearly 200 grape varieties to study which are better adapted to climate change impacts and their consequences.

Ensuring water and energy efficiency and adopting circular economy strategies are important targets for both wineries. In Company S, biomass from the by-products of destemming was used as soil compost. In Company E (who produces both wine and olive oil), olive stones were used to produce biomass for heating purposes.

Company E had a photovoltaic installation of 250 kW, but the owners aimed to increase it to a 500-kW installed capacity, which would provide 50% of energy needs during the peak season. Company E's managers were also considering using storage to supply energy during peak hours when it is more expensive. As one engineer explained, this would allow for switching off the grid connection in these periods and to use stored energy, as well as energy from the photovoltaic station. As storage costs decrease, this option will become increasingly viable.

When considering direct exchanges with other prosumers, the managers of the two wineries agree that this would be important to incentivise the adoption of RES technologies. Both wineries would be interested in developing business models that would promote a wider adoption of RES by other economic agents in the region, where RES prosumers would be able to directly exchange their surplus energy.

3.3. Questionnaire on Barriers and Opportunities

Of the 59 grapevine growers and wineries who responded to the questionnaire, only 5% did not plan to implement any RES production, 47.5% had installed RES systems, while 47.5% planned to integrate RES production in the future.

Regarding the reasons for installing (or not) RES systems, 20 respondents provided an answer (this was the only open question included in the form). The key words mostly used were "sustainability of the company", "lack of information" and "high investment". Responses to this open question stressed the value of RESs as part of a more environmentally and economically sustainable policy for the company. Four respondents pointed out the benefit of saving on energy costs, as well as the benefit of using RESs together with water-pumping systems. The need for more information and the high cost of initial investments was mentioned by three respondents. One respondent pointed out that the high cost was particularly restrictive given that there are other investment priorities related to the wineries' core businesses. Finally, the possibility of lowering dependency on conventional fuels and centralized systems was also mentioned by one respondent, and bureaucracy and administrative hurdles for installing solar panels were referred to as a problem by another.

Concerning the main energy source used, solar photovoltaics was the most common (36), followed by combinations of solar PV and thermal (18) (Figure 1). There were no responses indicating that biogas was used.

Up until 2019, the Portuguese self-consumption law had defined two types of small-scale RES installations, namely small production unit (UPP) (of up to 250 kW), in which all energy is injected to the grid (this type of unit does not allow self-consumption and is to be discontinued in 2020); and small self-consumption units (UPAC), which should be dimensioned according to local energy consumption needs. When asked about the regulatory regime of an existing or future RES installation in their winery or property (Figure 2), 23 responded they were covered (or planned to be) by the UPAC regime, 11 have had an UPP and 3 were using both types of units. Additionally, up until 2012, the regulatory framework for RES micro-generation included a feed-in-tariff (FIT), which has ended. Yet, 5 respondents (who must have installed their systems before 2012) were still benefiting from the previous legal regime, in combination or not with others. Finally, 2 respondents replied "None" and 15 replied "don't know". Those that "didn't know" corresponded to some of the respondents who did not yet have an installation but were planning to. Also, some respondents did not know which regime their installation fitted into.

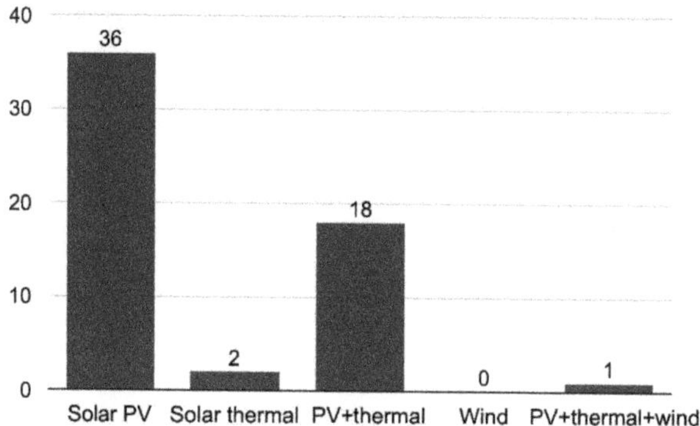

Figure 1. RESs used or planned to be used by the viticulture or winemaking companies (The (y) axis shows number of replies; and the (x) axis the response options).

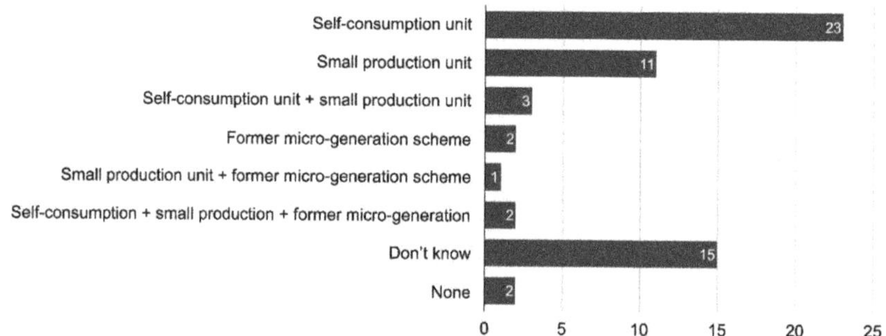

Figure 2. Regulatory regime of the current or future (planned) RES production unit (The (y) axis shows the response options; and the (x) axis the number of replies).

Overall, grape and wine producers were pretty (31%) or very interested (64%) in selling their surplus energy to another company, local entity or consumer (see Figure 3). Until October 2019, this was not allowed in Portugal, yet a law for collective self-consumption (Decree-Law 162/2019 for collective self-consumption and renewable energy communities) is coming into force in early 2020.

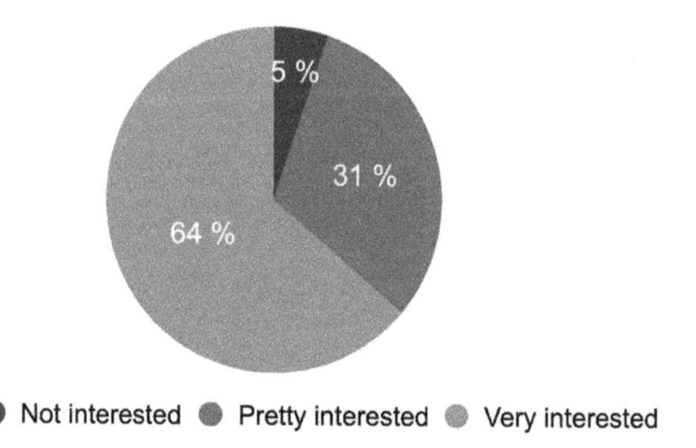

Figure 3. Interest in selling surplus energy to other companies or consumers (in percentage).

Concerning the challenges constraining the production and self-consumption of renewables, the top three options (of the 13 response options, plus the option "others") were the cost of the installation of photovoltaic systems (83.1%), followed by the cost of storage technologies (62.7%) and the low renumeration for surplus energy sold to the grid (49.2%) (Figure 4).

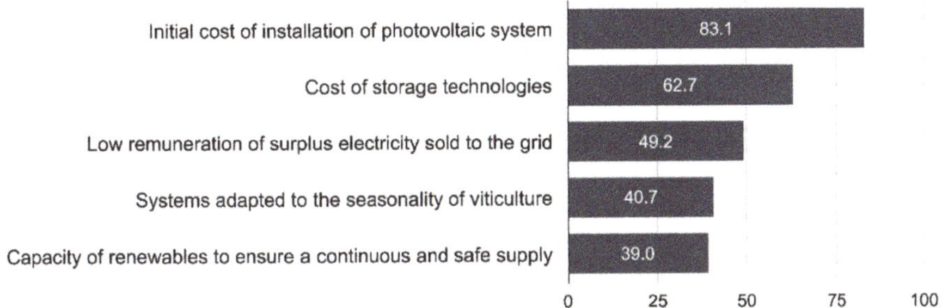

Figure 4. Top five challenges that constrain RES production and self-consumption (in percentages).

Regarding the opportunities motivating the production and self-consumption of renewables in grape and wine producers, the top three options (of the 13 response options, plus the option "others") were a good company image (71.2%) and meeting the expectations of society for a cleaner environment (71.2%), followed by a reduced dependency on future price fluctuations in the energy market (61.0%) (Figure 5).

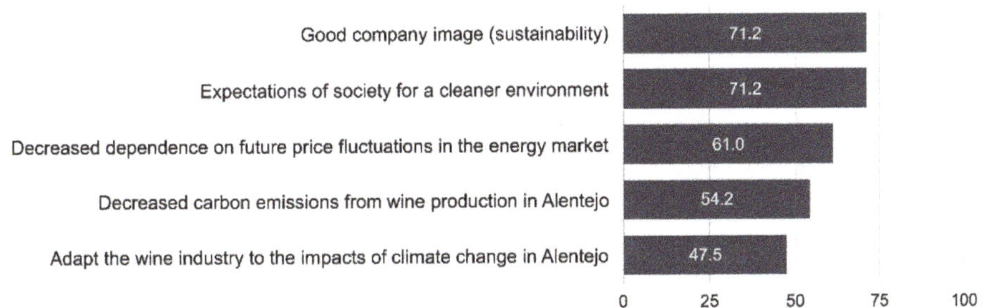

Figure 5. Top five opportunities that motivate RES production and self-consumption (in percentage).

3.4. Triangulation of Results

Results of the different methods applied show a strong agreement in the findings obtained between the workshop and the online questionnaire, while the field notes reinforced the environmental and cultural factors. The most relevant and correlated factors found in the workshop and questionnaire were the following:

The key constraining factors were the initial cost of installing PV systems, the high cost of storage technologies and a low remuneration of surplus electricity sold to the grid and/or not being possible to sell surplus energy to other consumers or prosumers. This last aspect was especially relevant since 95% of the questionnaire respondents seemed to be in favour of this possibility (see Figure 5). Another correlated and relevant legal challenge is the need for RES systems to be adapted to the seasonality of viticulture and winemaking practices (allowing for an optimized installed capacity). This result was also confirmed through the field notes.

The key enabling factors were gaining a good company's marketing image (sustainability); meeting society's expectations for environmental sustainability; reducing dependency on future energy price fluctuations and dependency on fossil fuels; development of RES technologies (including electric vehicles, which could provide a use to surplus energy from local installations); and the prospect of reducing carbon emissions from wine production.

While most constraints related to financial, regulatory and technological reasons (in this order), the key enabling factors were social, environmental and technological (also in this order).

The field notes provided some additional findings regarding the perspectives of winemakers related to a sense of responsibility to contribute to climate change mitigation, as they are aware of the need to reduce emissions and seem open to experiment and adopt new RES-based solutions.

4. Discussion

Portuguese vine-growers and winemakers are aware of the need to transform the current energy sub-system within the wine SES to increase its resilience. Moreover, these stakeholders expressed their views on the interlinkages between soil regeneration, irrigation needs and energy sources. RES prosumerism is seen as part of a broader strategy for preserving and restoring local ecosystem services, which includes climate adaptation measures (e.g., testing different grape varieties, choosing not to irrigate and reducing or eliminating the use of chemical pesticides and fertilizers). In this context, traditional local and ecological knowledge is a significant asset for viticulture realities, supporting the claim for the importance of different systems of knowledge in building SES resilience [49,50]. A holistic perspective of the social, economic and environmental dimensions seems prevalent, and beyond quantity, the quality of wine production is constantly measured against the consequences of inaction in the context of climate change.

When reflecting on the enabling and constraining factors for integrating a higher share of RES and becoming prosumers, winemakers considered different interrelated aspects.

RES prosumerism is seen as an economic benefit. Similarly, a study of Spanish wineries showed that the use of RES for energy production can reduce energy bills and increase energy efficiency in the sector [43]. Using RES is also perceived to be important for the marketing of a greener and more environmentally sustainable wine company, which is relevant for wine tourism. From the stakeholders' perspectives, there is a strong correlation between the companies' marketing images and wine tourism, which is consistent with the literature on wine tourism and sustainability [51,52].

In the context of Alentejo, by creating new "green jobs", introducing RES was also understood to prevent, or even reverse, land abandonment. This topic has been explored in detail by Zambon and colleagues who presented the case of agro-energy districts, such as in the Italian wine sector, arguing that these represent a new future for viticulture [53], creating new economic opportunities for local communities, whilst developing new viticulture practices.

Despite the high cost of initial investments, the environmental, economic and social benefits that may come with a wider update of renewables are important incentives. Related to this, the personal values of managers and winery owners could play an important role as an incentive to seek for more sustainable practices, which was also concluded in studies of wineries in the United States and New Zealand [54,55].

Thus, the transition to a new energy model in which viticulturists and winemakers become RES prosumers is perceived to be more than a decarbonization strategy, it is also seen as a business strategy (e.g., reducing energy costs), a marketing and innovation strategy (with benefits for wine tourism) and a strategy against land abandonment in the region (e.g., creating new jobs, involving local communities and developing new partnerships).

Technological and technical-related cultural aspects are equally relevant, and some winemakers were wary of the capacity of renewable technologies to respond to the high energy needs of wineries, in particular, during the harvest season, which is consistent with the study of Garcia-Casarejos and colleagues [11] who clearly identified a group of wineries in Spain with this perspective. Changing cultures was identified as a barrier since it requires managers to thoroughly explain to their teams the benefits of introducing RES technologies.

Financial and regulatory aspects appeared as key constraints for individual producers. Regulatory barriers were also found in the study of the adoption of RES in Spanish wineries, where "administrative obstacles" were thought to be reinforcing the reluctance of some wineries in not investing in renewables [11]. However, the regulatory barriers in the Portuguese case were linked to self-consumption policies. The impossibility of selling surplus energy to other consumers appeared as a top constraint, which strengthens the argument that collective action is important to further develop the integration of RESs in the sector. Legal frameworks for self-consumption are currently changing in Portugal to allow, for the first time, setting up collective self-consumption schemes and renewable energy communities (i.e., new Decree-Law 162/2019). These changes will open the way to innovative collaborative business schemes that are likely to provide more attractive options and promote investment of individual wineries in RES systems [44]. The issue of the seasonality of production activities is equally a key constraint which could be resolved by new legislations, allowing for seasonal capacity contracts for farmers who use RESs, or even specific tax benefits.

The results indicate there was a drive to work together in developing collective solutions that accelerate the adoption of RES in the sector. Despite being competitors in the market, Alentejo winemakers seemed to recognize a benefit in developing new partnerships and networks when it comes to adopting RES. Working collectively, prosumers could set up business models that would provide new opportunities to increase RES adoption [56], and would therefore increase the adaptability and transformability of the SES. New business models could include peer-to-peer schemes [7], setting up community energy services involving a nearby village or other wineries [57] or using excess energy during the non-peak harvest season to charge electrical vehicles [58]. Allowing wineries to offer new services, such as charging electric cars [59], could imply a collective self-consumption scheme where other actors participate (e.g., cooperatives, municipalities and villages) [60], with mutual benefits for those involved [61,62]. Finally, integrating RESs in winemaking could also offer a better final product for the increasing number of consumers who value environmentally friendly wines [51,63].

5. Conclusions

Sustainable viticulture is intertwined with the environmental, economic and cultural domains in which it develops. Viticulturists and winemakers in the Alentejo region are aware of the necessity to understand and manage vineyards as complex ecosystems in which environmental aspects, but also technical and cultural practices, shape crop features and wine quality. The practice of viticulture and winemaking can be part of a global transition to low-carbon and more sustainable production and consumption patterns through conferring winemakers with the role of RES prosumers. Regions, such as Alentejo, are extremely vulnerable to climate change, yet offer optimal geographic and climatic conditions for mainstreaming the use of RESs. Despite the focus on winemaking, the conclusions of this study offer equally relevant insights for research on the integration of RES prosumerism in other farming sectors.

The top constraining factors were financial and legal, while top enabling factors were related to environmental reasons and to meeting society's expectations for a cleaner and greener environment in relation to sustainable production. These were valuable incentives for RES prosumerism, more so in Alentejo where wine tourism is an increasingly important revenue source.

Given the key identified hindering factors, agriculture subsidies should help finance small-scale RES production and self-consumption schemes. New business models, involving the collaboration of different stakeholders and local communities, seem fundamental to incentivising a wider uptake of RES prosumerism in wineries and other farming activities. Legislation should be as flexible as possible to allow for bottom-up creativity and innovative solutions for adopting renewables. New energy and sustainability policies for the sector and for the region, as well as climate change mitigation and adaptation plans, need to seriously consider the need to support the development of new business models, which make it more attractive for winemakers and other farmers to invest in RES technologies as part of their business.

Future research and innovation studies should focus on developing winemaking business models (which are also applicable to other agro-industry sectors) that integrate new services through collective schemes (e.g., energy cooperatives) and new governance arrangements where companies collectively benefit from energy exchanges and provide new services, including sustainable tourism services. Research should consider the economic benefits of a higher share of RES prosumers in viticulture and winemaking activities, but also explore other social and cultural benefits, including the role of a more resilient winemaking SES in the fight against land abandonment in Alentejo and other Mediterranean viticulture regions.

Finally, the approach used in this study, namely Living Labs, offered a platform for collective action, and co-management and co-creation approaches, that are capable of addressing the complexity of an SES and contributing to build up resilience. Thus, Living Labs can be used as a tool to increase a systems' adaptability and resilience in other SES studies.

Supplementary Materials: The following are available online at http://www.mdpi.com/2071-1050/11/23/6781/s1, Table S1: Respostas do Formulário.

Author Contributions: The two first authors, I.C. and E.M.-G. have equally contributed to the investigation and to the final draft of the paper. Together, the two authors have conceptualized the study, developed and implemented the methodology, collected data and analysed the results. I.C. has coordinated the writing of the full final draft of the paper, to which E.M.-G. has contributed mainly in the methodology and results sections, including the production of the final graphs (i.e., Figures 1–5). The final draft was also extensively discussed by both first authors. G.L. has reviewed the manuscript and contributed to the design and implementation of the methodology, validation and investigation, including data collection. J.B. and N.O.'s contributions to the paper were related with the validation of some of the data presented, and resources, as well as reviewing and editing the paper's content regarding the Alentejo wine industry and its potential role regarding prosumerism.

Funding: This research was funded by the European Union's Horizon 2020 research and innovation programme under grant agreement no. 764056. The sole responsibility for the content of this document lies with the authors. It does not necessarily reflect the opinion of the funding authorities. The funding authorities are not responsible for any use that may be made of the information contained therein. The Centre for Ecology, Evolution and Environmental Changes to which the three authors are affiliated, received national funds through FCT (Fundação para a Ciência e a Tecnologia) in the frame of the project UID/BIA/00329/2019

Acknowledgments: The authors acknowledge all those who participated in the Living Lab work (wine companies, wine cooperatives, researchers and invited experts). A special thanks to CVRA (the Wines of Alentejo) for facilitating contacts with wine producers and other stakeholders, and for collaborating with our research team throughout the Living Lab process.

Conflicts of Interest: The authors declare no conflict of interest.

References

1. De Vries, G.W.; Boon, W.P.C.; Peine, A. User-led innovation in civic energy communities. *Environ. Innov. Soc. Transit.* **2016**, *19*, 51–65. [CrossRef]

2. Butenko, A. User-Centered Innovation and Regulatory Framework: Energy Prosumers' Market Access in EU Regulation. 2016. Available online: http://dx.doi.org/10.2139/ssrn.2797545 (accessed on 30 May 2019).

3. Bellekom, S.; Arentsen, M.; van Gorkum, K. Prosumption and the distribution and supply of electricity. *Energy Sustain. Soc.* **2016**, *6*, 22. [CrossRef]

4. CE Delft. *The Potential of Energy Citizens in the European Union*; CE Delft: Delft, The Netherlands, September 2016; Available online: https://www.cedelft.eu/publicatie/the_potential_of_energy_citizens_in_the_european_union/1845 (accessed on 25 August 2019).

5. Van der Schoor, T. Local Citizen Initiatives and Transitions to Energy Sustainability. Available online: https://research.hanze.nl/ws/portalfiles/portal/3444082/107.Local_citizen_initiatives_SB13_Oulu_vdSchoor_1_.pdf (accessed on 25 July 2019).

6. Newbery, D.M. Towards a green energy economy? The EU Energy Union's transition to a low-carbon zero subsidy electricity system—Lessons from the UK's Electricity Market Reform. *Appl. Energy* **2016**, *179*, 1321–1330. [CrossRef]

7. Morstyn, T.; Farrell, N.; Darby, S.J.; McCulloch, M.D. Using peer-to-peer energy-trading platforms to incentivize prosumers to form federated power plants. *Nat. Energy* **2018**, *3*, 94. [CrossRef]

8. Dóci, G.; Vasileiadou, E.; Petersen, A.C. Exploring the transition potential of renewable energy communities. *Futures* **2015**, *66*, 85–95. [CrossRef]

9. Croonenbroeck, C.; Lowitzsch, J. From Fossil to Renewable Energy Sources. In *Energy Transition: Financing Consumer Co-Ownership in Renewables*; Lowitzsch, J., Ed.; Springer International Publishing: Cham, Switzerland, 2019; pp. 29–58. ISBN 978-3-319-93518-8.

10. Smyth, M.; Russell, J. 'From graft to bottle'—Analysis of energy use in viticulture and wine production and the potential for solar renewable technologies. *Renew. Sustain. Energy Rev.* **2009**, *13*, 1985–1993. [CrossRef]

11. Garcia-Casarejos, N.; Gargallo, P.; Carroquino, J. Introduction of Renewable Energy in the Spanish Wine Sector. *Sustainability* **2018**, *10*, 3157. [CrossRef]

12. Galbreath, J.; Charles, D.; Oczkowski, E. The Drivers of Climate Change Innovations: Evidence from the Australian Wine Industry. *J. Bus. Ethics* **2016**, *135*, 217–231. [CrossRef]

13. Fraga, H.; García de Cortázar Atauri, I.; Malheiro, A.C.; Santos, J.A. Modelling climate change impacts on viticultural yield, phenology and stress conditions in Europe. *Glob. Change Biol.* **2016**, *22*, 3774–3788. [CrossRef]

14. Fraga, H.; Malheiro, A.C.; Moutinho-Pereira, J.; Santos, J.A. An overview of climate change impacts on European viticulture. *Food Energy Secur.* **2012**, *1*, 94–110. [CrossRef]

15. Malheiro, A.; Santos, J.; Fraga, H.; Pinto, J. Climate change scenarios applied to viticultural zoning in Europe. *Clim. Res.* **2010**, *43*, 163–177. [CrossRef]

16. Costa, J.M.; Vaz, M.; Escalona, J.; Egipto, R.; Lopes, C.; Medrano, H.; Chaves, M.M. Modern viticulture in southern Europe: Vulnerabilities and strategies for adaptation to water scarcity. *Agric. Water Manag.* **2016**, *164*, 5–18. [CrossRef]

17. Mozell, M.R.; Thach, L. The impact of climate change on the global wine industry: Challenges & solutions. *Wine Econ. Policy* **2014**, *3*, 81–89.

18. Galitsky, C.; Worrell, E.; Radspieler, A.; Healy, P.; Zechiel, S. *BEST Winery Guidebook: Benchmarking and Energy and Water Savings Tool for the Wine Industry*; Lawrence Berkeley National Laboratory: Berkeley, CA, USA, 2005.

19. FAO. *The State of Food and Agriculture. Climate Change, Agriculture and Food Security*; FAO: Rome, Italy, 2016.

20. Dressler, M. Prosumers in the wine market: An explorative study. *Wine Econ. Policy* **2016**, *5*, 24–32. [CrossRef]

21. Folke, C. Resilience: The emergence of a perspective for social-ecological systems analyses. *Glob. Environ. Change* **2006**, *16*, 253–267. [CrossRef]

22. Johnson, V.; Hall, S. Community energy and equity: The distributional implications of a transition to a decentralised electricity system. *People Place Policy Online* **2014**, *8*, 149–167. [CrossRef]

23. Haukipuro, L.; Väinämö, S.; Hyrkäs, P. Innovation Instruments to Co-Create Needs-Based Solutions in a Living Lab. *Technol. Innov. Manag. Rev.* **2018**, *8*, 22–35. [CrossRef]

24. Campos, I.S.; Alves, F.M.; Dinis, J.; Truninger, M.; Vizinho, A.; Penha-Lopes, G. Climate adaptation, transitions, and socially innovative action-research approaches. *Ecol. Soc.* **2016**, *21*. [CrossRef]

25. Schaffers, H.; Guzman, J.G.; Merz, C. An Action Research Approach to Rural Living Labs Innovation. 2008. Available online: https://www.semanticscholar.org/paper/An-Action-Research-Approach-to-Rural-Living-Labs-Schaffers-Guzm%C3%A1n/4c8d3706527ee4b0752e17c014da6362b7b25b26?citationIntent=background#citing-papers (accessed on 25 August 2019).

26. Schumacher, J.; Feurstein, K. Living Labs–the user as co-creator. In Proceedings of the 2007 IEEE International Technology Management Conference (ICE), Sophia-Antipolis, France, 4–6 June 2007; pp. 1–6.

27. Santonen, W.T.; Creazzo, L.; Griffon, A.; Bódi, Z.; Aversano, P. Cities as Living Labs–Increasing the Impact of Investment in the Circular Economy for Sustainable Cities. 2017. Available online: https://ec.europa.eu/research/openvision/pdf/rise/cities_as_living_labs.pdf (accessed on 26 August 2019).

28. Armitage, D.R.; Plummer, R.; Berkes, F.; Arthur, R.I.; Charles, A.T.; Davidson-Hunt, I.J.; Diduck, A.P.; Doubleday, N.C.; Johnson, D.S.; Marschke, M.; et al. Adaptive co-management for social–ecological complexity. *Front. Ecol. Environ.* **2009**, *7*, 95–102. [CrossRef]

29. Liu, J.; Dietz, T.; Carpenter, S.R.; Alberti, M.; Folke, C.; Moran, E.; Pell, A.N.; Deadman, P.; Kratz, T.; Lubchenco, J.; et al. Complexity of Coupled Human and Natural Systems. *Science* **2007**, *317*, 1513–1516. [CrossRef]

30. McGinnis, M.D.; Ostrom, E. Social-ecological system framework: initial changes and continuing challenges. *Ecol. Soc.* **2014**, *19*, art30. [CrossRef]

31. Ostrom, E. A General Framework for Analyzing Sustainability of Social-Ecological Systems. *Science* **2009**, *325*, 419–422. [CrossRef] [PubMed]

32. Folke, C.; Carpenter, S.R.; Walker, B.; Scheffer, M.; Chapin, T.; Rockström, J. Resilience Thinking: Integrating Resilience, Adaptability and Transformability. *Ecol. Soc.* **2010**, *15*. [CrossRef]

33. Garmestani, A.S.; Benson, M.H. A Framework for Resilience-based Governance of Social-Ecological Systems. *Ecol. Soc.* **2013**, *18*, art9. [CrossRef]

34. Correia, T.P. Threatened landscape in Alentejo, Portugal: the 'montado' and other 'agro-silvo-pastoral' systems. *Landsc. Urban Plan.* **1993**, *24*, 43–48. [CrossRef]

35. Guimarães, M.H.; Guiomar, N.; Surová, D.; Godinho, S.; Pinto Correia, T.; Sandberg, A.; Ravera, F.; Varanda, M. Structuring wicked problems in transdisciplinary research using the Social–Ecological systems framework: An application to the montado system, Alentejo, Portugal. *J. Clean. Prod.* **2018**, *191*, 417–428. [CrossRef]

36. Jones, N.; de Graaff, J.; Rodrigo, I.; Duarte, F. Historical review of land use changes in Portugal (before and after EU integration in 1986) and their implications for land degradation and conservation, with a focus on Centro and Alentejo regions. *Appl. Geogr.* **2011**, *31*, 1036–1048. [CrossRef]

37. Campos, I.; Vizinho, A.; Truninger, M.; Penha Lopes, G. Converging for deterring land abandonment: a systematization of experiences of a rural grassroots innovation. *Community Dev. J.* **2016**, *51*, 552–570. [CrossRef]

38. Huntjens, P.; Pahl-Wostl, C.; Rihoux, B.; Schlüter, M.; Flachner, Z.; Neto, S.; Koskova, R.; Dickens, C.; Nabide Kiti, I. Adaptive Water Management and Policy Learning in a Changing Climate: A Formal Comparative Analysis of Eight Water Management Regimes in Europe, Africa and Asia: Adaptive Water Management and Policy Learning in a Changing Climate. *Environ. Policy Gov.* **2011**, *21*, 145–163. [CrossRef]

39. Sadras, V.; Moran, M.; Petrie, P. Resilience of grapevine yield in response to warming. *OENO ONE* **2017**, *51*. [CrossRef]

40. Baêta, S.A. de A. Promoção Do Uso Eficiente De Água E Energia Em Unidades De Produção Vitivinícola: Estudos De Caso Da Adega Mayor E Granacer. Ph.D. Thesis, Lisbon University, Lisboa, Portugal, 2016.

41. Lopes, C.W. Promoção Do Uso Eficiente De Água E Energia Em Unidades De Produção Vitivinícola: Estudos De Caso Da Adega Cooperativa De Vidigueira, Cuba E Alvito, Herdade Das Servas E Roquevale. Ph.D. Thesis, Lisbon University, Lisboa, Portugal, 2017.

42. Mendonça, A.M.R. Promoção Do Uso Eficiente De Água E De Energia Em Unidades De Produção Vitivinícola: Estudo Dos Casos Da Herdade Dos Grous E Herdade Da Mingorra. Ph.D. Thesis, Lisbon University, Lisboa, Portugal, 2016.

43. Gómez-Lorente, D.; Rabaza, O.; Aznar-Dols, F.; Mercado-Vargas, M. Economic and Environmental Study of Wineries Powered by Grid-Connected Photovoltaic Systems in Spain. *Energies* **2017**, *10*, 222. [CrossRef]

44. Hall, S.; Roelich, K. Business model innovation in electricity supply markets: The role of complex value in the United Kingdom. *Energy Policy* **2016**, *92*, 286–298. [CrossRef]

45. Grimstad, S.; Burgess, J. Environmental sustainability and competitive advantage in a wine tourism micro-cluster. *Manag. Res. Rev.* **2014**, *37*, 553–573. [CrossRef]

46. Carrasco, I.; Castillo-Valero, J.-S.; Pérez-Luño, A. Wine Tourism and Wine Vacation as a Cultural and Creative Industry: The Case of the Bullas Wine Route. In *Cultural and Creative Industries*; Peris-Ortiz, M., Cabrera-Flores, M.R., Serrano-Santoyo, A., Eds.; Springer International Publishing: Cham, Switzerland, 2019; pp. 181–195. ISBN 978-3-319-99589-2. [CrossRef]

47. Olsson, P.; Folke, C. Local Ecological Knowledge and Institutional Dynamics for Ecosystem Management: A Study of Lake Racken Watershed, Sweden. *Ecosystems* **2001**, *4*, 85–104. [CrossRef]

48. Berkes, F.; Folke, C.; Gadgil, M. Traditional Ecological Knowledge, Biodiversity, Resilience and Sustainability. In *Biodiversity Conservation: Problems and Policies*; Perrings, C.A., Mäler, K.-G., Folke, C., Holling, C.S., Jansson, B.-O., Eds.; Springer: Dordrecht, The Netherlands, 1994; Volume 4, pp. 269–287. ISBN 978-0-7923-3140-7. [CrossRef]

49. Folke, C. Traditional Knowledge in Social–Ecological Systems. *Ecol. Soc.* **2004**, *9*, art7. [CrossRef]

50. Moller, H.; Berkes, F.; Lyver, P.O.; Kislalioglu, M. Combining Science and Traditional Ecological Knowledge: Monitoring Populations for Co-Management. *Ecol. Soc.* **2004**, *9*. [CrossRef]

51. Barber, N. Consumers' Intention to Purchase Environmentally Friendly Wines: A Segmentation Approach. *Int. J. Hosp. Tour. Adm.* **2012**, *13*, 26–47. [CrossRef]

52. Carmichael, B.A.; Senese, D.M. Competitiveness and Sustainability in Wine Tourism Regions: The Application of a Stage Model of Destination Development to Two Canadian Wine Regions. In *The Geography of Wine*; Dougherty, P.H., Ed.; Springer: Dordrecht, The Netherlands, 2012; pp. 159–178. ISBN 978-94-007-0463-3.

53. Zambon, I.; Colantoni, A.; Cecchini, M.; Mosconi, E. Rethinking Sustainability within the Viticulture Realities Integrating Economy, Landscape and Energy. *Sustainability* **2018**, *10*, 320. [CrossRef]

54. Marshall, R.S.; Akoorie, M.E.M.; Hamann, R.; Sinha, P. Environmental practices in the wine industry: An empirical application of the theory of reasoned action and stakeholder theory in the United States and New Zealand. *J. World Bus.* **2010**, *45*, 405–414. [CrossRef]

55. Cordano, M.; Marshall, R.S.; Silverman, M. How Do Small and Medium Enterprises Go "Green"? A Study of Environmental Management Programs in the U.S. Wine Industry. *J. Bus. Ethics* **2010**, *92*, 463–478. [CrossRef]

56. Boo, E.; Molinero, S.; Vicente, E.S. Novel Business Models and Main Barriers in the EU Energy System. 2016. Available online: https://zenodo.org/record/3479275#.Xd-httURXIU (accessed on 25 August 2019).

57. Dóci, G.K. Renewable Energy Communities. A Comprehensive study of Local Energy Initiatives in the Netherlands and Germany. Ph.D. Thesis, Vrije Universiteit Amsterdam, Amsterdam, The Netherlands, 2017.

58. Riedner, L.; Mair, C.; Zimek, M.; Brudermann, T.; Stern, T. E-mobility in agriculture: differences in perception between experienced and non-experienced electric vehicle users. *Clean Technol. Environ. Policy* **2019**, *21*, 55–67. [CrossRef]

59. Nigro, N.; Frades, M. Business Models for Financially Sustainable EV Charging Networks. 2015. Available online: https://www.c2es.org/site/assets/uploads/2015/03/business-models-ev-charging-infrastructure-03-15.pdf (accessed on 25 August 2019).

60. Ford, R.; Stephenson, J.; Whitaker, J. *Prosumer Collectives: A Review*; New Zealand's Smart Grid Forum; Centre for Sustainability, University of Otago: Dunedin, New Zealand, 2016; pp. 1–28.

61. Koirala, B.P.; Koliou, E.; Friege, J.; Hakvoort, R.A.; Herder, P.M. Energetic communities for community energy: A review of key issues and trends shaping integrated community energy systems. *Renew. Sustain. Energy Rev.* **2016**, *56*, 722–744. [CrossRef]

62. Herbes, C.; Brummer, V.; Rognli, J.; Blazejewski, S.; Gericke, N. Responding to policy change: New business models for renewable energy cooperatives – Barriers perceived by cooperatives' members. *Energy Policy* **2017**, *109*, 82–95. [CrossRef]

63. Schäufele, I.; Hamm, U. Consumers' perceptions, preferences and willingness-to-pay for wine with sustainability characteristics: A review. *J. Clean. Prod.* **2017**, *147*, 379–394. [CrossRef]

 © 2019 by the authors. Licensee MDPI, Basel, Switzerland. This article is an open access article distributed under the terms and conditions of the Creative Commons Attribution (CC BY) license (http://creativecommons.org/licenses/by/4.0/).

MDPI

St. Alban-Anlage 66

4052 Basel

Switzerland

Tel. +41 61 683 77 34

Fax +41 61 302 89 18

www.mdpi.com

Sustainability Editorial Office

E-mail: sustainability@mdpi.com

www.mdpi.com/journal/sustainability

www.ingramcontent.com/pod-product-compliance
Lightning Source LLC
LaVergne TN
LVHW071242170726
843515LV00006BA/1310